New Materialisms and Environmental Educati

Edited by
David A. G. Clarke and Jamie Mcphie

Routledge
Taylor & Francis Group
LONDON AND NEW YORK

First published 2024
by Routledge
4 Park Square, Milton Park, Abingdon, Oxon OX14 4RN

and by Routledge
605 Third Avenue, New York, NY 10158

Routledge is an imprint of the Taylor & Francis Group, an informa business

British Library Cataloguing in Publication Data
A catalogue record for this book is available from the British Library

ISBN13: 978-1-032-46162-5 (hbk)
ISBN13: 978-1-032-46164-9 (pbk)
ISBN13: 978-1-003-38033-7 (ebk)

DOI: 10.4324/9781003380337

Typeset in Myriad Pro
by Newgen Publishing UK

Publisher's Note
The publisher accepts responsibility for any inconsistencies that may have arisen during the conversion of this book from journal articles to book chapters, namely the inclusion of journal terminology.

Contents

Citation Information

The chapters in this book were originally published in the journal *Environmental Education Research*, volume 26, issue 9–10 (2020). When citing this material, please use the original page numbering for each article, as follows:

Preface

New materialisms and environmental education: editorial
David A. G. Clarke and Jamie Mcphie
Environmental Education Research, volume 26, issue 9–10 (2020), pp. 1255–1265

Chapter 1

Tensions, knots, and lines of flight: themes and directions of travel for new materialisms and environmental education
David A. G. Clarke and Jamie Mcphie
Environmental Education Research, volume 26, issue 9–10 (2020), pp. 1231–1254

Chapter 2

From action to intra-action? Agency, identity and 'goals' in a relational approach to climate change education
Blanche Verlie and CCR 15
Environmental Education Research, volume 26, issue 9–10 (2020), pp. 1266–1280

Chapter 3

Entangled threads and crafted meanings – students' learning for sustainability in remake activities
Hanna Hofverberg
Environmental Education Research, volume 26, issue 9–10 (2020), pp. 1281–1293

Chapter 4

More-than-human stories: experimental co-productions in outdoor environmental education pedagogy
Scott Jukes and Ya Reeves
Environmental Education Research, volume 26, issue 9–10 (2020), pp. 1294–1312

Chapter 5

Informal environmental learning: the sustaining nature of daily child/water/dirt relations
Sarah Crinall and Margaret Somerville
Environmental Education Research, volume 26, issue 9–10 (2020), pp. 1313–1324

Chapter 6

What if schools were lively more-than-human agencements all along? Troubling environmental education with moldschools
Tuure Tammi
Environmental Education Research, volume 26, issue 9–10 (2020), pp. 1325–1340

Chapter 7

'An atmosphere, an air, a life:' Deleuze, elemental media, and more-than-human environmental subjectification and education
Marcelina Piotrowski
Environmental Education Research, volume 26, issue 9–10 (2020), pp. 1341–1352

Chapter 8

Re-assembling environmental and sustainability education: orientations from New Materialism
Greg Mannion
Environmental Education Research, volume 26, issue 9–10 (2020), pp. 1353–1372

Chapter 9

Fieldnotes and situational analysis in environmental education research: experiments in new materialism
Andy Ruck and Greg Mannion
Environmental Education Research, volume 26, issue 9–10 (2020), pp. 1373–1390

Chapter 10

Doing little justices: speculative propositions for an immanent environmental ethics
David Rousell
Environmental Education Research, volume 26, issue 9–10 (2020), pp. 1391–1405

Chapter 11

Painting trees in the wind: socio-material ambiguity and sustainability politics in early childhood education with refugee children in Denmark
Nanna Jordt Jørgensen and Asger Martiny-Bruun
Environmental Education Research, volume 26, issue 9–10 (2020), pp. 1406–1419

Chapter 12

Challenging amnesias: re-collecting feminist new materialism/ecofeminism/climate/education
Annette Gough and Hilary Whitehouse
Environmental Education Research, volume 26, issue 9–10 (2020), pp. 1420–1434

Chapter 13

Anthropocentrism's fluid binary
Ramsey Affifi
Environmental Education Research, volume 26, issue 9–10 (2020), pp. 1435–1452

Chapter 14

Dark pedagogy – speculative realism and environmental and sustainability education
Jonas Andreasen Lysgaard and Stefan Bengtsson
Environmental Education Research, volume 26, issue 9–10 (2020), pp. 1453–1465

Chapter 15

Dark places: environmental education research in a world of hyperobjects
Antti Saari and John Mullen
Environmental Education Research, volume 26, issue 9–10 (2020), pp. 1466–1478

Chapter 16

Environmental end game: ontos
Chris Beeman and Sean Blenkinsop
Environmental Education Research, volume 26, issue 9–10 (2020), pp. 1479–1490

Chapter 17

Words (are) matter: generating material-semiotic lines of flight in environmental education research assemblages (with a little help from SF)
Noel Gough and Chessa Adsit-Morris
Environmental Education Research, volume 26, issue 9–10 (2020), pp. 1491–1508

Chapter 18

Nature matters: diffracting a keystone concept of environmental education research – just for kicks
Jamie Mcphie and David A. G. Clarke
Environmental Education Research, volume 26, issue 9–10 (2020), pp. 1509–1526

Notes on Contributors

CCR 15, RMIT University, Melbourne, Victoria, Australia.

Chessa Adsit-Morris, Department of the History of Art and Visual Culture, University of California, Santa Cruz, CA, USA.

Ramsey Affifi, Institute for Education, Teaching, and Leadership, The University of Edinburgh, Edinburgh, UK.

Chris Beeman, Faculty of Education, Brandon University, Brandon, Manitoba, Canada.

Stefan Bengtsson, SWEDESD, Uppsala University, Uppsala, Sweden.

Sean Blenkinsop, Faculty of Education, Simon Fraser University, Burnaby, British Columbia, Canada.

David A. G. Clarke, Moray House School of Education and Sport, The University of Edinburgh, Edinburgh, UK.

Sarah Crinall, Melbourne Graduate School of Education, University of Melbourne, Melbourne, Victoria, Australia.

Annette Gough, School of Education, RMIT University, Melbourne, Australia.

Noel Gough, School of Education, La Trobe University, Melbourne, VIC, Australia.

Hanna Hofverberg, Department of Creative Studies, Umeå University, Umeå, Sweden.

Nanna Jordt Jørgensen, University College Capital, Copenhagen, Denmark.

Scott Jukes, Institute of Education, Arts and Community, Berwick Campus, Federation University Australia, Victoria, Australia.

Jonas Andreasen Lysgaard, Danish School of Education, Aarhus University, Copenhagen, Denmark.

Greg Mannion, Faculty of Social Sciences, University of Stirling, Stirling, Scotland.

Asger Martiny-Bruun, Department of Social Education, University College Copenhagen, Copenhagen, Denmark.

Jamie Mcphie, MA Outdoor and Experiential Learning, University of Cumbria, Ambleside Campus, Ambleside, UK.

John Mullen, College of Education, Eastern Michigan University, Ypsilanti, Michigan.

Marcelina Piotrowski, Milieux Institute for Arts, Culture, and Technology, Concordia University, Canada.

Ya Reeves, Outdoor Environmental Education, La Trobe University, Bendigo, Australia.

David Rousell, School of Education, RMIT University, Melbourne, Victoria, Australia.

Andy Ruck, Swedish University of Agricultural Sciences (SLU), Uppsala, Sweden.

Antti Saari, Faculty of Education, University of Tampere, Tampere, Finland.

Margaret Somerville, School of Education, Western Sydney University, Sydney, NSW, Australia.

Tuure Tammi, Faculty of Education, University of Oulu, Hämeenlinna, Finland.

Blanche Verlie, RMIT University, Melbourne, Victoria, Australia.

Hilary Whitehouse, College of Arts, James Cook University, Cairns, Australia.

Preface

David A. G. Clarke and Jamie Mcphie

ABSTRACT

In this editorial, we introduce the 17 papers that make up this special issue of *Environmental Education Research* on new materialisms and environmental education, and discuss themes of significance arising from this collection. The papers are grouped in the special issue under the following headings: (1) pedagogical/research orientations; (2) critiques; (3) dark endings?; and (4) concepts mattering; but as we show here and in our introduction to the collection, they can also be connected in multiple ways to any other and a range of other relevant topics. As with our Introduction to the collection, we trust this commentary provokes and inspires, if not energises and materialises the furthering of axiological pathways away from dominant onto-epistemologies of environmental education research.

Introducing the papers

The 17 papers within the special issue are uniquely diverse in their articulations and interpretations of new materialisms when thought alongside environmental education and its research. Other than the obvious links to scholarship on the new materialisms and environmental education, all of these papers are creative in what we see as their Deleuzian inspired turn to *thinking with* and the fashioning of new concepts, in order to wrestle with contemporary environmental and educational challenges. As guest editors of this collection, we have ordered the papers into four groupings: pedagogical/research orientations; critiques; dark endings?; and concepts mattering. In truth, each paper has elements of each of these labels within them and could be connected in multiple ways to any other. But in order to meet the reader half way, in this editorial we briefly introduce each paper through these themes in order to comment on their particular offerings.

Pedagogical/research orientations

We begin with papers which we regard as most obviously applying, thinking with, or otherwise living with concepts from the new materialisms in more traditional 'pedagogical' or 'research' focused realities. Such papers help illustrate some of the broader themes of the new materialisms we commented on in our Introduction to the collection, in that they provide illustrations of more-than-human entanglements of pedagogical, empirical and speculative research.

There are nine pedagogically themed environmental education papers that ruminate with the new materialisms in educational establishments, written by authors attempting to do empirical research differently by focusing on relations that 'matter'. Extracting the concept 'pedagogy' from more static or mechanical versions, these papers illustrate how new materialist strategies might lead to various forms of play with educational concepts, in order to break free from traditional essentialist boundaries or what might be seen as staid pointillist practice. For example, via 'relationally attuned [...] climate pedagogies' (Verlie and CCR15, 2018), 'remaking pedagogies' (Hofverberg, 2019), 'pedagogical strategies' (Jukes and Reeves, 2019), 'water and art as philosophical muse' and 'relational pedagogies' (Crinall and Somerville, 2019), thinking and 'hesitating' 'with microbes' as 'more-than-human agencements' in 'moldschools' (Tammi, 2019) we witness scholars from around the world re-working pedagogical possibilities by emphasising the relational qualities of the matter that co-create distributed agencies in environmental education. Similarly, articles on 'thinking with the elements' (Piotrowski, 2019) 'Assemblage Pedagogy' (Mannion, 2019), 'research assemblage' (Ruck and Mannion, 2019), and 'immanent environmental ethics' (Rousell, 2018) all endeavour to muddy arbitrary humanistic boundaries and binaries, and instead include the idea that we are always already 'of' the world, not simply in it or on it.

In 'From action to intra-action? Agency, identity and 'goals' in a relational approach to climate change education', Verlie and CCR15 (2018) explore how and why they should become 'climate killjoys' in an attempt to decentre the human from climate pedagogies where a more response-able agency is revealed to emerge through relationships. Thinking with Karen Barad's concept of intra-action, Blanche as tutor and CCR15 as class, demonstrate the impossibility of dualistic analysis of people's impacts on the climate within the confines and affordances of a unit on an Australian university environmental education course. They note 'that these processes of change might be indicative of a more mutual enmeshment, where neither 'entity' would exist in its current form without the other', and that this understanding 'often seems to elude our full appreciation'. Verlie and CCR15 posit that a focus on climate intra-action enacts approaches that are not as accessible to more humanistic interpretations, namely: intrasectionality, acting-with, and diffraction. Responding to the particular problem of the location of ethical responsibility within a nondualist interpretation, Verlie and CCR15 suggest that 'intrasectionality navigates this [critique] by accounting for the dynamic processes of connecting-excluding that compose situated-but-dynamic hierarchies'. In our reading, Verlie and CCR15 are making novel contributions that both circumvent and reinvent the problem of the location of ethical obligation in a post-dualistic 'higher' environmental education. This essential work pushes at the margins of what it is currently possible to think in environmental education practices in university courses, forcing deep reconsideration of the manner in which we formulate participation in ethical research and pedagogy.

Hanna Hofverberg (2019) explores secondary school students' encounters and correspondence with crafting materials in 'Entangled threads and crafted meanings – students' learning for sustainability in remake activities' to reveal the stories that emerge and the educational possibilities for remaking pedagogy. Hofverberg shifts the scale of the sustainability at play to the everyday materialiality of recycling clothes in the classroom. Researching remake activities through Ingoldian correspondence, the practice of making can be witnessed as a co-constitutive activity of material-body relationality. The form is not imposed on the materials by the young adults in this research, but is rather formulated *with* them through experiments in learning. For us, Hofverberg's offering speaks to both the practical understandings of students' engagement with remake activities as activities of and for sustainability, but also pushes at conceptions of the human maker as the master in creative activities. Here we see the dual possibilities of a new materialist take, which on the one hand, concerns itself with more traditional ethical orientations towards reusing, remaking and recycling, while on the other, also helps demonstrate an ontological reorientation of the relationship between (presumed) intentionality and human action. Here, we believe, we see an application of the critique of the hylomorphic model which Ingold (following Deleuze and Guattari who in turn follow Paul Klee) describes as the realisation that 'the essential relation, in a world of life, is not between matter and form but between *materials* and *forces*' (Ingold, 2010, pp.92-93). Form (as either a plan to turn old clothing into a new accessory or, for instance, a curricula or pedagogical plan or policy) is not conceived outside of relations with the world, in a perceived realm of abstraction, but instead always (cor)responds to/with it as force and action in movement. This understanding should curtail overly universalising pedagogies and educational policies, as the precautionary principle leads away from the potential ethical blindness of grand, untethered gestures, and points instead to minor interventions, cautious experiments, and tiny changes. We could regard this not only as an ethical orientation, but one that is wrapped up with the very nature of the possible. Reading Hofverberg's contribution helps us better perceive these challenges.

Scott Jukes and Reeves (2019) offer 'methodological and pedagogical insight into using new materialist theories in creative and productive ways' in their paper, 'More-than-human stories: Experimental co-productions in outdoor environmental education pedagogy'. They introduce a number of concepts at work within new materialist discourses (including material agencies; mutual entangled relationships-assemblages; and becoming part of a more-than-human world) to build on storying approaches and place-responsive pedagogies to develop their concept of *more-than-human stories*. In this instance, the more-than-human story being offered is *One Single Moment*, a picturebook written by Ya when she was a pre-service teacher participating in a ski-touring journey in the Australian Alps, for which Scott was an educator. The authors offer *One Single Moment* as an example of creative empirical materials, rather than as 'data', which might suggest a 'brute fact gathered'. In offering further exposition on *One Single Moment,* the authors reveal insights into student-educator-place-theory-picturebook relationships at work in the presence of the aforementioned concepts, and demonstrate the temporal nature of the education continuing to be at work. Their paper offers guidance for experiments in practice, but also many provocations for our understanding of places as becoming educational. This has pedagogical potential. For example, one such provocation reminds us of something a senior scholar said to us at the Cambridge University PhD summer school organised by fellow early career researcher, Elsa Lee. We met Professor Marcia McKenzie, there as a 'keynote', a chance encounter with whom was the initial provocation for this special issue. After we performed a paper we were working on about distributed agency in a park, we were told that things cannot have agency without humans being present, as part of that particular agential assemblage. We disagreed. We reason(ed) it's rather like that age-old question about the tree falling in the forest. The sound of the tree falling in the forest is only ever a sound for those temporal anomalies gifted with ears to translate changing vibrations in the air into sound. The resulting sound waves/particles are a unique experience and yet also a shared experience for the multiple material phenomena involved in the processual event. But - acknowledging the agency of the 'more-than-human world' which Scott has 'been playing with' - places are lively actors in themselves and have their own narratives, regardless of any human intra-actions (not forgetting that humans are places too for many species).

More broadly, these *it-narratives* could be communities of microbiota continually pulsating in multiple directions rather than simply responding to things outside of their perceived boxes of taxonomic status, just as humans are not place-responsive in any cause-and-effect linearity. Life is not dot-to-dot. And so, places are not educational of themselves in a unidirectional manner. They cannot teach us as if we were a *tabula rasa* waiting to receive wisdom from some highly gendered romantically conceived Nature, 'out there'. We are the 'there' itself. We are place/s. We are made of place/s. So what does it really mean to know place/s? There will always be secrets to/in places that hide in differing timelines, outside of our evolutionary heritage, our sensory bandwidth. There are countless narratives happening all the time, that perform at different scales, and temporal, spatial and sensorial frequencies, that we simply cannot perceive or even conceive (even with extra-sensory cyborgian additions like radar or ultraviolet detectors).

Even then, another spiral and perturbation is that conceptions themselves perform ecologically in more ways than we can fathom. If we, as assemblages of multiple complex materials, can be said to have this concept called 'agency', then so can other places of assembled multiple complex materials, including forests, shopping malls and even the world, and beyond. While we often think of agency as *intent*, as with Verlie and CCR15, the conjoined nature of actions or reciprocal performance doesn't really have a starting point which we could pinpoint to state with accuracy that that's where and when a decision was made. Starting points have always already begun in a multidirectional torrent of events. These accumulate other events and physical processes over time to eventually form what many people now 'think' as agency. Perhaps intent, or agency, might be better rephrased or simply understood differently to the Cartesian ghost or Freudian ego - as something like an assemblage or conglomeration (bearing in mind that each of these concepts would perform differently). The human is not so special in this multispecies/multilithic interpretation (fabric? If we return to Hofverberg), as we are just another conglomeration of events and physical processes - anthropoglomerate. This is where provocations from Jukes and Reeves, and inevitably others, might take *us,* anyway.

Crinall and Somerville (2019) also move us to consider binaries of solid and fluid states and the muddying of these, in how they problematise sustainability. Thinking with Deleuze, they draw on Grosz and 'ontologies of water that do post-qualitative research playfully' in their paper 'Informal environmental learning: the sustaining nature of daily child/water/dirt relations'. Post-inquiry, or post-qualitative inquiry, argues that qualitative inquiry within the social sciences has become overly procedural and that opportunities for thinking with theory can be sought in an attempt to produce new thoughts, and indeed new worlds, as researchers grapple with empirical materials (St. Pierre, 2019). Crinall and Somerville offer their understanding of art, derived from Deleuze and Grosz, as bodily affective, to an everyday artfulness of walking with early years children *as* sustaining. They invite their paper to be read 'as an experience of sensation' in what we perceive as another attempt to reinterpret the ethical location of sustaining, and sustainability. This is away from or enveloping parcelled notions of the environment, or nature, demarcated by discourses of lack, to instead create/explore sustaining material relationships both in the events of everyday walking with children, and with the very act of reading their paper. In the first sense, this notion of sustainability as sustenance helps us conceive a broadening of environmental education, one in which the dichotomy between health, wellbeing, environments, and bodies of all kinds is disrupted. We intuit both radical possibilities and dangers in this approach. In the second sense, in which sustenance might be gleaned from the reading of their paper, we see affinities with the powerful potential of evocative research approaches, but also wonder at the distance (spatial and temporal, but also affectual and conceptual) involved in attempts to affect (or is it infect? or inflect? ...) readers from/with new materialist orientations. We find ourselves both hoping for their success, but understanding that, as with all educational endeavours, the ends to which we travel can never be guaranteed.

Next, in the paper, 'What if schools were lively more-than-human agencements all along? Troubling environmental education with moldschools', Tuure Tammi (2019) deconstructs 'the animal turn with a turn to microbes' by asking 'what moldschools might teach us or ask from us in terms of environmental education and ethics'. Tammi deploys the concept of agencement, more commonly and perhaps misleadingly translated from Deleuze and Guattari as 'assemblage'. For Tammi, agencement indexes the movement in bodies 'becoming-with', and shows 'how the phenomenon of moldschools' emerges 'in the intersections of building and cleaning, practices, materials, technologies, microbes, policies, and more'. Combining this conceptual work with empirical materials gathered from a moldschool, Tammi emphasises the material-discursive enactment of making moldschools.

Tammi's paper also demonstrates how the practice of education is always already environmental, regardless of location or subject matter. We are all becoming-with microbes and 'world-making practices'. Indeed, Tammi suggests that, 'through these encounters the school is no longer what it was thought to be, and its human inhabitants are not quite what was anticipated either. In other words, material-discursive mold troubles anthropocentric epistemologies and ontologies: the commonly held dichotomies (e.g. nature/culture, body/mind), school routines, relations with materials, and more'. As with the previous papers, the exciting perspective taken here suggests to us that environmental education researchers might gain similarly deep insights into the material-discursive nature of educational contexts by following surprising or under acknowledged intra-actants where they are. Tammi demonstrates the diligence required for this kind of work, which necessitates a careful mapping of affects across boundaries which can often appear sedimented in thought and conventional humanist research practices.

Another entry point to these aspects can be found in, ''An atmosphere, an air, a life:' Deleuze, elemental media, and more-than-human environmental subjectification and education', a conceptual piece from Marcelina Piotrowski. Piotrowski (2019) articulates 'ways of engaging with more-than-human forms of subjectification in environmental movements' by 'thinking alongside an 'elemental Deleuze'' to include the 'classical elements: air, water, earth, fire'. Piotrowski points to the importance of environmental movements as locations for public pedagogy and political education, and thus for subjectification or forming ways of being. In attempting to interpret subjectification from a posthuman, rather than humanist orientation, Piotrowski flattens the perceived peak of human intentionality to locations dispersed among relational entanglements with matter. The elements perform as the media for these experiments, and Piotrowski shows how the philosophy of Deleuze renders philosophy itself as always already a worldly act. By, for instance, focusing on how the air, water, earth, and fire create the realities of which we are manifest, Piotrowski argues that the political subjectivities of adults 'cease to function as beings, and need to be examined as composites of folds, as process-ontologies that are vital because of elements' process-power'. The upshot is a resistance to depictions of human sovereignty in environmental movement education research, typically understood via traditional educational institutions like schools, through experiments in (re)thinking matters of educational subjectification with Deleuze. The Deleuzian term 'a life' is significant for the manner in which it depicts a posthuman politics of locations and timings of 'life'. We agree with Piotrowski as she calls for environmental education researchers to understand the imbrication of subjectivities with materialities, rather than focusing necessarily on the formation of beings (through, for instance, capacity building in orthodox humanist subjectivities as 'pupils' in schools). However, we also wonder at the complexities and practicalities of calls of this nature and urge researchers to further explore the 'challenge' that Piotrowski identifies.

Perhaps a way of exploring is neither from within to without, or from outside to in? Greg Mannion (2019) provides advice on how to entangle 'from the middle', 'drawing on data from a place-responsive heritage education project' and 'employing theory from Deleuze and Guattari' in the paper, 'Re-assembling environmental and sustainability education: orientations from new materialism'. Mannion expresses how 'humanistic approaches to education in general appear to have failed us to date in many ways to adequately address issues such as climate change and wider global precarity,' and so reconnoitres how some ESE has engaged with this critique by turning to the new materialisms. Mannion's concerns lay with what might be thought of as the 'practical', 'how to do it', orientations to questions about research and pedagogy with new materialisms and investigates 'assemblage theory' as a potent(ial) toolbox for what he terms 'assemblage research' and 'assemblage pedagogy'. These concepts are demonstrated through the project, *Stories in the Land*, a 'place-responsive, arts-based, outdoor environmental education programme that connected participants from different generations from schools, youth groups and their communities with local places through journeys on the widespread network of 'drove roads' across Scotland'. Both 'assemblage pedagogy' and 'assemblage research' are argued to operate along ethical lines, where the 'good' is viewed as attempting to attain an immanent ethics of increased connection, and capacity to affect and be affected, to the middles we find ourselves of. In our reading, we sense Mannion pushing at several tensions within the turn to new materialisms in environmental education research and pedagogy. By orientating his contributions towards 'orientations', and in effect advising directions of travel for both researchers and pedagogues, we see Mannion as doing essential work in facilitating access. While in providing a 'tool box' and set of recommendations, Mannion offers a conceptual map that he acknowledges should not be seen as comprehensive. The tension we see is between this essential work, and the non-methodological (post-qualitative) and often anti-prescriptive challenge brought by some who sense deep implications for the very nature of research within the social sciences brought by the disruptions of the new materialisms. This tension remains practically and philosophically unresolved, bringing further burdens to those attempting to do good research whilst operating within the limitations of contemporary higher education institutional and research models, particularly those in precarious research positions.

To elaborate, in 'Fieldnotes and situational analysis in environmental education research: experiments in new materialism', Ruck and Mannion (2019) explore 'the constant tensions that arose when attempting to use existing qualitative research methods in combination with new materialist theories'. Ruck and Mannion attempt to provide 'an insight into the tensions, synergies and on-the-ground methodological struggles within this site of experimentation'. In so doing, they explore how their sympathy for new materialist theory, and its (non)methodological post-qualitative implications, conflicted with the instrumental requirements of a time-bound, funded, structured, evaluation of the *Polli:Nation* initiative - a UK wide project which 'engaged young people from 260 primary and secondary schools in the transformation of their school grounds into pollinator-friendly habitats'. In particular, Andy Ruck explains how his research practices within his doctoral studies were altered as he came to the discourses surrounding new materialisms: fieldnotes become 'theoretically sensitive' whilst 'performative privilege' became an ongoing concern. Ruck and Mannion try to provide insight and encouragement to researchers by highlighting how they negotiated the space between what they see as 'demonstrable impact' and 'high theory'. Whilst there is hope in this example, we also wonder at the types of compromises qualitative researchers have to make in demonstrating a certain perception of 'impact'? This includes the ways these compromises help reinforce the neoliberal research impact agenda (Laing, Mazzoli Smith, & Todd, 2018) and limit the potential for alternative ethical research. These struggles are not 'new' either, but are perhaps further exacerbated as the distance between conventional humanist orientated research paradigms, funding regimes and quality criteria and the contemporary moment are widened by the popularity of new materialisms? However, what the 'ethical' is, is also contested within this moment, necessitating experimentations with ethics itself within inquiry and further complexifying any claim to 'impact'.

In another example of such ethically concerned research, David Rousell (2018) suggests that 'immanent ethics cannot be reduced to a set of predetermined values or prescriptions for environmental education, but should proceed through a speculative process of creative experimentation and negotiation in the pursuit of unforeseen openings'. In the paper, 'Doing little justices: speculative propositions for an immanent environmental ethics', Rousell, draws on the *States and Territories* project in which a series of artworks were created across an Australian university campus. Thinking with immanence as the conceptual orientation, Roussell shows how an ethics inspired by Deleuze and Whitehead operates in the present as it encounters the problems of new realities, such as the Anthropocene or techno-scientific creations, to call for renewed notions of environmental justice brought on by the Anthropocene. In an aversion to moral principles, 'doing little justices' is thus an experiment in the 'performative, speculative, collective, and minor fluctuations of a micropolitics'. An immanent ethics produces new forms of rights, where rights are no longer seen as a characteristic which humans possess, but rather 'there is a continuous becoming, distribution, and re-assembling of rights in relation to the mutual construction of an ethical problem'. Rousell's exploration of the ethical within environmental education research prompts us to consider the manner of *be(com)ing* implied in pedagogic practices, and how potentially new conceptions of ethics might be made more manifest within, and as a result of, altered practices.

So, with their working of methodological issues and Deleuzian inspired explorations of ethics and subjectification, the papers in this section might be thought of as performing critical endorsement of the material turn. In contrast, the following section illustrates a shift to critical caution, with authors contributing generative appraisal in historical, empirical, and philosophical terms.

Critiques

Whilst each of the contributions across the special issue offers some variety of critical perspective, it is within three papers that we sense the most elaborated productive critiques for how we conceptualise new materialism within environmental education and environmental education research.

First, in 'Painting trees in the wind: Socio-material ambiguity and sustainability politics in early childhood education with refugee children in Denmark', Jørgensen and Martiny-Bruun (2019) reiterate the intractability of social justice and environmental sustainability issues which much, though not all, new materialist writing also acknowledges. It is to 'common world' approaches within environmental education in particular that they offer a Merleau-Pontian critique, emphasising simultaneous 'immersion and detachment' as they consider the features of sustainability education at work for refugee children in Danish day care institutes. In particular, Jørgensen and Martiny-Bruun explore how 'human relations with the nonhuman cannot be understood independently of social relations' and 'argue against completely erasing or overlooking the experiences of being separated from the world which have been seen as a characteristic of modern Western thought'. In reading Jørgensen and Martiny-Bruun's critique we are reminded that there exists some scholarship which, whilst struggling with or even embracing the material turn, can rush to overly simplistic application, describing material interactions as inevitably an assemblage of some form or other, as if that is all that can or should be said. What is jettisoned in such an uncritical approach is any focus on the complexities of contested concepts and ethics, and the different ways of thinking they might imply, including within or against new materialisms. The upshot is that, without heeding the lessons of critical theory - which are well acknowledged in much new materialist scholarship - environmental education scholarship may fail in its egalitarian aims.

To illustrate, in their generative and subtle critique, Gough and Whitehouse (2020) apply the concept of 'amnesias' in tracing the implications of forgetting the relationships between feminist new materialisms and ecofeminism, conceptual and anthropogenic climate disruption, and 'the impact of climate disruption on women'. In their paper, 'Challenging amnesias: Re-collecting feminist new materialism/ecofeminism/climate/education', Whitehouse and Gough are less concerned with definitions of categories and more with the challenge to 'denaturalize nature and deculturalize culture', acknowledging that 'the climate emergency is forcing a realisation there is not one conceptual model that is going to see us through'. Whitehouse and Gough argue for the importance of acknowledging genealogical flows in thought, not so much to critique the present, but so that 'the past and the present can be drawn together to provide meaningful impetus for research'. This form of generative or immanent critique orientates thinking towards the task at hand by recollecting that different approaches have their conceptual strengths when considering the intractability of gender inequality and the climate emergency. Whilst we see more strident accusations of 'amnesia' levelled at those exploring new materialisms in environmental education from elsewhere, Whitehouse and Gough demonstrate the skill required in thinking and writing with immanent care - fostering understanding with what we perceive as a commitment to what Latour (2004) calls 'the matters of concern', rather than the matters of fact.

Rounding out this section, in 'Anthropocentrism's fluid binary', Ramsey Affifi (2019) delves deeper into environmentalisms' attempts to de-anthropocentrise environmental education and calls for 'pedagogies that help students imagine and engage with what [...] conceptualisations actually *do* when believed'. With perhaps this special issue's most critical commentary, Affifi recommends employing 'humble bafflement' in order to navigate the '(non)Anthropocene'. Affifi takes aim at what he sees as an abstracted project of de-anthropocentrism which he detects within new materialist discourses, and counter offers variations on centering and decentering which he argues are always at play within 'anthropocentrism's fluid binary'. We see Affifi's paper as similarly offering a survey of amnesias, but also an attention to the material performativity of concepts when he asks:

> 'what does a given deanthropocentrism do? Does it actually lead us to attend to the field of more-than-human presences around us? Does it invite us to relate to the birch trees and magpies around us, and to recreate our environments to enable such relationships? Does it actually compel us to stand in front of the bulldozers? Does it create beauty? Or, despite its claim to radical nonhumanness, does it instead merely invite us into more arguments, more time in front of computers, and more wandering around stuffy conference rooms?'

The answer, we think, is that it may do many things, including each of these. The multitude of directions that might be considered to be contributing to the material turn can each be evoked or practiced in multiple ways and to multiple degrees. They may both replicate the present and inspire action within and beyond the habitus of Western academia. Nonetheless it seems less of a choice we have, and more of a struggle we are in, and with which we must contend. Affifi's provocations add significant insight for those in these struggles.

In sum then, the papers in this section encourage us to pause and reflect on what the new materialisms make im/possible for us as researchers and practitioners of environmental education. Questions about how we understand the nature of critique also become relevant to how we might respond, and might well be caught up with our ontological proclivities (e.g. Latour, 2004). In the next section we see these ontological proclivities shift to darkened renderings and readings, in an attempt to activate ourselves in/of the world to ontological ends.

Dark endings?

The penultimate papers in this special issue take a darker turn, employing fictional and speculative conceptual apparatus in their dark ecologies and place-based pedagogies. For example, Lysgaard and Bengtsson's (2020) 'Dark pedagogy – speculative realism and environmental and sustainability education' discusses the 'inherent withdrawnness of the objects at play' and makes 'explicit the potential educational implications of speculative realisms' in environmental and sustainability education. They utilise Timothy Morton's *Dark Ecology*, concepts from Object Oriented Ontology (OOO) - hyperobjects, undermining, overmining, duomining, correlationism - and the author H.P. Lovecraft's speculative horror fiction to 'interrogate, substantiate and, thereby, to open up for discussions central concepts to the field'. Lysgaard and Bengtsson, for instance, explore how a dark pedagogy would endeavour to ask questions about the nature of self and attempt to account for educationally unaccounted for objects. Taking Morton's notion of the hyperobject, with its near yet massively distributed effects, Lysgaard and Bengtsson speculate on an environmental education which is less interested in subjectification or place, and more concerned with affective states of never fully grasping, never fully apprehending: 'as the hyperobject is both not-here (nonlocal) and very near (viscous), it undermines the very identity of the self engaging with it. The displacement of the self takes place in the emotional resonance of the hyperobject'. Accordingly, what OOO might offer are strange new renderings of how the self always develops as a result of an unknowable world. The argument here is that our notions of democracy, and (environmental and sustainability) education's role in this, are altered as a result.

In 'Dark places: environmental education research in a world of hyperobjects', Saari and Mullen (2018) offer further specific pedagogical conceptualisation as they take up Timothy Morton's OOO inspired concept of 'hyperobjects' as a 'springboard' for thinking 'through' strategies for living in 'dark places that are haunted' by hyperobjects such as global warming. Theorising with place specifically, Saari and Mullen explore how Morton's concepts might further 'the post-stucturalist and new materialist critique of bounded, stable places through introducing an ontology of finitude, withdrawal, and excess'. Within this approach, our expectations of comprehending places are shifted by 'working-through' the uncanniness of both our relationships to places, and the possibilities for straightforward environmental education in the local. According to Saari and Mullen, 'If our affects, desires, hopes and fears are projected into material and more-than-human environments, working-through can also take place in the form of recognizing sites of mourning, ambiguous affects and the corresponding senses of injustices. And all the while recognizing these emotions as uncanny and ambivalent rather than splitting them into good and bad or local and global'. Dark places, then, force us to reconsider the parochial or romantic in place-based pedagogies and invite understanding of places as multiple, alien, and only ever partially knowable.

Another invitation awaits in the next paper. Beeman and Blenkinsop (2019) sum up their narrative of, 'Environmental end game: ontos' by asking us to '[entangle] ourselves differently' through diffractive readings and experiences, 'shifting conceptualizing towards a different ontological position compatible with an immanent materialist position'. They instruct the reader to think with their paper *whilst* enacting various activities that the reader might perform. We second this call to heterogeneity in modes of literacy. We also acknowledge the challenges calls of this nature can create. The ability to respond to such calls for action is not evenly distributed, including in terms of literacies, competencies and intentionalities. Social and environmental (in)equities are often barriers to many (assemblages), where lines of flight are clipped at the wing and made inaccessible via powerful binaries themselves (as highly active performative concepts - concepts *mattering*). As we note in the introduction to the collection, these include privilege/poverty or human/other-than-human (even though the human is also the other-than-human too). It is to this idea, that concepts matter in the world, that we now turn with the final papers of the special issue.

Concepts mattering

Deleuze and Guattari played with the idea 'Monism = Pluralism', and in so doing brought forth a seemingly necessary contradiction. The unity of monism is co-produced out of the multiplicity of pluralism, just as transcendence is a product of immanence, mind is a product of body, theory is a product of practice, and culture is a product of nature (and/or *vice versa*). This could also be said for binary existence as a product of both monism and pluralism. That Cartesian dualisms exist isn't in debate here, it's what these dualisms *do* and produce or the resulting effects of how they perform that is contentious. Several papers throughout this special issue urge a move to performing new practices and creative experimentation with new approaches, and this is just what the last two papers attempt to do with the performativity of words – concepts mattering.

Gough and Adsit-Morris' (2019) contribution, 'Words (are) matter: generating material-semiotic lines of flight in environmental education research assemblages (with a little help from SF)' speculates on the performative capacity of concepts derived from Deleuze and Guattari and from SF (science/speculative fiction). Gough and Adsit-Morris offer *cognitive estrangement* and *object-oriented thought experiments* as 'little machines' that 'can be plugged into an environmental education research assemblage to produce new material-semiotic conjunctions and configurations of researchers, empirical materials, methods and contexts'. With a series of 'ruptures' the authors explore how 'critical thinking about materiality is apparent in many areas of cultural production and practice relevant to environmental education'. With their focus on SF, we think that Gough and Adsit-Morris make a compelling case that the way we imagine human-nature-technological futures in many ways condition what is possible for us in the political and educational present, and that this has implications for all our futures. Indeed, as they state, 'SF and material worlds are now so entwined that they cannot be understood in isolation'.

Finally, with our own offering 'Nature matters: diffracting a keystone concept of environmental education research – just for kicks' (Mcphie and Clarke, 2018), we speculate on the performative mattering of various concepts of 'Nature' which we perceive as occurring to various degrees within the environmental education research literature. In so doing, we operationalise the new materialist concept 'diffraction' to see what thinking through nature(s) might do. In an attempt to overcome the problem of the plasticity of the term, we articulate nature, and the related notions of landscape and environment, as concepts always in process, and so offer natur*ing*, landscap*ing*, and environ*ing*. Indeed, we speculate that envisioning concepts as pedagogical playmates might reinforce their ecological physicality as performative agents, bringing to life forceful energetic processes that are still perceived by many as inert linguistic phenomena - *dead matter.*

Summary

The four themes we have explored in this editorial are ones that came through for us in our reading of the contributions to the special issue, but should by no means be thought of as describing the sole contributions of each paper. Each is, and has to be, multifaceted as it tackles multiple issues and makes multiple contributions. That being the case, we urge you to not take our word for what these papers are communicating, but to instead delve into what they make im/possible for you within environmental education and beyond.

The editing journey has been a long road. We have been exhausted and elated, challenged and confronted, exasperated and educated, carrying what at times felt like a heavy burden of collating and marrying a hugely varying diversity of styles and expressions whilst also carrying on with work, life, and young families. As is by now evident, there are many issues at stake and many voices involved in thinking about the in/significance of new materialisms and environmental education. There are, no doubt, important points left unsaid or underexplored in this special issue but we feel the collection offers some heavy lifting, exploring various avenues a little further than before, and in so doing provides many novel provocations for environmental education and its research. As with our Introduction, caught in the middle and mire of creating and curating a collection, it still/must/will/always already leaves us with a sense that the work is more open than closed to further inquiries, even as it highlights the potential of new materialisms to forge axiological pathways away from dominant onto-epistemologies of environmental education research. And thus, we conclude our introductory and editorialising comments with a final invitation: for further contributions on the topics, that join us in engaging with what the collection does and does not offer, and to rework it.

Acknowledgment

We have thought with many scholars (and things) along this journey, some of whom have helped us beyond their remit, and we thank you wholeheartedly for that. We'd like to thank the contributing authors, the reviewers, members of the editorial board, and the journal production team. We extend particular thanks to Alan Reid for sharing with us his keen professional insight, enthusiasm, and scholarly interest in a truly collaborative spirit.

Disclosure statement

No potential conflict of interest was reported by the authors.

ORCID

David A. G. Clarke http://orcid.org/0000-0002-4330-0818
Jamie Mcphie http://orcid.org/0000-0001-5290-1685

References

Affifi, R. 2019. "Anthropocentrism's Fluid Binary." *Environmental Education Research* 1–18. doi:10.1080/13504622.2019.1707484.

Beeman, C., and S. Blenkinsop. 2019. "Environmental End Game: Ontos." *Environmental Education Research* 1–12. doi:10.1080/13504622.2019.1663792.

Crinall, S., and M. Somerville. 2019. "Informal Environmental Learning: The Sustaining Nature of Daily Child/Water/Dirt Relations." *Environmental Education Research* 1–12. doi:10.1080/13504622.2019.1577953.

Gough, A., and H. Whitehouse. 2020. "Challenging Amnesias: Re-Collecting Feminist New Materialism/Ecofeminism/Climate/Education." *Environmental Education Research* 1–15. doi:10.1080/13504622.2020.1727858.

Gough, N., and C. Adsit-Morris. 2019. "Words (Are) Matter: Generating Material-Semiotic Lines of Flight in Environmental Education Research Assemblages (with a Little Help from SF)." *Environmental Education Research* 1–18. doi:10.1080/13504622.2019.1663793.

Hofverberg, H. 2019. "Entangled Threads and Crafted Meanings–Students' Learning for Sustainability in Remake Activities." *Environmental Education Research* 1–13. doi:10.1080/13504622.2019.1664414.

Ingold, T. 2010. "The Textility of Making." *Cambridge Journal of Economics* 34 (1): 91–102. doi:10.1093/cje/bep042.

Jørgensen, N. J., and A. Martiny-Bruun. 2019. "Painting Trees in the Wind: Socio-Material Ambiguity and Sustainability Politics in Early Childhood Education with Refugee Children in Denmark." *Environmental Education Research* 1–14. doi:10.1080/13504622.2019.1602755.

Jukes, S., and Y. Reeves. 2019. "More-than-Human Stories: Experimental co-Productions in Outdoor Environmental Education Pedagogy." *Environmental Education Research* 1–19. doi:10.1080/13504622.2019.1699027.

Laing, K., L. Mazzoli Smith, and L. Todd. 2018. "The Impact Agenda and Critical Social Research in Education: Hitting the Target but Missing the Spot?" *Policy Futures in Education* 16 (2): 169–184. doi:10.1177/1478210317742214.

Latour, B. 2004. "Why Has Critique Run out of Steam? from Matters of Fact to Matters of Concern." *Critical Inquiry* 30 (2): 225–248. doi:10.1086/421123.

Lysgaard, J. A., and S. Bengtsson. 2020. "Dark Pedagogy–Speculative Realism and Environmental and Sustainability Education." *Environmental Education Research* 1–13. doi:10.1080/13504622.2020.1739230.

Mannion, G. 2019. "Re-Assembling Environmental and Sustainability Education: Orientations from New Materialism." *Environmental Education Research* 1–20. doi:10.1080/13504622.2018.1536926.

Mcphie, J., and D. A. G. Clarke. 2018. "Nature Matters: Diffracting a Keystone Concept of Environmental Education Research–Just for Kicks." *Environmental Education Research* 1–18. doi:10.1080/13504622.2018.1531387.

Piotrowski, M. 2019. "An Atmosphere, an Air, a Life:' Deleuze, Elemental Media, and More-than-Human Environmental Subjectification and Education." *Environmental Education Research* 1–12. doi:10.1080/13504622.2018.1485134.

Rousell, D. 2018. "Doing Little Justices: Speculative Propositions for an Immanent Environmental Ethics." *Environmental Education Research* 1–15. doi:10.1080/13504622.2018.1517408.

Ruck, A., and G. Mannion. 2019. "Fieldnotes and Situational Analysis in Environmental Education Research: Experiments in New Materialism." *Environmental Education Research* 1–18. doi:10.1080/13504622.2019.1594172.

Saari, A., and J. Mullen. 2018. "Dark Places: Environmental Education Research in a World of Hyperobjects." *Environmental Education Research* 1–13. doi:10.1080/13504622.2020.1739230.

St. Pierre, E. A. 2019. "Post Qualitative Inquiry, the Refusal of Method, and the Risk of the New." *Qualitative Inquiry* 1–7. 107780041986300. doi:10.1177/1077800419863005.

Tammi, T. 2019. "What If Schools Were Lively More-than-Human Agencements All along? Troubling Environmental Education with Moldschools." *Environmental Education Research* 1–16. doi:10.1080/13504622.2019.1584881.

Verlie, B., &. CCR 15 2018. "From Action to Intra-Action? Agency, Identity and 'Goals' in a Relational Approach to Climate Change Education." *Environmental Education Research* 1–15. doi:10.1080/13504622.2018.1497147.

Introduction—Tensions, knots, and lines of flight: themes and directions of travel for new materialisms and environmental education

David A. G. Clarke and Jamie Mcphie

ABSTRACT

In this introduction to a special issue of *Environmental Education Research* on New Materialisms and Environmental Education, we begin with a brief overview to this publishing project and to scholarship on new materialisms and environmental education. Against this backdrop, we then discuss various themes of significance arising from the broader tumult of thought that occurs in the 17 papers that bring these areas into conversation. In brief, papers gathered in this collection illustrate a series of engagements with: (1) new empiricism and post-qualitative inquiry; (2) the meeting of politics, ethics, and decolonial theory with new materialisms; (3) conceptions of nature, environment, sustainability and the human subject; and (4) new materialisms as environmental pedagogy. We recognise that readers will imagine, detect and respond in diverse ways to the papers and the thematics on their own terms too, noting that other inclusions in the collection would likely have generated different patterns and affordances for insight, challenge and debate. Thus, we argue that in some senses, the collection must remain open rather than closed, while we also invite further contributions on the topics, that engage with what the collection does and does not offer, and to rework it. In other words, we trust our introduction underscores the immanent performativity expected of many of the new materialisms, and highlights their potential to forge axiological pathways away from dominant onto-epistemologies of environmental education research.

Decoupling endings as a curious place to start

When we issued a Call for Papers for this special issue of *Environmental Education Research* in April 2017, we were acutely aware of currents of work in environmental education and its research that had ebbed and flowed on the conception and experience of matter, and the 're/turn' to the material. But it was often unclear to us what further ramifications, implications and performativities might be of the associated materialisms which were being invoked, criticised or offered as generative in this work. Some of this work seemed to rely on a continuum of uncritical-critical-postcritical materialisms, and were prone to position their authors or interests in a vanguard accordingly. Others sought ways to supplant the 'old' with the 'post' or the 'new', perhaps in terms of research design or questions, but increasingly through categories and preferences in theory and theorists, that sought to destabilise and re-imag(in)e what or who was appealing or authoritative in governing the field of practice and inquiry. If we were struggling to make sense of this, we reflected, we were probably not alone. So, we looked for opportunities to invite our colleagues to join us on a series of departures and sense-making journeys, reckoning more deeply, we hoped, with the aforementioned pulses and debates, and helping clarify some of this proliferation and its possible confusions. In short, our intention was to better engage the contributions and challenges new materialisms might offer, as one of the latest incarnations of engagement with new theory and praxis in environmental education.

In our wider work, we had already started examining some of the specific issues and implications that the new materialisms might raise for the field of environmental education (see Clarke 2017; Clarke and Mcphie 2014, 2016; Mcphie 2018; Mcphie and Clarke 2015, 2018; Clarke 2019). While with the encouragement and support of the journal's board, we proposed to guest edit a special issue that would broaden and deepen such conversations. Looking back, we had hoped that the special issue would act as a site that brought about both a tightening and loosening of various potential tensions and entanglements, 'knotted' or/and 'fabricated' as they seemed to us, around the ethical, the practical, the methodological, the historical, and, of course, the pedagogical. And in this introduction, and the accompanying editorial essay, we illustrate how we and others have wrestled with these challenges.

But first, to further contextualise our interests in broadening the conversation, if another truth were told, these were spurred by our own research challenges. We had each failed, in a similar manner to St. Pierre (2014), to reconcile and integrate prevailing humanist qualitative research approaches with a burgeoning new materialist and posthuman literature. Our stumblings were further complicated in that neither stable notions of the environment, ecocentric perspectives *nor* discursive approaches seemed to make sense for us anymore. At the time of the call, we had already identified a broad movement within environmental and sustainability education (ESE) which appeared similarly engaged with considering the implications of a (re)turn to the material (e.g. Adsit-Morris 2017; Gough 2016; Malone 2016; Malone, Truong, and Gray 2017; Payne 2016; Tuck and McKenzie 2015; Van Poeck and Lysgaard 2016). These attempts, we felt, aligned with a rich history of environmental education scholarship that has engaged, and continues to contemplate, fundamental questions about being, knowing, and the axiological for research and practice development. In this regard, recent moves to engage new materialist theory with environmental education research/policy/practices were subsequently discussed with colleagues at a PhD summer school of the European Educational Research Association Network 30 on ESE research, organised by Elsa Lee at Homerton College, University of Cambridge in 2016. And if there was a singular moment of crytallisation, it was somewhere there, in between the lecture halls and libations, that the idea for creating a platform for further exploration, deliberation and debate via a special issue took first form.

After various negotiations and iterations, the call was published (a copy can be found at https://eerjournal.wordpress.com/tag/new-materialism/). But in this introduction to the collection, rather than rehearse that, we rather flag that in tracing the origins of this particular upwelling of interest in the ESE community, readers will detect strong echoes in the argumentation and citation patterns of the turn to new materialisms in the broader academy, primarily in relation to the cultural studies, humanities, and the social sciences. These concrete practices of scholarship might be regarded as evidence of inquiry more broadly becoming more closely aligned with long-lived, but sometimes buried, questions of fundamental significance to environmental education and the related fields of environmental philosophy and environmental ethics, and more recently, the environmental humanities. Indeed, one recurring argument for the prefix 'new' has been that we are in new times, politically, technologically, and environmentally (Coole and Frost 2010). Unpacking this, politically, many Westernised cultures have seen shifts to populism along with an increasing far-right presence that preys on fear, sorrow, and loss to force an agenda of isolationist propaganda. In terms of the Overton Window – the range of policies politically acceptable to the mainstream population at a given time – this particular 'window of discourse' has shifted so much in socio-political spheres that what had once seemed abhorrent (to many liberal progressives) has too often become the norm. In short, the orthodoxy of liberal democracy is threatened in this authoritarian age, and that includes its cultural institutions (Applebaum 2020; Gessen 2020), such as public education and publicly-funded research and inquiry. However, amid this shifting political milieu there is also an invigorated youth and antiracist movement for environmental and political justice, challenging both contemporary alt-right as well as institutionalised ways of being. Questioning what schools are really for, and being used for, in times of deepening climate emergency, has become a key challenge for society from these social movements and activists, as well as to ESE, iconized in the life and trials of Greta Thunberg. Inseparable from these developments are the decentralisation of 'news' and media consumption in the turn away from legacy media. As seen – or more to the point, not seen – in the activities and coverage given to the work and networks rallying around Extinction Rebellion, there is potential but also a cloaking nature of 'organic democratisation', providing opportunities to the grassroots whilst allowing the powerful to subvert new and alternative media platforms, 'faking' what is going on, be that on the ground, and in the airwaves, so to speak.

These intersecting themes make their presence felt and recombine within the flux of the material in nonconventional ways, as in 'advances' in human gene editing and events such as the birth of a three parented baby. Commentators note we have not only become cyborgs – where our presence is now stored in our phones and dispersed in vast data storage centres – we are also becoming outmoded and *re*moded. As an increase in robotization and artificial intelligence leads to 'breakthroughs' at the interface of the 'human' and the 'technological', so too will it lead to increasing income inequality as well as increases in xenophobia (Gamez-Djokic and Waytz 2020). These developments produce a contestation in the manner in which we understand 'the human condition'; including as either transhuman or posthuman: 'the former signifies an intensification of Enlightenment humanist thought, while by contrast the latter typically denotes normative distancing from the canons of violence and subjugation associated with the humanist project' (MacFarlane 2020, 2). These vastly divergent visions of what it is to be and become human can also produce political movements and shifts in the public imaginary: each has direct and divaricating implications for how the environmentally concerned might act, and how critical perspectives on these matters are formulated or prosecuted.

Equally, in what might generally be thought of as a more 'earthly' register, we continue to see seismic shifts that have dramatically altered ways of life which had once seemed stable – at least for the more privileged, who make-up only a minority of the world's human population. The least privileged, the majority, battle ever increasing temperatures, air pollution, rising waters, and depleted lands, and have been doing so for some time – or more specifically, increasingly do so since the onset of the most recent incarnations and centuries of industrialisation and globalisation. For the most privileged of the now, these changes might sometimes strike up close with the lived experience of extreme weather events, but also with an uncanny presence/distance as we see/hear rumours of a new 'record' reported in the news – instruments in Furnace Creek in California documenting a smothering 54.4 °C in August, 2020. The spectre and data of catastrophic environmental crises can seep into our lives indirectly, rendering us ontologically disoriented as we grapple with felt ethical imperatives whilst being imbricated in a globalised carbon dependent economy. Of course, worlds have always ended, and it is not necessarily the speed or extent of this change that marks it apart. It might instead be the fact that we are the ones witnessing this change, and this pain, now. Despite the wide effects, we only register changes as *crises* when they are ours to feel.

Pulling this deeper into the current moment by adding epidemiological and epistemological dimensions to the mix, the coronavirus pandemic alters thoughts and behaviours, whilst revealing and reinforcing certain economic, social, and political tensions of our times. Some argue the Coronavirus has 'piggybacked on racism' (Hattenstone 2020) and contrary to received wisdom, it does discriminate, because it becomes a part of us, inseparable from our cultural constructions.

Similarly, contemporary scientific insights reveal that the human biome depicts 'us' as more symbiotically alien than traditional biological or cultural narratives of the human would have us believe – as a singular, skin-bounded unit of flesh and bone. We, human and other-than-human, are a conglomeration of multiple species from mites, bacteria and fungal mycelium to water, minerals, electricity, chemicals, and viruses, but *also* material-discursive drives and performative concepts, furthering the material and posthuman perspective. Similar to viral contagions, micro and nano-plastics have now entered the troposphere, hydrosphere, lithosphere (plastiglomerate rocks) and biosphere (recently found in human organs, adding plastic – and oil – to the conglomerate human biome – the anthropo-glomerate), causing the potential for unmitigated environmental disasters as we find ourselves ever more dispersed in the Plastocene (Carrington 2020; Mcphie 2019; Preston 2017; Wright and Kelly 2017). These changes have collectively kick-started a worldwide depression – economic, medical, spatial, temporal, and (environ)mental – all new, all material.

This 'newness' to our times is signalled particularly, we feel, by the emergence and prominence of the term 'anthropocene'. It can be further witnessed in the ways this term creates novel jumping off points for (and even the purported 'death' of) environmental education (Le Grange 2019) due to its performative effects and the contemporary conditions it is meant to describe. In regard to wider debates, we might imagine the new materialisms as a sort of 'ecologising' of social science (see Murdoch 2001; Affifi et al. 2017), but an ecologising which, at its most complex/sticky/critical, wrestles deeply to extricate core binaries which remain alive and (un)well in the idea of the 'ecological'. In fact, we might also be drawn to ponder the potential for 'ecologis*ing*', where the weight is placed on *doing*, to demonstrate the always ongoing nature of deconstructing the world as understood as a correlate of human thought.

Wherever we find ourselves with such twists and turns though (including the possibility of profound disorientation), together, and with others, we wondered what work might be being done by those we knew, and those we would like to know, who might treat these topics as, in some way or other, matters that could be addressed by bringing environmental education into closer proximity, even co-mingling, with the new materialisms. To that end, our call for scholarly papers on these matters has been a success. But while a special issue can never be considered as a snapshot of the state of affairs of a topic, we were both pleased and surprised at the extent of the response to the call, both from established and early career scholars. So abundant was the response, that we have been able to accept two issues worth of papers for publication. And as such, we trust this collection affords a wider range of perspectives on new materialisms employed, and thus runs a fuller gamut of criticism, caution, application, and wholehearted endorsement, than initially anticipated.

We begin the special issue with this introduction to the collection. It offers a brief overview of new materialisms, with a specific focus on feminist new materialisms, before going on to discuss the ways in which some environmental education research has been moving with this 'turn'. We close this introduction with a reflection on various tensions, knots, and lines of flight which, whilst not derived directly from the papers that make up this special issue, we nonetheless see as potential themes and directions of travel for new materialisms and environmental education research. In our second paper, an editorial essay, we introduce the 17 papers that make up the special issue, outlining our sequencing of them in relation to themes we see emerging from their contributions.

New materialisms

The terms 'new materialism' and 'neomaterialism' were coined by both Rosi Braidotti and Manuel DeLanda independently of each other in the second half of the 1990s (Van der Tuin and Dolphijn 2010). However, a move (back) to materialist-process thinking in the social sciences can perhaps be traced to, for instance, Latour and Woolgar (1979), as well as having links to other process-oriented philosophies of the 20th century, i.e. the work of Alfred North Whitehead and Gregory Bateson. But we could go further, rooting out other influences and influencers of new materialisms, be that travelling back to Heraclitus, animistic Celts and Baruch Spinoza, the historical materialism of Friedrich Engels and Karl Marx, and certain (post)phenomenological perspectives in Western thought, or looking to more contemporary similarities from other cultures that might align more ontologically, epistemologically or axiologically with new materialist sensibilities, such as certain forms of Buddhism (Nagarjuna's or Ambedkar's) or contemporary animisms (Mcphie and Clarke 2015). In this sense and reckoning, new materialisms are not really new, in that many First Nations ontologies have been appropriated on the way to constructing 'the Material Turn', a point we will revisit presently.

The 'New Materialisms' has thus become something of an umbrella term used to encapsulate a more general turn in academic circles towards revisiting ontology in light of contemporary political and environmental events and their historical antecedents. Often referred to as 'the Material Turn', it has now spread, in tentacular and rhizomatic fashion, across many disciplines. These include, but are by no means limited to: children's geographies and education (e.g. Crinall 2019; Cutter-Mackenzie-Knowles, Malone, and Barratt Hacking 2020; Diaz-Diaz and Semenec 2020; Malone, Tesar, and Arndt 2020; Merewether 2019; Myers 2019; Rautio and Stenvall 2019), human geography (e.g. Anderson and Wylie 2009; Forman 2020; Whatmore 2006), drama and performance studies (e.g. Schneider 2015), philosophy (e.g. Latour 2005), medical practice (e.g. Mol 2002), archaeology (e.g. Witmore 2014), sport, exercise and health studies (e.g. Thorpe and Clark 2020), mental health (e.g. Duff 2014; Fox 2016; McLeod 2017; Mcphie 2019), critical studies of men and masculinities (e.g. Garlick 2019), quantum physics (e.g. Barad 2007), international relations (e.g. Lundborg and Vaughan-Williams 2015), politics (e.g. Bennett 2010; Braun and Whatmore 2010; Coole and Frost 2010; Latour 2004), research (e.g. Fox and Alldred 2015; St. Pierre 2015), and decolonial studies (e.g. Rosiek, Snyder, and Pratt 2020; Smith 2017). Thus, this special issue of *Environmental Education Research* also provides a space and place to acknowledge the potentially deep implications that new materialisms prompt for theory, research and practice in environmental education.

But what is a new materialism? Whilst encapsulating a broad move toward ontology and the materially real, there is neither a singular nor settled definition, or encapsulating philosophy. As with the brief sketch above, there is, of course, an enduring though varying focus on materiality in the history of Western thought, despite materiality sometimes appearing marginalised (Coole and Frost 2010). While it is not our intention to provide an overview of this here, for the purposes of this special issue, we detect that the main thrust of the new materialisms is an attempt to (re)turn to, renew, or create better philosophies of matter. Such philosophies are an attempt to offer a response to contemporary scientific revelations, the intractability of the social and environmental, and a perceived over-reliance on discourse and language in the so-called post-modern era of philosophy and social science. In short, 'new materialists' posit that, philosophically speaking, inquiry in the social sciences and cultural studies in the last several decades has paid too much attention to theorising and interrogating subjectivity, at the expense of similar but different considerations of matter, owing to a perceived inaccessibility of the material world:

> There is an apparent paradox in thinking about matter: as soon as we do, we seem to distance ourselves from it, and within the space that opens up a host of immaterial things seems to emerge: language, consciousness, subjectivity, agency, mind soul; also imagination, emotions, values, meaning, and so on. These have typically been presented as idealities fundamentally different from matter and valorized as superior to the baser desires of biological material or inertia of physical stuff. It is true that over the past three decades or so theorists have radicalized the way they understand subjectivity, discovering its efficacy in constructing even the most apparently natural phenomena while insisting upon its embeddedness in dense networks of power that outrun its control and constitute its wilfulness. Yet it is on subjectivity that

> their gaze has focused. Our motivation ... has been a conviction that it is now time to subject objectivity and material reality to a similarly radical reappraisal (Coole and Frost 2010, 2)

Whilst there are disparate strands in the new materialisms, affinities between philosophies and theories that have attempted to move beyond discursive or linguistic accounts have been identified. Connolly (2013) for instance, acknowledges that whilst escaping discourse seems difficult, we have never been able to truly escape matter, and thus new materialists offer various forms of contestable metaphysics. Their work is often characterised by an interest in the implications of ontological positions for first order concerns such as agency, time, subjectivity, objectivity, and knowledge. But it can be more or less than this. Specifically, new materialist theorists 'find it essential to bring such a cosmology into play in concrete explorations of ethics, state politics and global politics, exposing by contrast and comparison as we do so conventional cosmologies now tacitly in play in the human sciences' (Connolly 2013, 400).

The political potential of the new materialisms has been particularly attractive to feminist scholarship of the last few decades. First, the political effects of troubling pregiven ontologies is perhaps one of the fundamental characteristics of the various new materialisms. Ethical concerns both inform and spring from a general troubling of those concepts that are often taken, ontologically, as relatively stable in developing policy, theory and research approaches. New materialisms explore the possibility of post-dualistic conceptions of agency and body (and/or mind). They can be deployed in the ongoing questioning of axiomatic distinctions between what is 'natural' and what is human or human derived, and thus trouble common conceptions of life as deriving solely in the organic, as in Jane Bennett's *Vibrant Materialism* and materially informed contemporary animism (Harvey 2013).

Together, these attempts can be read and used to question the nature of the political, and the location of ethics and culture. More concretely, troubling established dualisms by way of a new materialism might be spearheaded by attempts to articulate forms of protean monism (Connolly 2013). For instance, in reference to the nature culture dualism, Van der Tuin and Dolphijn (2010, np) offer that:

> New materialism is a cultural theory that does not privilege culture, but focuses on what Haraway would call 'naturecultures'. It explores a monist perspective of the human being, disposed of the dualisms that have dominated the humanities until today, by giving special attention to matter, as it has been so much neglected by dualist thought. New materialism, a cultural theory inspired by the thoughts of Deleuze, that spurs a renewed interest in philosophers such as Spinoza and Leibniz, shows how cultured humans are always already in nature, and how nature is necessarily cultured, how the mind is always already material, and how matter is necessarily something of the mind. New materialism opposes the transcendental and humanist (dualist) traditions that are haunting a cultural theory that is standing on the brink of both the modern and the post-postmodern era.

As an example of the ties to many First Nations ontologies, in keeping with the promise of respectful dialogue and acknowledgement, Standing Rock Sioux author Vine Deloria (1999) emphasised 'Indians do not talk about nature as some kind of concept or something "out there." They talk about the immediate environment in which they live' (p. 233). This relational, non-dualistic perspective has been, and is still, prominent in animistic thought, and has seeped into the material turn via a host of prominent scholars (and environmentalists), influential to new materialist writing (e.g. Spinoza, Bateson, Deleuze, Haraway, Barad, Bennett, Latour) and who have held on to its ethico-onto-epistemological potentials (see Mcphie and Clarke 2015).

However, new materialisms refuse to be one thing – yet! (although it may seem they are rapidly becoming so?). For instance, speculative realism (often in the form of object oriented ontology, or OOO), feminist new materialisms, new empiricisms, the ontological turn, PhEmaterialism, as well as an emerging material ecocriticism are prominent labels that have arisen, amongst others, to engage within/beyond the spirit of new materialisms and otherwise, with some of the binaries highlighted above. Each of these trajectory-tributaries have been discussed widely, and it is not our intention to explain the similarities and differences in this editorial, although you can find evidence of doses of each of these concepts throughout this special issue. However, we feel the need to further introduce the prefix 'feminist' as it performs some-'thing' effectively (and affectively) in an equitable direction of travel.

Feminist new materialisms

Feminist new materialisms focus on 'how the forces of matter and the processes of organic [and in-organic] life contribute to the play of power or provide elements or modes of resistance to it' (Frost 2011, 70). In this way feminist new materialisms are positioned as fundamentally political. Material feminists acknowledge how feminist theory and practice have been significantly enriched by the exploration of power, discourse and performativity of postmodern feminism. However, acknowledging the work of, for instance, Donna Haraway and Clare Colebrook, material feminists argue for a return to matter so as to allow exploration of the effects and affects of bodies and the myriad material conditions of power:

> The retreat from materiality has had serious consequences for feminist theory and practice. Defining materiality, the body, and nature as products of discourse has skewed a tremendous outpouring of scholarship on 'the body' in the last 20 years, nearly all of the work in this area has been confined to analysis of discourse about the body (Alaimo and Hekman 2008, p.3).

Further, and significantly, the environment, non-human, more-than-human, or other-than-human are each positioned as upfront concerns in feminist new materialisms. This concern arises both as a result of the challenging environmental conditions of the 21st century and in response to the need to critically address 'nature' from feminist positions. Yet despite the critical work of Carol Merchant, Val Plumwood and others, the concern here is that mainstream feminist theory:

> relegated ecofeminism to the backwoods, fearing that any alliance between feminism and environmentalism could only be founded upon a naïve, romantic account of reality ... The problem with this approach, however, is that the more feminist theories distance themselves from 'nature', the more that very 'nature' is implicitly or explicitly reconfirmed as the treacherous quicksand of misogyny. Clearly, feminists who are also environmentalists cannot be content with theories that replicate the very nature/culture dualism that has been so injurious—not only to nonhuman nature but to various women, Third World peoples, indigenous peoples, people of color, and other marked groups. Rather than perpetuate the nature/culture dualism, which imagines nature to be the inert ground for the exploits of Man, we must reconceptualize nature itself. Nature can no longer be imagined as a pliable resource for industrial production or social construction. Nature is agentic—it acts, and those actions have consequences for both the human and nonhuman world (Alaimo and Hekman 2008, 4–5).

Feminist new materialisms, then, reassert the natural as more than 'Nature' but must also acknowledge the ecofeminist histories with which they might align. Within this special issue, Gough and Whitehouse (2020) offer a deeper exploration of this point. In this way feminist new materialists including but by no means limited to the work of Karen Barad, Donna Haraway and Rosi Braidotti, have drawn on poststructuralist theory, posthuman theory and developments in the biological and quantum sciences, to be at the forefront of critical discussion surrounding the nature of nature/culture debates and the return to ontological matters. Haraway has revolutionised feminist politics over decades with her various fabulated offerings of, for instance, the feminist cyborg, the modest witness, onco mouse, and situated knowledges. Within feminist new materialisms there is also/often a particular taking up of the combined work of 20th century continental philosophers, Gilles Deleuze and Felix Guattari, who over the course of 30 years of collaborative and solo work, developed a politically critical metaphysics of immanence to contrast with more prevailing transcendent metaphysics. Transcendent here implies the essential existence of categories or forms which transcend the mattered universe, whereas immanence implies no outside, essential essence or referent for being. Deleuze, for example, argues that transcendence is in immanence rather than in opposition to it. We would concur, as any 'thought' about transcendence is derived from the material, physical and so 'of' this world. Therefore, any transcendent and/or binary conceptions are firstly and always immanent. Transcendences exist, but only ever as a matter of immanence.

Within feminist new materialisms, as with Deleuze and Guattari, another key feature we might detect is the ways in which a posthuman understanding is proposed and propelled: knowledge/world production is neither anti-biological nor posited on biology as the master plan, dismissive of other philosophical perspectives (Ahmed 2008; Hinton and van der Tuin 2014; Sullivan 2012). Rather, Taylor and Ivinson (2013) highlight how ''new' material feminisms displace the human as the principal ground for knowledge [...] and accepts that matter is alive' (p. 666). Here 'priority [is] given to difference, entanglement and undecidability' as it confronts 'the distance, separation and categorical assurance that shores up the self-mastery of the oedipal (male) subject of humanism' (Maclure 2015, 5).

In her book, *Posthuman Knowledge*, Rosi Braidotti (2019) posits that 'the field of new materialism is especially strong in education in general and in feminist educational practices in particular' where 'the growth and high quality of this research field is so intense as to deserve a study of its own' (2019, p.141). Indeed, the contemporary blossoming of attempts to prioritise the differences, entanglements and undecidability in *education,* Maclure (2015) attributes to feminist new materialisms. Other instances are documented in, *Feminist Posthumanisms, New Materialisms and Education,* edited by Ringrose, Warfield, and Zarabadi (2020). The editors highlight how contributions to their collection represent a particular inflection: of the 'PhEmaterialism' movement, 'where the *Ph* refers to posthuman, the *Fem* from feminisms, and materialism from the new materialist movement' (p.1, original emphasis). Such scholarship brings together 'previously published work that [*Ringrose, Warfield and Zarabadi*] feel moves the field of educational research somewhere different than before' (p.1). The *feminist* prefix to new materialisms is deployed here to encourage the material turn to perform *differently.* To borrow from Haraway, it is perhaps this difference that makes the difference, a difference that may be needed more acutely at this ecological precipice.

We also recognise that a healthy bibliography of feminist environmental education research continues to grow at a steady pace, challenging the normative dominance of gendered environmental binaries (for more on this, see the ever-expanding reference list on the *Environmental Education Intersectional Feminist Caucus* Facebook page, compiled by Connie Russell). And yet, beyond a series of special issues (for example, in *The Journal of Environmental Education*) this literature appears to have had little purchase within broader environmental education research discourses to date. Nonetheless, we see this literature as already ahead of the curve in considering any turn to the material within the social sciences (see, for instance, Gough and Whitehouse 2020). Indeed, while this special issue is not specifically advocating feminist new materialisms, it most certainly embraces them and highlights their potential as one of many pedagogical possibilities that new materialisms can help explore and develop as an ethically orientated imperative within environmental education and its research.

Considering the need, then, for more equitably positioned epistemological engagements with human-environment assemblages, 'environmental' education and research is perhaps uniquely positioned to engage various playfully serious intentions and axiological possibilities of the material turn. In fact, this work is already well underway, as we now discuss.

New materialisms and environmental education

Developing and redeveloping materialist theories have already been identified as an emerging and important 'route' for environmental education (e.g. Howles, Reader, and Hodson 2018). In *Environmental Education Research* for instance, Katrien Van Poeck, and Jonas Lysgaard consider the potentials for materially-informed theories to explore policy on environmental and sustainability education and their capacity to

> challenge certain ideas and perspectives in philosophy of education, (social) constructivism and critical theory by revealing the need to move beyond an exclusive focus on discourse and language and to seriously take the materiality of context of sustainability debates into account (Van Poeck and Lysgaard 2016, p.313, emphasis in original).

Van Poeck and Lysgaard (2016, p.314) further articulate how, amongst other approaches, claims of new materialists to operate beyond the strictly discursive may 'offer relevant and inspiring ideas, concepts, frameworks and findings to ESE policy research as well as the broader field of educational research'. Concurrently the new materialisms have been characterised as a new movement in thought for outdoor environmental education research (Gough 2016) as well as a theoretical area that might hold potential for interrogating various 'absences and silences' within environmental education research (Payne 2016). While we also note there has been a genealogical thread of engagement with poststructural and/or materialist thought in outdoor and environmental education from the preceding decades (e.g. Barrett 2005; Bell and Russell 2000; Gough 1999; Gough and Price 2004; Hart 2005; McKenzie 2005; Russell 2005; Stables 2007; Payne 2005; Payne 2019), as well as work that has drawn on the materialist philosophy of Deleuze and Guattari within environmental education specifically (e.g. Cole 2019; Gough and Sellers 2004; Stewart 2008), that offer conversation partners for locating and historicising any contemporary 'new' turn to materiality in environmental education.

Examples of recent work in environmental education that reference authors we perceive as aligning themselves with new materialist theory include Pauliina Rautio (2014) and Malone's (2016) uptake of the agential realism of feminist new materialist Karen Barad in considering children's geographies, Lysgaard and Fjeldsted's (2015) examination of speculative realism and nature education 'between discourse and matter', and our own use of broad feminist new materialist, Deleuzo-Guatarrian and animist literature to consider environmental education as materially immanent and 'becoming' (Clarke and Mcphie 2014, 2016; Mcphie and Clarke 2015). Examples of creative engagements include Fawcett's (2009) *Feral Sociality and (Un)Natural Histories* in which she thinks environmental education, salmon, and, among other things, earthworms rhizomatically to consider the (nomadic) ethical practice of environmental education scholarship. While Adsit-Morris' (2017) book *Restorying Environmental Education*, in which she thinks with the feminist new materialisms of Rosi Braidotti, Donna Haraway, and Karen Barad, is yet another example of new theory being put to work to reconfigure more contemporary approaches to environmental education.

Drawing further on Haraway, there are scholars articulating 'common worlds' approaches, particularly in research on education in the early years (e.g. Nxumalo and Pacini-Ketchabaw 2017). Within this journal, for instance, we see Duhn and Galvez's (2020) creative exploration of Haraway's notion of tentacularity to speculate on how children animate their relationality as an example of these. Further, within *Environmental Education Research*'s special issue on 'Troubling the intersections of urban/nature/childhood in environmental education', edited by Duhn, Malone, and Tesar (2017), several articles explore concepts from common worlds ideas as well as new materialisms.

We also see commonalities being articulated between new materialist theory and Indigenous and decolonial approaches. As in the Land Education special issue of this journal, these have tended to focus on cultural perspectives, ontology and critiques of dominant ways of knowing and being, whilst acknowledging incommensurabilities which require attention and which we return to in our discussion at the end of this article (Tuck, McKenzie, and McCoy 2014; Tuck and McKenzie 2015). Some recent attempts to integrate these perspectives within environmental and place education include work by Chris Beeman and Laura Sims (2020) and Elmarie Kotzé, Kathie Crocket and Cheri Waititi (2016). Furthering this attempt, Debbie Sonu and Nathan Snaza's (2015, p.258) interrogation of ecological pedagogy with new materialisms suggests a need to engage in creative practice to 'disrupt theories of vertical domination and conquest' which may appear within environmental education practice-research.

Additionally, a series of environmental education authors have acknowledged the related field of non-representation theory (see Thrift 2008), and a movement in anthropology to move to sociomaterial theory beyond nature/culture duality via a focus on ontology (spearheaded by, for example, Tim Ingold 2000, 2007, 2011, 2013, 2015) as significant for environmental education research (e.g. Carolan 2007; Gannon 2017; Le Grange 2018; Lynch and Mannion 2016; Rautio 2013; Rooney 2016; Ross and Mannion 2012).

Lastly, we are aware of the diverse terminology in play in turns towards materiality and the particular move within educational discourse towards 'sociomateriality'. Sociomateriality has been referenced in research works with a particular focus on the conception of place-based environmental education (e.g. Mannion, Fenwick, and Lynch 2013; Nxumalo 2016). While Marcia McKenzie and Andrew Bieler, for instance, offer a series of examples of educational place-oriented projects interpreted through a sociomaterial lens, drawing from many authors aligned with the new materialisms to argue that 'categorizing some concerns as 'social justice' or 'critical' issues and others as 'environmental' becomes increasingly untenable' (McKenzie and Bieler 2016, p.5).

There are, no doubt, important omissions in this brief survey and we acknowledge that some may perceive the new materialisms in much broader, or even narrower, terms than we do here. Nonetheless, we see this special issue as appealing to diverse approaches and terminology within the move to materiality in social science and are excited by the opportunities this diversity presents. Indeed, as we introduce the 17 papers that make up the special issue in the following article, we can see this diversity at work. We underscore too, that the contents of this special issue arise from an *open* call, and so the submissions do not necessarily respond to specific gaps or questions of the convergence of new materialisms of environmental education. Below we sketch out four themes that we think speak to this convergence and provide, perhaps, starting places for those concerned with furthering scholarship and practice development in these areas.

Tensions, knots, and lines of flight in new materialisms and environmental education: themes and directions of travel across and beyond the collection

The themes described below are not drawn directly from the articles in the special issue, but are rather areas we see as being of concern for environmental educators and researchers interested in the new materialisms. We see each theme as offering problems and directions for travel. These themes are: (1) new empiricism and post-qualitative inquiry; (2) the meeting of politics, ethics, and decolonial theory with new materialisms; (3) conceptions of nature, environment, sustainability and the human subject; and (4) new materialisms as environmental pedagogy. Below we lay out these problems, highlighting how some, including through reference to articles contained in the special issue, have responded to them.

New empiricism and the post-qualitative

There is a particular move in the social sciences to consider the new materialisms as, at the very least, informing the manner in which we (re)think methodology. More forcefully there is an emerging movement that seeks to take new materialisms and the material-thinking they result in as a serious contestation to prevailing forms of qualitative methodology. Within the special issue, examples of these include Ruck and Mannion (2019), Jukes and Reeves (2019), and Crinall and Somerville (2019). Works that advocate this 'post-method' thinking include Law's (2004) *After Method*, Coleman and Ringrose (2013) *Deleuze and Research Methodologies*, Jackson and Mazzei (2011) *Thinking with Theory*, Snaza and Weaver's (2015) *Posthumanism and Educational Research*, Taylor and Hughes' (2016) *Posthuman Research Practices in Education* and various special issues in the journals *Qualitative Inquiry* and *Cultural Studies ↔ Critical Methodologies*, which explore what Patti Lather and Elizabeth St. Pierre have referred to as the new empiricisms and/or post-qualitative research. Thomas and Bellingham's (2020) edited collection, *Post-Qualitative Research and Innovative Methodologies* is another recent addition to this post-method movement.

Within the field of environmental education, edited collections interested in new theory, troubling/creating research methods concerned with materiality, place, the Anthropocene and environmental education practice have begun to appear (e.g. Somerville, Davies, Power, Gannon and de Cartet, 2011; Tuck and McKenzie; 2015; Reinertsen 2016; Malone, Suon and Gray, 2017; Cole and Malone 2019). Much of this work describes authors experimenting with or revisiting their own research, and we recognise in Noel Gough's articulation that 'it is much too soon to be making definitive, conclusive or prescriptive recommendations for practice' (Gough 2016, 13), a reflection that applies to post-qualitative inquiry generally. Indeed, prescription is exactly what not to do, as the approach is about creating the new, which can never be prescribed (Massumi 2010). However, conversations about the significance of new materialisms within environmental education research are well underway (e.g. Hart et al. 2018; Mcphie and Clarke 2019).

What, then, does research become for environmental education with a new materialism? New materialisms are informing environmental education research through approaches such as thinking with concepts (e.g. Hart 2017), post autoethnography or immanent life writing (Clarke 2019), cartographic and diffractive storytelling (Riley 2019), engaging with the messy entanglements of theory/practice (Pleasants and Stewart 2019) as well as through other myriad examples within and outside this special issue. We are interested in seeing work aligned with environmental education but which take up the research issues appearing in these broad debates including the troubling of the human 'subject', 'data' and 'analysis' as well understood features of social research and praxis. Within this special issue we see authors take up these issues and provide further methodological innovation and provocation for environmental education research and practice development.

Politics, ethics and decolonial theory

New materialist scholars often situate their work as in some way presenting directions forward for understanding and addressing broad environmental and social problems. How ethics is conceived, whether 'environmental' or otherwise, within new materialism is, however, not well explored in relation to the ethics of environmental education practice and research, nor necessarily the direction of movement implied, Cartesianally or otherwise framed. There are some meditations on modes and energies of directionality articulated within the special issue. Whilst Rousell (2018) makes the most obvious contribution to a discussion of ethics with his immanentist perspective, the ethical and political is alive within each article. Verlie and CCR15 (2018) further contribute with their application of intrasectionality, acting-with, and diffraction, which attune our attention to the constant flux and rearrangement of boundaries and hierarchies in attempts at posthuman practices of pedagogy and research. Even with the most critical offering, Affifi (2019) points to the importance of considering what our concepts do with the world, and how they come to matter. Experiments in the direction of concepts matterings are most forcefully endorsed by Gough and Adsit-Morris (2019), and Mcphie and Clarke (2018). However, there are many ethical issues still at play within the convergence of new materialism and environmental education, and cautionary voices have already begun to emerge. Where new materialist approaches have been taken up methodologically, Tuck and McKenzie (2015, 17) note:

> The increasing influence of Indigenous and decolonizing scholarship, spatial theories, and new materialism on theories, methodologies, and methods of social science cannot be disputed. However, although one might suppose that such innovations and recalibrations might prompt a more robust discussion of place in social sciences, this is not often the case. In many cases, flattened ontological or materialist frameworks de-emphasize the agency of people and politics in attempting to better attend the interconnected 'networks' or 'mangles' of practice in researching social life.

Tuck and McKenzie (2015) point instead to Indigenous methodologies as seldom failing to account for issues of people, land, and place. Indeed, the vision of an ontological landscape where there is no distinction between one thing and the next, no fulcrum upon which to gain purchase in either tackling the problems of the world or, indeed, identifying what those problems are, will be of great concern to those who have worked so hard for decolonial social and environmental justice – and education. We see some reference to Indigenous people within the special issue, and yet we know more needs to be done to decolonise new materialisms.

Hinton, Mehrabi, and Barla (2015, p.4) are particularly interested in this issue drawing as they do on Karen Barad's agential realist materialism to ask 'how might the concerns of postcolonial politics meet with a posthumanist emphasis on the non-human, and what might it mean to undertake postcolonial inquiry that takes the non-human as its point of departure?'. Further to this, Zoe Todd (2016) urges that Indigenous authors be cited more prominently and generously given their articulation of millenia old cosmologies which the new materialisms resonate with. For example, in academia there is an ongoing danger of misappropriating and anthropologising under-represented voices – speaking from a more privileged perspective *in place* of others – rather than listening more carefully to what those voices might be saying, and why. Perspectives which do just this have been forwarded in this journal, in a special issue on *Land education: Indigenous, postcolonial, and decolonizing perspectives on place and environmental education research*, edited by Eve Tuck, Marcia McKenzie & Kate McCoy (2014).

The recognition that academic endeavours that do not acknowledge and actively resist settler colonialism are implicitly invested in settler futurities (Tuck and Gaztambide-Fernández 2013) must be taken seriously by new materialisms. This, combined with the issue that new materialisms itself may be seen as a majority white academic field located within privileged institutions (Hinton, Mehrabi, and Barla 2015) produce a cocktail of ethical issues around place/race/materialist theorising and praxis. For instance, a question of the basis of knowledge/matter generation must be considered both in terms of whether a conflict exists between prioritising new materialist/decolonial onto-epistemologies but also in terms of a critical understanding of how knowledge generated from within each onto-epistemology matters the world in 'ecological', or colonial terms. Hinton, Mehrabi, and Barla (2015, 13) demonstrate the potential iterative movement that may be required in this critical discussion:

> we would ask whether it is possible to both acknowledge the importance of these critiques of new materialism, to say 'yes' to their concerns, while also performing that doubled gesture of asking how new materialist onto-epistemologies might perform such erasures or elisions of race.

There are many seeming overlaps, such as distributed agencies among other-than-human things. Rosiek, Snyder, and Pratt (2020) suggest that there has been 'relatively little articulation' between the two as yet, with perhaps the most voiceful criticism coming from Zoe Todd (2016) who suggests ontology is just another word for colonialism, or perhaps Eve Tuck's (2014) aptly named paper, *A turn to where we already were*? These tensions must be acknowledged and the appropriate respect must be articulated within the theorising process itself, whilst the misappropriations must be identified and torn down, like the oppressive colonialist statues of historicised endeavours. This is a difficult task for many people due to the implicit nature of colonial bias. How can we know what we don't know? To a certain degree, in academia at least, reflection and reflexivity were supposed to provide a suitable empathetic tool to counter this trend but new materialist scholars have found that these (often) self-referential concepts only produce 'more of the same' and so have had to trial new concepts with which to think with, such as 'diffraction' (see Haraway 1997; and Barad 2007), a term that you will see put to use in many new materialist inquiries, including in this special issue. So far, we think, diffraction seems to perform a little more equitably than reflections and reflexivities. Perhaps this is partially because it was designed for this specific purpose in the attempt to make life more tolerable, fair and healthy for those – humans and other-than-humans – with little power or privilege; the subaltern, or perhaps more appropriately, the minoritarian. Yet, many voices are still largely omitted from, or ignored within, academia, even though they may be written 'about', as if anthropologizing the Indigenous voice will somehow make amends.

What then, are the pedagogical ethics implicated in the new materialisms? Davies (2016, p.9) considers the genealogy of ethics in the new materialisms through Foucault, Butler, and Deleuze, and guides the reader through understanding how ethics is reconfigured in poststructuralism and specifically the new materialist thought of Karen Barad:

> It is worth noting ... that in none of these ways of making sense of the world, of our place in it, and our ethical obligations, whether that be poststructuralist or post-humanist new materialism, is ethics a matter of separate individuals following a set of rules. Ethical practice requires thinking beyond the already known, being open in the moment of the encounter, pausing at thresholds and crossing over (De Schauwer & Davies, 2015; Wyatt, 2014). Ethical practice is emergent in encounters with others, in emergent listening with others, including non-human others (Davies, 2014b). It is a matter of questioning what is being made to matter and how that mattering affects what it is possible to do and to think. Ethics is emergent in the intra-active encounters in which knowing, being and doing (epistemology, ontology and ethics) are inextricably entangled (Barad 2007). (Davies 2016, p.9)

This form of immanent ethics is not so much a form of guiding morality as an intra-relational knowing-becoming-doing (the use of hyphens here is to demonstrate the indissolubility of these acts). Within immanent materiality we are all already always in this mode, but our material conception of matter/s influence the manner in which we participate with the participation. The implication of this for environmental ethics and our modes of practicing research and pedagogy remain, perhaps perpetually, unclear. At the very least the forms of both humanist and environmental ethics that have developed in the postmodern era are not straightforwardly transposed to new materialist thinking (Johns-Putra 2013). However, similar to the debate that raged over whether an anthropocentric or ecocentric ethic mattered as long as we got the job done, concerns over decolonialism, epistemology, accessibility, anthropocentricity, semantics, transcendent ethics and agential distribution, and so forth, need a response from the new materialisms to clarify issues of its performance (Todd 2016). In a sense we see this as an area in which 'critical' environmental education may attempt to 'give back', and inform discourse with developing new materialist approaches which seek ethical practices.

'Nature', 'sustainability', 'the environment' and 'the subject undone'

Along with questions raised by the meta-ethics of new materialisms come a number of concerns around 'sustainability' or 'environmental care'. Within the special issue, Affifi (2019) infers the question: *What is to be sustained if the nature/culture boundary is blurred to an extent where we appreciate the vibrant materiality of the Great Pacific Garbage Patch as much as, say, a coral reef?* Varying forms of this question are appearing in the literature and require an urgent response. For our part, we wonder if this kind of flattening might retain the notion of ethics as something outside of the world, rather than immanent to/of it. Immanent materialism includes ethics, which is lived and practiced in existence, rather than expunged from a flattened world. Our moral concerns are present, whether we like it or not. We *feel* when they are transgressed by thinking. Indeed, this process is much more like *thinking-feeling* in which the body 'expands its repertory of dynamic postures by mixing, matching and alloying them, explores its own living potential, strikes new postures – invents new ways of affording itself of the world, in collaboration with the world, with what the world throws before it' (Massumi 2008, p.11). Indeed, Affifi (2019) articulates this very point, though through different language, and we agree that situatedness must be a core concern for environmental education researchers.

Traditionally the 'flip-side' to 'the environment' has been the human, the subject, or/and the individual who in some way is *the* source of agency that environmental education aims to reach. The picture no longer holds with a new materialist lens, where 'no primacy of the human actor

– individual or collective – over the nonhuman actor can be accepted on a priori grounds' (Knappett and Malafouris 2008, xi). Latour (1999) suggested agency 'resides in the blind spot in which society and matter exchange properties' (p. 190), yet society is itself matter-ed, and so agency can be thought of as co-constructed and distributed in time and space, regardless of the human actor – as the human actor is also a co-constructed multiplicity of other so-called actants – bacteria, virus, fungal mycelium, water, mites, plastic, and concepts (see Mcphie 2018). This point is made particularly well within this special issue by Tammi (2019) with their exploration of how 'mold-schools' emerge as sites of interaction between practices and beings both human and other-than-human.

As Knappett and Malafouris (2008, xi–xii) recognised, 'agency need not be coterminous with intentionality, which releases nonhumans into the process of agency'. Even so, intentionality in humans is still contested (see Libet 2006), especially if we exorcise the Cartesian ghost from such shells. Hofverberg (2019) demonstrates how intentionality may conceptually spill into and out of re-making practices with her paper in this special issue. Thus, perhaps agency might be(come) thought of as 'a dance' (Pickering 2010), a 'dance of animacy' (Ingold 2013) or just 'living lines of environing' (Mcphie and Clarke 2018). Acknowledging a new materialist theoretical position, St. Pierre (2004) suggests educational research needs to consider the potential for a 'subject undone', but it appears the repercussions of this thinking for environmental education remain under explored.

The immanent ethics of the material turn described by Davies (2016) and Rousell (2019), imply a letting go of both the 'subject' and the 'environment', when used in a categorical sense. We perceive the turn to materiality as in no way, or at least not necessarily or straightforwardly, a turn away from thinking 'environmental' problems, but rather an(other) understanding. As Donna Haraway (2014), referring to Marilyn Strathern, states; 'it matters what stories tell stories. It matters what thoughts think thoughts. It matters what worlds world worlds'. Haraway (unpaginated) continues:

> we need to take seriously the acquisition of that kind of skill, emotional, intellectual, material skill, to destabilise our own stories to retell them with other stories and vice versa. A kind of serious de-normalisation of that which is normally held still in order to do that which one thinks one is doing. It matters to destabilise worlds of thinking with other worlds of thinking. It matters to be less parochial, if ever there was a time to need to be worldly it is surely now.

In other words, within the special issue, we can see some of the ways in which both the 'subject' as learners and educators, as well as the 'environment' or 'nature', which may have been held too still within environmental education practice and research, might be thought differently, and how the various ethico-onto-epistemological (Barad 2007) implications of the various new materialisms might *matter* in our work. However, doing this work is by no means straightforward.

When authors – many White, many globally northern, many middle-class – are applying new materialisms to environmental education, what exactly are they applying it *to*? Perhaps we could work our way back from current perceptions of romanticised White landscapes to a particular zeitgeist where the foundations of a privileged epistemological accessibility to so-called green scenery was formed in Europe. There are many references to how environmental framing first began to alter how privileged White Westerners optically distanced themselves from so-called scenic views of romanticised 'natural' landscapes (Antrop 2013; Ellison 2013; Ingold 2000; Schama 2004; Thompson 2010), or the eco-fascistic origins of the modern environmental movement (Bookchin 1987; Forchtner 2019; Smyth 2019; Staudenmaier 2011). But there is not much on how this might manifest its White supremacist ideologies or epistemological hierarchies 'implicitly' in the conceptual apparatus of contemporary environmental scholarship, including on environmental education. In keeping with the theme of this special issue, we'd like to further articulate how a particular ontology and therefore conception of 'the environment' might have huge ramifications for any new turn attempting to get past the barrier of privileged implicit Whiteness.

For example, along the inquirous path to this introduction we (White ... Westernised ... Male ...) wondered at the lingering impression in the burgeoning literature – across publications – that the concerns of environmental education should be for one certain group of things/phenomena rather than another group of things/phenomena. In some attempts to link environmental education and new materialist thinking, common things that are often exampled as the object of concern remain distinctly 'natural', aligning themselves on one side of a divide they purport to circumvent and undermine (e.g. flora, fauna, water, outdoor spaces, etc). We hardly ever noticed examples of human-made phenomena being conceived as the thing to sustain or educate for/about. Certainly no thoughts or concepts (unless of course, we include the particular manmade concept of green/White 'nature'). It seemed to us that those who wished to embrace new materialisms were trying really hard not to romanticise or bifurcate the world, but often fell back into the Cartesian trap by focusing purely on what they have been conditioned to think of as 'natural'.

Indeed, we believe this is the most troublesome issue in thinking together environmental education and new materialisms, a potential barrier that prevents contemporary conceptual environmental rebellions from achieving their goals. If other perceptions of environments are omitted from thought, or other ontologies relegated to the sidelines of history by assuming what you are saying is 'the' truth as opposed to 'a' truth amongst others competing/converging/coalescing/diffracting... what potential futures disappear? We believe this White landscape ontology is implicitly institutionalised in much environmental scholarship by the very linguistic assumptions utilised. Even in employing a purportedly more equitable ontological turn, it seems almost too difficult for many scholars to weed out the implicit bias that has rooted itself into the very fabric of our lexicon and thought. However, time is of the essence – what, with the rise of the Far Right, the 6th Mass Extinction looming, along with climate catastrophe, pandemics, and the ethically obscure post/transhuman future we find ourselves in – perhaps we, at least some of us, need to explore every avenue, and push these boundaries as far as they will stretch, by working harder to both imagine and understand how they never really existed in the first place?

New materialisms as 'environmental' pedagogy

Lastly there are a range of approaches that advocate the taking up of new materialist positions as potentially generative for educational practice. We recognise these follow a lineage of feminist, embodied, and situated approaches in broader educational practices as well as environmental education (see, e.g. Warren 1990; Lather and Ellsworth 1996; Bell and Russell 2000; Gough 2004; Le Grange 2004; Fawcett 2009). From outwith environmental education, Hickey-Moody and Page's (2016, 12) edited collection offers perspectives of pedagogies of matter for political resistance in the arts, stating that:

> bodies and the process of embodiment are core to our ways of knowing-being. However, they are also fundamental to the entanglement of *matter and learning and teaching (pedagogy)*. This embodied entanglement of matter and teaching as pedagogy – the moments when materials and spaces impact on bodies and bodies impact ideas – is our primary interest... through bodies and with matter, we are always making, performing and learning. Therefore we posit that new materialist pedagogy is embodied and is intra-action between bodies and matter.

This vision articulates an 'environmental education' in the broadest sense. Within the special issue we can find pedagogical experiments offered by Jukes and Reeves (2019), Mannion (2019), and Crinall and Somerville (2019) too. However, in noting these examples we wish to stay aware of the potential of reifying a practice/theory divide, and instead want to remain open to the ways in which our readings, as bodily practices, seep through and confront our other bodily practices, and vice versa. Within this broad view, environmental educations are occurring constantly in the process of living.

New materialisms are also helping scholars reconsider pedagogy in ways that might resonate with environmental educators, particularly those accustomed to thinking with place and embodiment in their practices. And yet there is something decidedly (re)new(ed) about the scope of pedagogic discussions sweeping the social sciences. Pedagogy, irrespective of the environment as the originating concern, is being described as 'thingy' and slimy (Sojot 2019) and as a process of placemaking, where the 'who' is entangled with the 'where' (Page 2020). Geerts (2019, 128) suggests that new materialist informed ideas can infer a 'pedagogy that is centred on critique and creativity, situatedness, geopolitical (self-)awareness, accountability, and an immanent ethical attitude that takes current-day political constellations and complications into account'. Again, these descriptions appear outwith environmental education discourses. In these ways, new materialisms impel a pedagogy that is inherently critical, cautious, and environmental in the broadest sense of the term. The political potential of new materialist informed pedagogy is sketched particularly effectively by Juelskjaer (2020, 66) who concludes that:

> In moving forward with feminist new materialist analyses of pedagogy … it would be fruitful to think about natureculture pedagogies of the sensorium and to begin to develop frameworks through which these pedagogies may be researched. Policy agendas that facilitate such research would also be welcomed, with a view to reversing climate change and building sustainable educational futures.

Irrespective of the warmth (or/and lack thereof) with which the new materialisms might be met by pedagogues and pedagogical scholars within environmental education, there is a sense in which the material, placeful, relational, situated, crisis aware, and ultimately environmental are no longer reserved for a small clique of scholars on the pages of this and related journals (as if they ever were). Rather, we wonder how might they come centre stage in the present and developing theoretical milieu of the social sciences and humanities and, indeed, education and pedagogy? There are questions about the extent to which this broadening engages with past philosophical debates within environmental ethics, philosophy and education. Nonetheless, environmental education, we would contend, has already become something quite other than it once was through this broadening. At the very least, new materialisms renew questions of what is actually going on when we are interacting with learners in the hope of expanding what is possible for 'practice'.

Present thoughts

Across new materialist discourses there is a recognition that key themes of the material, post-dualistic accounts, distributed agencies, and relational ontologies have been visited before, but also that this (re)turn may offer something new to our current and pressing predicaments in education. For instance, in their edited collection *Pedagogical Matters*, Snaza et al. (2016, xxix) argue:

> new materialisms gather together a range of ideas that have been taken up before, both in and outside of curriculum studies, but we think this gathering offers us a crucial chance to look at our practices as educators again, differently, more closely perhaps, as gesture toward the emergence of new political action.

Within the special issue we see example after example of how this new political action may be conceived in the present.

The present. Pandemic: lockdown/exposure; mass extinction: extinction rebellion; climate catastrophe: the youth climate movement; scandals of patriarchy: #MeToo; the rise of hate (again): moments of empathy and rolling Black Lives Matter protests. Where and what is the remit of environmental education and research in such instances of the present? Should it, *can* it even, be severed from broader renderings of reality that feed and filter our daily lives, which themselves dictate what is environmental, what is *natural*, and what is not? Might these framings be altered so as to increase the plurality with which we can approach and understand 'the problem'? Besides, are these creations, these concepts, not also environmental? Is this not the rejection of any advocation for the ethical desert of the featureless plain, and rather a call to widen our understanding of methodologies as always already environmentally performative? As always already pedagogical in the way they condition our ways of being, regardless of their (non)paradigmatic allegiances – as always already political and ethical?

So, are the potential problems and solutions of environmental education research best articulated and tackled axiologically, ontologically, epistemologically, methodologically or by queer(y)-ing the historicised concepts themselves? Language always seems to *get in the way*. Yet, as 'language' is just as real, ecological and *of* this world as, say, 'experience' – just as 'theory' is as physical as, say, 'practice' – then these (often conceived of as) 'abstract' phenomena will always *get in the way*. To continue our return to where we began, there is no outside to this path we travel to be able to pit one boundaried thing against another, such as nature and culture or mind and body. Abstraction remains immanent.

Put differently, if concepts perform ecologically (Mcphie and Clarke 2018), can we relearn, discover, invent, or put to use concepts to differently explore and challenge inequitable socio-ecologizing conditions – concepts that might diffract these/their stories and purposefully *get in the way* of less helpful extremes of thought/behaviour? These are hefty responsibilities for scholars of pedagogy. And why are we even writing about this stuff – in this way? Is it a futile attempt to 'make a difference'? Do we even have the agency to control or shape the direction of travel? Because, let's face it, much of the new materialisms discourse appears too radical, inaccessible, incomprehensible or controversial even for some of the most acclaimed environmental academics, some of whom seem to have developed *new materialophobia*. As with any movement of thought there are, as ever, critics. Some scholars suggest new materialists and/or posthumanists contradict themselves by continuing to use the very dualisms they reject (James 2017). This may be so in some cases, but just because a concept is critiqued doesn't mean it cannot be used to effect, in the same way that if you never speak another dualism again it doesn't mean you have erased them from the realms of thinking or behaviour. Regarding the problem of subjectification, Deleuze and Guattari (2004) suggested, 'conversely, you can keep on saying 'I', just for kicks, and already be in another regime in which personal pronouns function only as fictions' (p. 152). So, the key is not necessarily to reject dualisms, but to acknowledge the inherent problems with their performativities. Carry on using them if you wish, with the newly acquired knowledge of how they can perform inequitably if left unsupervised?

Critical consideration is essential in these ongoing movements as we grapple with the tumult created by fast moving theory (for example, Rodrigues et al. 2020). However, one response to the *Call for Papers* for this special issue suggested that it read like 'babble'. The term 'babble', with its onomatopoetic origins in child's speech might be interpreted as derogatory. This is not an uncommon reaction, rather like the angry, defensive responses first encountered by the returned prisoner in the Plato's Cave analogy. The description of new worlds to others yet to encounter them, or with those trying to make sense of them, invariably seems to require new and seemingly impenetrable terminology to the uninitiated, or critic. Further, the challenges to received ontologies can be seen as so counterintuitive as to seem ridiculous, or cause confusion and vexation. Nonetheless, from the perspective described within this introduction, we can understand the perception of something as 'babble' as always also materially performative, adding to the tumult of converging ways of thinking and to the changing nature of the topics at hand. Conservative approaches might hope to pull in one direction, maintain the current order of things, or look to the past for orientation. Yet to others, a future of this field is already upon and among us: the emergence of 'critical, poststructural and post-qualitative-based new materialist, posthumanist, speculative ontologies and research practices has already changed EE and EE research' (Hart and Hart 2019, 280). This special issue illustrates some of the ways in which this is the case; and in reality, it may even prove a turning point? Put differently, perhaps as the contributors to this collection show in their various ways, it's in the *attempt* and in the ever-shifting *moments, movements and turns* where healthier trajectories may emerge, and travel. For *attempting* is also acting. So, let's attempt and adjust to the little changes that may be provoked, as we cautiously feel our way through a much needed, but not necessarily 'new', material turn in environmental education and its research.

Disclosure statement

No potential conflict of interest was reported by the authors.

ORCID

David A. G. Clarke http://orcid.org/0000-0002-4330-0818
Jamie Mcphie https://orcid.org/0000-0001-5290-1685

References

Adsit-Morris, C. 2017. *Restorying Environmental Education*. London: Palgrave Macmillan. doi:10.1007/978-3-319-48796-0.

Affifi, R. 2019. "Anthropocentrism's Fluid Binary." *Environmental Education Research*, 1–18. doi:10.1080/13504622.2019.1707484.

Affifi, R., S. Blenkinsop, C. Humphreys, and C. W. Joldersma. 2017. "Introduction to Ecologizing Philosophy of Education." *Studies in Philosophy and Education* 36 (3): 229–241. doi:10.1007/s11217-017-9574-3.

Ahmed, S. 2008. "Imaginary Prohibitions: Some Preliminary Remarks on the Founding Gestures of the 'New Materialism'." *European Journal of Women's Studies* 15 (1): 23–29. doi:10.1177/1350506807084854.

Alaimo, S., and S. Hekman. 2008. *Material Feminisms*. Chicago: Indiana University Press.

Anderson, B., and J. Wylie. 2009. "On Geography and Materiality." *Environment and Planning A: Economy and Space* 41 (2): 318–335. doi:10.1068/a3940.

Antrop, M. 2013. "A Brief History of Landscape Research." In *The Routledge Companion to Landscape Studies*, edited by P. Howard, I. Thompson, and E. Waterton, 12–22. London: Routledge. doi:10.4324/9780203096925.ch1.

Applebaum, A. 2020. *Twilight of Democracy: The Seductive Lure of Authoritarianism*. London: Penguin Random House.

Barad, K. 2007. *Meeting the Universe Halfway: Quantum Physics and the Entanglement of Matter and Meaning*. Durham: Duke University Press. doi:10.1215/9780822388128.

Barrett, M. J. 2005. "Making (Some) Sense of Feminist Poststructuralism in Environmental Education Research and Practice." *Canadian Journal of Environmental Education* 10 (1): 79–93.

Beeman, C., and L. Sims. 2019. "From Relationship to Something More: Environmental and Sustainability Education and a New Ontological Position." In *Environmental and Sustainability Education in Teacher Education*, edited by D. Karrow and M. DiGuiseppe, 193–208. Cham: Springer. doi:10.1007/978-3-030-25016-4_12.

Bell, A. C., and C. L. Russell. 2000. "Beyond Human, beyond Words: Anthropocentrism, Critical Pedagogy, and the Poststructuralist Turn." *Canadian Journal of Education / Revue Canadienne de L'éducation* 25 (3): 188–203. doi:10.2307/1585953.

Bennett, J. 2010. *Vibrant Matter: A Political Ecology of Things*. Durham: Duke University Press. doi:10.1215/9780822391623.

Bookchin, M. 1987. "Social Ecology versus Deep Ecology: A Challenge for the Ecology Movement." *Green Perspectives: Newsletter of the Green Program Project*, 4–5. http://theanarchistlibrary.org/library/murray-bookchin-social-ecology-versus-deep-ecology-a-challenge-for-the-ecology-movement

Braidotti, R. 2019. *Posthuman Knowledge*. Hoboken, NJ: John Wiley & Sons.

Braun, B., and S. Whatmore, eds. 2010. *Political Matter: Technoscience, Democracy and Public Life*. Minneapolis: University of Minnesota Press.

Carolan, M. S. 2007. "Introducing the Concept of Tactile Space: Creating Lasting Social and Environmental Commitments." *Geoforum* 38 (6): 1264–1275. doi:10.1016/j.geoforum.2007.03.013.

Carrington, D. 2020, August 17. "Microplastic Particles Now Discoverable in Human Organs." *The Guardian Newspaper*. https://www.theguardian.com/environment/2020/aug/17/microplastic-particles-discovered-in-human-organs

Clarke, D. A. G. 2017. "Educating beyond the Cultural and the Natural: (Re)Framing the Limits of the Possible in Environmental Education." In *Reimagining Sustainability Education in Precarious Times*, edited by K. Malone, S. Truong, and T. Gray, 305–319. London: Springer. doi:10.1007/978-981-10-2550-1_21.

Clarke, D. A. G. 2019. "Practising Immanence:(Still) Becoming an Environmental Education Academic." Unpublished Doctoral thesis, University of Edinburgh. https://era.ed.ac.uk/bitstream/handle/1842/35533/Clarke2019.pdf?sequence=1&isAllowed=y
Clarke, D. A. G., and J. Mcphie. 2014. "Becoming Animate in Education: Immanent Materiality and Outdoor Learning for Sustainability." *Journal of Adventure Education & Outdoor Learning* 14 (3): 198–216. doi:10.1080/14729679.2014.919866.
Clarke, D. A. G., and J. Mcphie. 2016. "From Places to Paths: Learning for Sustainability, Teacher Education and a Philosophy of Becoming." *Environmental Education Research* 22 (7): 1002–1024. doi:10.1080/13504622.2015.1057554.
Cole, D. R. 2019. "The Designation of a Deleuzian Philosophy for Environmental Education and Its Consequences … ." *Australian Journal of Environmental Education* 35 (3): 173–182. doi:10.1017/aee.2019.16.
Cole, D. R., and K. Malone. 2019. "Environmental Education and Philosophy in the Anthropocene." *Australian Journal of Environmental Education* 35 (3): 157–162. doi:10.1017/aee.2020.5.
Coleman, R., and J. Ringrose, eds. 2013. *Deleuze and Research Methodologies*. Edinburgh: Edinburgh University Press.
Connolly, W. E. 2013. "The 'New Materialism' and the Fragility of Things." *Millennium: Journal of International Studies* 41 (3): 399–412. doi:10.1177/0305829813486849.
Coole, D., and S. Frost, eds. 2010. *New Materialisms*. Durham: Duke University Press Books. doi:10.1215/9780822392996.
Crinall, S. 2019. *Sustaining Childhood Natures: The Art of Becoming with Water*. Singapore: Springer. doi:10.1007/978-981-13-3007-0.
Crinall, S., and M. Somerville. 2019. "Informal Environmental Learning: The Sustaining Nature of Daily Child/Water/Dirt Relations." *Environmental Education Research*, 1–12. doi:10.1080/13504622.2019.1577953.
Cutter-Mackenzie-Knowles, A., K. Malone, and E. Barratt Hacking, eds. 2020. *Research Handbook on Childhoodnature: Assemblages of Childhood and Nature Research*. Singapore: Springer. doi:10.1007/978-3-319-51949-4.
Davies, B. 2016. "Ethics and the New Materialism: A Brief Genealogy of the 'Post'philosophies in the Social Sciences." *Discourse: Studies in the Cultural Politics of Education* 39 (1): 113–127. doi:10.1080/01596306.2016.1234682.
Deleuze, G., and F. Guattari. 2004. *A Thousand Plateaus*. London: Bloomsbury.
Deloria, V. 1999. *Spirit and Reason: The Vine Deloria, Jr., Reader*. Golden, CO: Fulcrum.
Diaz-Diaz, C., and P. Semenec. 2020. *Posthumanist and New Materialist Methodologies: Research after the Child*. Singapore: Springer. doi:10.1007/978-981-15-2708-1.
Duff, C. 2014. *Assemblages of Health: Deleuze's Empiricism and the Ethology of Life*. Dordrecht: Springer. doi:10.1007/978-94-017-8893-9.
Duhn, I., and S. Galvez. 2020. "Doing Curious Research to Cultivate Tentacular Becomings." *Environmental Education Research* 26 (5): 731–741. doi:10.1080/13504622.2020.1748176.
Duhn, I., K. Malone, and M. Tesar. 2017. "Troubling the Intersections of Urban/Nature/Childhood in Environmental Education." *Environmental Education Research* 23 (10): 1357–1368. doi:10.1080/13504622.2017.1390884.
Ellison, A. M. 2013. "The Suffocating Embrace of Landscape and the Picturesque Conditioning of Ecology." *Landscape Journal* 32 (1): 79–94. doi:10.3368/lj.32.1.79.
Fawcett, L. 2009. "Feral Sociality and (un)Natural Histories. On Nomadic Ethics and Embodied Learning." In *Fields of Green. Restorying Culture, Environment, and Education*, edited by M. McKenzie, P. Hart, H. Bai, and B. Jickling, 227–236. Cresskill, NJ: Hampton Press.
Forchtner, B., ed. 2019. *The Far Right and the Environment: Politics, Discourse and Communication*. London: Routledge.
Forman, P. J. 2020. "Materiality, New Materialisms." In *International Encyclopaedia of Human Geography*, edited by A. Kobayashi, 2nd ed. https://www.sciencedirect.com/referencework/9780081022962/international-encyclopedia-of-human-geography#book-description
Fox, N. J. 2016. "Health Sociology from Post-Structuralism to the New Materialisms." *Health* 20 (1): 62–74. doi:10.1177/1363459315615393.
Fox, N. J., and P. Alldred. 2015. "New Materialist Social Inquiry: Designs, Methods and the Research-Assemblage." *International Journal of Social Research Methodology* 18 (4): 399–414. doi:10.1080/13645579.2014.921458.
Frost, S. 2011. "The Implications of the New Materialisms for Feminist Epistemology." In *Feminist Epistemology and Philosophy of Science: Power in Knowledge*, edited by H. E. Grasswick, 69–83. Dordrecht: Springer. doi:10.1007/978-1-4020-6835-5.
Gamez-Djokic, M., and A. Waytz. 2020. "Concerns about Automation and Negative Sentiment toward Immigration." *Psychological Science* 31 (8): 987–1000. doi:10.1177/0956797620929977.
Gannon, S. 2017. "Saving Squawk? Animal and Human Entanglement at the Edge of the Lagoon." *Environmental Education Research* 23 (1): 91–110. doi:10.1080/13504622.2015.1101752.
Garlick, S. 2019. "The Return of Nature: Feminism, Hegemonic Masculinities, and New Materialisms." *Men and Masculinities* 22 (2): 380–403. doi:10.1177/1097184X17725128.

Geerts, E. 2019. "Re-Vitalizing the American Feminist-Philosophical Classroom: Transformative Academic Experimentations with Diffractive Pedagogies." In *Posthumanism and Higher Education*, 123–140. Cham: Palgrave Macmillan. doi:10.1007/978-3-030-14672-6_7.

Gessen, M. 2020. *Surviving Autocracy*. New York: Penguin Random House.

Gough, A. 2004. "Blurring Boundaries: Embodying Cyborg Subjectivity and Methodology." In *Educational Research: Difference and Diversity*, edited by H Piper and I Stronach, 113–127. Farnham: Ashgate.

Gough, A., and H. Whitehouse. 2020. "Challenging Amnesias: Re-Collecting Feminist New Materialism/Ecofeminism/ Climate/Education." *Environmental Education Research*, 1–15. doi:10.1080/13504622.2020.1727858.

Gough, N. 1999. "Rethinking the Subject:(de) Constructing Human Agency in Environmental Education Research." *Environmental Education Research* 5 (1): 35–48. doi:10.1080/1350462990050102.

Gough, N. 2016. "Postparadigmatic Materialisms: A "New Movement of Thought" for Outdoor Environmental Education Research?" *Journal of Outdoor and Environmental Education* 19 (2): 51–65. doi:10.1007/BF03400994.

Gough, N., and C. Adsit-Morris. 2019. "Words (Are) Matter: Generating Material-Semiotic Lines of Flight in Environmental Education Research Assemblages (with a Little Help from SF)." *Environmental Education Research*, 1–18. doi:10.1080/13504622.2019.1663793.

Gough, N., and L. Price. 2004. "Rewording the World: Poststructuralism, Deconstruction and the 'Real' in Environmental Education." *Southern African Journal of Environmental Education* 21: 23–36.

Gough, N., and W. Sellers. 2004. "Re/de/Signing the World: Poststructuralism, Deconstruction and 'Reality' in Outdoor Environmental Education Research." Paper presented at 'Connections and Disconnections: Examining the Reality and Rhetoric'. International Outdoor Education Research Conference, July 6–9. Bendigo, VIC: La Trobe University.

Haraway, D. 2014. "Anthropocene, Capitalocene, Chthulucene: Staying with the Trouble." https://vimeo.com/ 97663518

Haraway, D. J. 1997. *Modest_Witness@Second_Millennium.FemaleMan_Meets_OncoMouse: Feminism and Technoscience*. New York: Routledge.

Hart, C. 2017. "En-Gendering the Material in Environmental Education Research: Reassembling Otherwise." *The Journal of Environmental Education* 48 (1): 46–55. doi:10.1080/00958964.2016.1249328.

Hart, P. 2005. "Transitions in Thought and Practice: Links, Divergences and Contradictions in Post-Critical Inquiry." *Environmental Education Research* 11 (4): 391–400. doi:10.1080/13504620500169270.

Hart, P., and C. Hart. 2019. "Changing Mindsets: Becoming Planetary." *The Journal of Environmental Education* 50 (4-6): 270–288. doi:10.1080/00958964.2019.1687410.

Hart, P., C. Hart, C. Aguayo, and F. T. Thiemann. 2018. "Theoretical and Methodological Trends in Environmental Education Research." *Pesquisa em Educação Ambiental* 13: 75–92. doi:10.18675/2177-580X.vol13.Especial.p75-91.

Harvey, G., ed. 2013. *The Handbook of Contemporary Animism. Handbooks on Contemporary Religion*. Durham, NC: Acumen Publishing. doi:10.4324/9781315728964.

Hattenstone, S. 2020, August 8. "The Virus Piggybacked on Racism': Why Did Covid-19 Hit BAME Families so Hard?" *Guardian Newspaper*. https://www.theguardian.com/world/2020/aug/08/the-virus-piggybacked-on-racism-why-did-covid-19-hit-bame-families-so-hard

Hickey-Moody, A., and T. Page. 2016. *Arts, Pedagogy and Cultural Resistance: New Materialisms*. London: Rowman & Littlefield International

Hinton, P., T. Mehrabi, and J. Barla. 2015. "New Materialisms/New Colonialisms." https://www.academia.edu/ 37684644/New_materialisms_New_colonialisms_2015_

Hinton, P., and I. van der Tuin. 2014. "Preface." *Women: A Cultural Review* 25 (1): 1–8. doi:10.1080/09574042.2014. 903781.

Hofverberg, H. 2019. "Entangled Threads and Crafted Meanings–Students' Learning for Sustainability in Remake Activities." *Environmental Education Research*, 1–13. doi:10.1080/13504622.2019.1664414.

Howles, T., J. Reader, and M. J. Hodson. 2018. "Creating an Ecological Citizenship': Philosophical and Theological Perspectives on the Role of Contemporary Environmental Education." *The Heythrop Journal* 59 (6): 997–1008. doi: 10.1111/heyj.13015.

Ingold, T. 2000. *The Perception of the Environment: Essays on Livelihood, Dwelling and Skill*. London: Routledge. doi: 10.4324/9780203466025.

Ingold, T. 2007. *Lines: A Brief History*. Oxford: Routledge. doi:10.4324/9780203961155.

Ingold, T. 2011. *Being Alive: Essays on Movement, Knowledge and Description*. Oxford: Routledge. doi:10.4324/ 9780203818336.

Ingold, T. 2013. *Making: Anthropology, Archaeology, Art and Architecture*. Oxford: Routledge. doi:10.4324/ 9780203559055.

Ingold, T. 2015. *The Life of Lines*. Oxford: Routledge. doi:10.4324/9781315727240.

Jackson, A. Y., and L. A. Mazzei. 2011. *Thinking with Theory in Qualitative Research: Viewing Data across Multiple Perspectives*. London: Routledge. doi:10.4324/9780203148037.

James, P. 2017. "Alternative Paradigms for Sustainability: Decentring the Human without Becoming Posthuman." In *Reimagining Sustainability in Precarious Times*, edited by K. Malone, S Truong, and T. Gray, 29–44. Singapore: Springer. doi:10.1007/978-981-10-2550-1_3.

Johns-Putra, A. 2013. "Environmental Care Ethics: Notes toward a New Materialist Critique." *Symploke* 21 (1): 125–135. doi:10.5250/symploke.21.1-2.0125.

Juelskjaer, M. 2020. "Mattering Pedagogy in Precarious Times of (un) Learning." *Matter: Journal of New Materialist Research* 1 (1): 52–79. https://revistes.ub.edu/index.php/matter/article/view/30067 doi:10.1344/jnmr.v1i1.30067.

Jukes, S., and Y. Reeves. 2019. "More-than-Human Stories: Experimental co-Productions in Outdoor Environmental Education Pedagogy." *Environmental Education Research*, 1–19. doi:10.1080/13504622.2019.1699027.

Knappett, C., and L. Malafouris. 2008. "Material and Nonhuman Agency: An Introduction." In *Material Agency: Towards a Non-Anthropocentric Approach*, edited by C. Knappett and L. Malafouris, x–xix. New York: Springer. doi:10.1007/978-0-387-74711-8.

Kotzé, E., K. Crocket, and C. Waititi. 2016. "Meeting the Wharenui Shapes Teaching/Learning Therapy: Place Pedagogy, New Materialism, and Whakaaro Māori." *Australian and New Zealand Journal of Family Therapy* 37 (3): 317–326. doi:10.1002/anzf.1172.

Lather, P., and E. Ellsworth. 1996. "This Issue: Situated Pedagogies – Classroom Practices in Postmodern Times." *Theory into Practice* 35 (2): 70–71. doi:10.1080/00405849609543704.

Latour, B. 1999. *Pandora's Hope: Essays on the Reality of Science Studies*. Cambridge, MA: Harvard University Press.

Latour, B. 2004. *Politics of Nature*. Cambridge: Cambridge University Press.

Latour, B. 2005. *Reassembling the Social*. Oxford: Oxford University Press.

Latour, B., and S. Woolgar. 1979. *Laboratory Life: The Construction of Scientific Facts*. Princeton, NJ: Princeton University Press.

Law, J. 2004. *After Method: Mess in Social Science Research*. London: Routledge. doi:10.4324/9780203481141.

Le Grange, L. 2004. "Embodiment, Social Praxis and Environmental Education: Some Thoughts." *Environmental Education Research* 10 (3): 387–399. doi:10.1080/1350462042000258206.

Le Grange, L. 2018. "Spinoza, Deep Ecology and Education Informed by a (Post) Human Sensibility." *Educational Philosophy and Theory* 50 (9): 878–887. doi:10.1080/00131857.2017.1384723.

Le Grange, L. 2019. "The Anthropocene: Becoming-Imperceptible of (Environmental) Education." *On Education: Journal for Research and Debate* 2 (4): 1–6. doi:10.17899/on_ed.2019.4.4.

Libet, B. 2006. "Reflections on the Interaction of the Mind and Brain." *Progress in Neurobiology* 78 (3-5): 322–326. doi:10.1016/j.pneurobio.2006.02.003.

Lundborg, T., and N. Vaughan-Williams. 2015. "New Materialisms, Discourse Analysis, and International Relations: A Radical Intertextual Approach." *Review of International Studies* 41 (1): 3–25. doi:10.1017/S0260210514000163.

Lynch, J., and G. Mannion. 2016. "Enacting a Place-Responsive Research Methodology: Walking Interviews with Educators." *Journal of Adventure Education and Outdoor Learning* 16 (4): 330–345. doi:10.1080/14729679.2016.1163271.

Lysgaard, J. A., and K. L. Fjeldsted. 2015. "Education between Discourse and Matter." In *Nature in Education*, edited by P. Kemp. Zürich: LIT Verlag.

MacFarlane, J. M. 2020. *Transhumanism as a New Social Movement: The Techno-Centred Imagination*. Cham, Switzerland: Palgrave Macmillan doi:10.1007/978-3-030-40090-3.

Maclure, M. 2015. "The 'New Materialisms': A Thorn in the Flesh of Critical Qualitative Inquiry?" In *Critical Qualitative Inquiry: Foundations and Futures*, edited by G. Cannella, M. S. Perez, and P. Pasque. Walnut Creek, CA: Left Coast Press.

Malone, K. 2016. "Theorizing a Child–Dog Encounter in the slums of La Paz Using Post-Humanistic Approaches in Order to Disrupt Universalisms in Current 'Child in Nature' Debates." *Children's Geographies* 14 (4): 390–407. doi:10.1080/14733285.2015.1077369.

Malone, K., M. Tesar, and S. Arndt. 2020. *Theorising Posthuman Childhood Studies*. Singapore: Springer. doi:10.1007/978-981-15-8175-5.

Malone, K., S. Truong, and T. Gray. 2017. *Reimagining Sustainability in Precarious Times*. Singapore: Springer. doi:10.1007/978-981-10-2550-1.

Mannion, G. 2019. "Re-Assembling Environmental and Sustainability Education: Orientations from New Materialism." *Environmental Education Research*, 1–20. doi:10.1080/13504622.2018.1536926.

Mannion, G., A. Fenwick, and J. Lynch. 2013. "Place-Responsive Pedagogy: Learning from Teachers' Experiences of Excursions in Nature." *Environmental Education Research* 19 (6): 792–809. doi:10.1080/13504622.2012.749980.

Massumi, B. 2008. "The Thinking-Feeling of What Happens: A Semblance of a Conversation." *Inflexions*, 1, 1–40. http://inflexions.org/n1_The-Thinking-Feeling-of-What-Happens-by-Brian-Massumi.pdf

Massumi, B. 2010. "What Concepts Do: Preface to the Chinese Translation of a Thousand Plateaus." *Deleuze Studies* 4 (1): 1–15. doi:10.3366/E1750224110000772.

McKenzie, M. 2005. "The 'Post-Post Period' and Environmental Education Research." *Environmental Education Research* 11 (4): 401–412. doi:10.1080/13504620500169361.

McKenzie, M., and A. Bieler. 2016. *Critical Education and Sociomaterial Practice: Narration, Place, and the Social*. New York: Peter Lang. doi:10.3726/978-1-4539-1715-2.

McLeod, K. 2017. *Wellbeing Machine: How Health Emerges from the Assemblages of Everyday Life*. Durham, NC: Carolina Academic Press.

Mcphie, J. 2017. "The Accidental Death of Mr. Happy: A Post-Qualitative Rhizoanalysis of Mental Health and Wellbeing." Unpublished Doctoral thesis. Lancaster University.

Mcphie, J. 2018. "I Knock at the Stone's Front Door: Performative Pedagogies beyond the Human Story." *Parallax* 24 (3): 306–323. doi:10.1080/13534645.2018.1496581.

Mcphie, J. 2019. *Mental Health and Wellbeing in the Anthropocene: A Posthuman Inquiry*. Basingstoke, UK: Palgrave Macmillan. doi:10.1007/978-981-13-3326-2.

Mcphie, J., and D. A. G. Clarke. 2015. "A Walk in the Park: Considering Practice for Outdoor Environmental Education through an Immanent Take on the Material Turn." *The Journal of Environmental Education* 46 (4): 230–250. doi:10.1080/00958964.2015.1069250.

Mcphie, J., and D. A. G. Clarke. 2018. "Nature Matters: Diffracting a Keystone Concept of Environmental Education Research–Just for Kicks." *Environmental Education Research*, 1–18. doi:10.1080/13504622.2018.1531387.

Mcphie, J., and D. A. G. Clarke. 2019. "Post-Qualitative Inquiry in Outdoor Studies." In *Research Methods in Outdoor Studies*, edited by B. Humberstone and H. Prince, 186–195. London: Routledge. doi:10.4324/9780429199004-18.

Merewether, J. 2019. "New Materialisms and Children's Outdoor Environments: Murmurative Diffractions." *Children's Geographies* 17 (1): 105–117. doi:10.1080/14733285.2018.1471449.

Mol, A. 2002. *The Body Multiple*. Durham, NC: Duke University Press. doi:10.1215/9780822384151.

Murdoch, J. 2001. "Ecologising Sociology: Actor-Network Theory, Co-Construction and the Problem of Human Exemptionalism." *Sociology* 35 (1): 111–133. doi:10.1177/0038038501035001008.

Myers, C. Y. 2019. *Children and Materialities: The Force of the More-than-Human in Children's Classroom Lives*. Singapore: Springer. doi:10.1007/978-981-13-8168-3.

Nxumalo, F. 2016. "Touching Place in Childhood Studies: Situated Encounters with a Community Garden." In *Youth Work, Early Education, and Psychology*, edited by H. Skott-Myhre, V. Pacini-Ketchabaw, and K. Skott-Myhre, 131–158. New York: Palgrave Macmillan. doi:10.1057/9781137480040_8.

Nxumalo, F., and V. Pacini-Ketchabaw. 2017. "Staying with the Trouble' in Child-Insect-Educator Common Worlds." *Environmental Education Research* 23 (10): 1414–1426. doi:10.1080/13504622.2017.1325447.

Page, T. 2020. *Placemaking. A New Materialist Theory of Pedagogy*. Edinburgh: Edinburgh University Press. doi:10.3366/j.ctv1453k6t.

Payne, P. 2005. "Lifeworld and Textualism: Reassembling the Researcher/ed and 'Others." *Environmental Education Research* 11 (4): 413–431. doi:10.1080/13504620500169411.

Payne, P. 2016. "What Next? Post-Critical Materialisms in Environmental Education." *The Journal of Environmental Education* 47 (2): 169–178. doi:10.1080/00958964.2015.1127201.

Payne, P. 2019. "Performative Abstractionism in Environmental Education: A Critical Theory of Theory." The Journal of Environmental Education 50 (4-6): 289–320. doi:10.1080/00958964.2019.1687405.

Pickering, A. 2010. "Material Culture and the Dance of Agency." In *The Oxford Handbook of Material Culture Studies*, edited by D. Hicks and M. C. Beaudry, 191–208. Oxford: Oxford University Press. doi:10.1093/oxfordhb/9780199218714.013.0007.

Pleasants, K., and A. Stewart. 2019. "Entangled Philosophical and Methodological Dimensions of Research in Outdoor Studies?" In Barbara Humberstone and Heather Prince (Ed.), *Research Methods in Outdoor Studies*. London: Routledge, p. 9–20. doi:10.4324/9780429199004-2.

Preston, C. J. 2017. "From Atom to Atmosphere." Blog. https://plastocene.com/2017/08/18/from-atom-to-atmosphere/

Rautio, P. 2013. "Children Who Carry Stones in Their Pockets: On Autotelic Material Practices in Everyday Life." *Children's Geographies* 11 (4): 394–408. doi:10.1080/14733285.2013.812278.

Rautio, P. 2014. "Mingling and Imitating in Producing Spaces for Knowing and Being: Insights from a Finnish Study of Child–Matter Intra-Action." *Childhood* 21 (4): 461–474. doi:10.1177/0907568213496653.

Rautio, P., and E. Stenvall. 2019. *Social, Material and Political Constructs of Arctic Childhoods: An Everyday Life Perspective*. Singapore: Springer. doi:10.1007/978-981-13-3161-9.

Reinertsen, A., ed. 2016. *Becoming Earth: A Post Human Turn in Educational Discourse Collapsing Nature/Culture Divides*. Singapore: Springer. doi:10.1007/978-94-6300-429-9.

Riley, K. 2019. "(Re) Storying Human/Nonhuman Relationships: Posthumanist Possibilities in Researcher/Teacher/Environmental Education Worlds." Unpublished Doctoral thesis, Deakin University.

Ringrose, J., K. Warfield, and S. Zarabadi, eds. 2020. *Feminist Posthumanisms, New Materialisms and Education*. Abingdon, UK: Routledge. doi:10.4324/9781351186674.

Rodrigues, C., P. G. Payne, L. L. Grange, I. C. Carvalho, C. A. Steil, H. Lotz-Sisitka, and H. Linde-Loubser. 2020. "Introduction: "New" Theory,"Post" North-South Representations, Praxis." *The Journal of Environmental Education* 51 (2): 97–112. doi:10.1080/00958964.2020.1726265.

Rooney, T. 2016. "Weather Worlding: Learning with the Elements in Early Childhood." *Environmental Education Research* 1 (24): 1–12. doi:10.1080/13504622.2016.1217398.

Rosiek, J. L., J. Snyder, and S. L. Pratt. 2020. "The New Materialisms and Indigenous Theories of Non-Human Agency: Making the Case for Respectful anti-Colonial Engagement." *Qualitative Inquiry* 26 (3-4): 331–346. doi:10.1177/1077800419830135.

Ross, H., and G. Mannion. 2012. "Curriculum Making as the Enactment of Dwelling in Places." *Studies in Philosophy and Education* 31 (3): 303–313. doi:10.1007/s11217-012-9295-6.

Rousell, D. 2018. "Doing Little Justices: Speculative Propositions for an Immanent Environmental Ethics." *Environmental Education Research*, 1–15. doi:10.1080/13504622.2018.1517408.

Ruck, A., and G. Mannion. 2019. "Fieldnotes and Situational Analysis in Environmental Education Research: Experiments in New Materialism." *Environmental Education Research*, 1–18. doi:10.1080/13504622.2019.1594172.

Russell, C. 2005. "'Whoever Does Not Write Is Written': The Role of 'Nature' in Post-Post Approaches to Environmental Education Research." Environmental Education Research 11 (4): 433–443. doi:10.1080/13504620500169569.

Schama, S. 2004. *Landscape and Memory*. London: Harper Perennial.

Schneider, R. 2015. "New Materialisms and Performance Studies." *TDR/the Drama Review* 59 (4): 7–17. doi:10.1162/DRAM_a_00493.

Smith, J. L. 2017. "I, River? New Materialism, Riparian Nonhuman Agency and the Scale of Democratic Reform." *Asia Pacific Viewpoint* 58 (1): 99–111. doi:10.1111/apv.12140.

Smyth, R. 2019, April 3. "Nature Writing's Fascist Roots." New Statesman. https://www.newstatesman.com/culture/books/2019/04/eco-facism-nature-writing-nazi-far-right-nostalgia-england

Snaza, N., D. Sonu, S. E. Truman, and Z. Zaliwska. 2016. *Pedagogical Matters: New Materialisms and Curriculum Studies*. New York: Peter Lang Publishing, Inc. doi:10.3726/978-1-4539-1648-3.

Snaza, N., and J. Weaver. 2014. *Posthumanism and Educational Research*. Abingdon: Routledge. doi:10.4324/9781315769165.

Sojot, A. N. 2019. "Slimer, Slime, and the Thingness of Pedagogy." *Educational Perspectives* 51 (1): 8–12.

Somerville, M., B. Davies, K. Power, S. Gannon, and P. de Carteret. 2012. *Place Pedagogy Change*. Rotterdam: Sense. doi:10.1007/978-94-6091-615-1.

Sonu, D., and N. Snaza. 2015. "The Fragility of Ecological Pedagogy: Elementary Social Studies Standards and Possibilities of New Materialism." *Journal of Curriculum and Pedagogy* 12 (3): 258–277. doi:10.1080/15505170.2015.1103671.

St. Pierre, E. A. 2004. "Deleuzian Concepts for Education: The Subject Undone." *Educational Philosophy and Theory* 36 (3): 283–296. doi:10.1111/j.1469-5812.2004.00068.x.

St. Pierre, E. A. 2014. "A Brief and Personal History of Post Qualitative Research: Toward "Post Inquiry." *Journal of Curriculum Theorizing* 30 (2): 2–19. https://journal.jctonline.org/index.php/jct/article/view/521

St. Pierre, E. A. 2015. "Practices for the "New" in the New Empiricisms, the New Materialisms, and Post Qualitative Inquiry." In N. Denzin and M. Giardina (Eds) *Qualitative Inquiry and the Politics of Research*, 75–96. London and New York: Routledge.

Stables, A. 2007. "Is Nature Immaterial? The Possibilities for Environmental Education without an Environment." *Canadian Journal of Environmental Education* 12 (1): 55–67.

Staudenmaier, P. 2011. "Fascist Ecology: The "Green Wing" of the Nazi Party and Its Historical Antecedents." In *Ecofascism Revisited: Lessons from the German Experience*, edited by J. Biehl and P. Staudenmaier, 13–42. Porsgrunn: New Compass Press.

Stewart, A. 2008. "Whose Place, Whose History? Outdoor Environmental Education Pedagogy as 'Reading' the Landscape." *Journal of Adventure Education & Outdoor Learning* 8 (2): 79–98. doi:10.1080/14729670801906125.

Sullivan, N. 2012. "The Somatechnics of Perception and the Matter of the Non/Human: A Critical Response to the New Materialism." *European Journal of Women's Studies* 19 (3): 299–313. doi:10.1177/1350506812443477.

Tammi, T. 2019. "What If Schools Were Lively More-than-Human Agencements All along? Troubling Environmental Education with Moldschools." *Environmental Education Research*, 1–16. doi:10.1080/13504622.2019.1584881.

Taylor, C. A., and C. Hughes, eds. 2016. *Posthuman Research Practices in Education*. Basingstoke, UK: Palgrave Macmillan. doi:10.1057/9781137453082.

Taylor, C. A., and G. Ivinson. 2013. "Material Feminisms: New Directions for Education." *Gender and Education* 25 (6): 665–670. doi:10.1080/09540253.2013.834617.

Thomas, M., and R. Bellingham. 2020. *Post-Qualitative Research and Innovative Methodologies*. New York: Bloomsbury. doi:10.5040/9781350062078.

Thompson, I. 2010. *The English Lakes: A History*. London: Bloomsbury.

Thorpe, H., and M. Clark. 2020. "Gut Feminism, New Materialisms and Sportwomen's Embodied Health: The Case of RED-S in Endurance Athletes." *Qualitative Research in Sport, Exercise and Health* 12 (1): 1–17. doi:10.1080/2159676X.2019.1631879.

Thrift, N. 2008. *Non-Representational Theory: Space, Politics, Affect*. London: Routledge. doi:10.4324/9780203946565.

Todd, Z. 2016. "An Indigenous Feminist's Take on the Ontological Turn: 'Ontology' is Just Another Word for Colonialism." *Journal of Historical Sociology* 29 (1): 4–22. doi:10.1111/johs.12124.

Tuck, E. 2014, April. "A Turn to Where We Already Were? Settler Inquiry, Indigenous Philosophy, and the Ontological Turn." Paper presented to the Annual Meeting of the American Educational Research Association, Philadelphia, PA.

Tuck, E., and R. A. Gaztambide-Fernández. 2013. "Curriculum, Replacement, and Settler Futurity." *Journal of Curriculum Theorizing* 29 (1): 72–89. https://journal.jctonline.org/index.php/jct/article/viewFile/411/pdf

Tuck, E., and M. McKenzie. 2015. *Place in Research*. New York: Routledge. doi:10.4324/9781315764849.

Tuck, E., M. McKenzie, and K. McCoy. 2014. "Land Education: Indigenous, Post-Colonial, and Decolonizing Perspectives on Place and Environmental Education Research." *Environmental Education Research* 20 (1): 1–23. doi:10.1080/13504622.2013.877708.

Van der Tuin, I., and R. Dolphijn. 2010. "The Transversality of New Materialism." *Women: A Cultural Review* 21 (2): 153–171. doi:10.1080/09574042.2010.488377.

Van Poeck, K., and J. Lysgaard. 2016. "The Roots and Routes of Environmental and Sustainability Education Policy Research." *Environmental Education Research* 22 (3): 305–318. doi:10.1080/13504622.2015.1108393.

Verlie, B., and CCR15. 2018. "From Action to Intra-Action? Agency, Identity and 'Goals' in a Relational Approach to Climate Change Education." *Environmental Education Research*, 1–15. doi:10.1080/13504622.2018.1497147.

Warren, K. 1990. "The Power and Promise of Ecological Feminism." *Environmental Ethics* 12 (2): 125–146. doi:10.5840/enviroethics199012221.

Whatmore, S. 2006. "Materialist Returns: Practising Cultural Geography in and for a More-than-Human World." *Cultural Geographies* 13 (4): 600–609. doi:10.1191/1474474006cgj377oa.

Witmore, C. 2014. "Archaeology and the New Materialisms." *Journal of Contemporary Archaeology* 1 (2): 203–246. doi:10.1558/jca.v1i2.16661.

Wright, S. L., and F. J. Kelly. 2017. "Plastic and Human Health: A Micro Issue?" *Environmental Science & Technology* 51 (12): 6634–6647. doi:10.1021/acs.est.7b00423.

From action to intra-action? Agency, identity and 'goals' in a relational approach to climate change education

Blanche Verlie and CCR 15

ABSTRACT

Engaging with new materialist/posthuman approaches to agency, in this paper I explore what might happen to the goal of cultivating climate action if we decentre the human from our climate pedagogies. Specifically, I engage with Karen Barad's concept of intra-action which argues that agency is not possessed by individual things or beings, but emerges through relationships. I work with experiences and occurrences from Climate Change Responses, an undergraduate social science course that I tutored in 2015 (CCR15). I explore how in CCR15, while trying to learn to better mitigate climate change, we became climate killjoys, resisting, challenging and disrupting pleasurable carbon intensive practices. Through these empirical examples, I show that 'climate intra-action' can enable us to attend to how human and more-than-human identities change through engagement with climate change; how our human capacities to affect climate emerge through acting-with more-than-human entanglements; and thus how unanticipated, different actions can emerge in climate change education. I therefore suggest that an intra-active approach to climate change education research and practice might enable less anthropocentric and more relationally attuned climate change 'response-abilities', for both teachers and students.

Introduction

In 2013, Australia elected Tony Abbott as our Prime Minister, apparently providing him with a mandate to repeal Australia's recently introduced price on carbon. So I went to the rallies that ensued, appalled, clutching placard after placard declaring 'I want climate action'. This was about the same time as I was applying to do a PhD, and I now realise that Abbott's election was a very influential participant in my decision to research climate change education. A year later I was reading Jane Bennett's *Vibrant Matter* (Bennett 2010), and Karen Barad's *Meeting the Universe Halfway* (Barad 2007). Bennett (2010) had me contemplating the 'thing-power' of carbon, because at the end of the day climate change 'all comes down to the carbon cycle; such a small amount of carbon has such a large impact' (CCR 2015). Perhaps climate change was carbon induced, not human induced? Barad (2007) pushed me to understand climate change as a co-produced outcome of the 'intra-action' between carbon and humans (and all the planet's more-than-human forces and entities). Maybe what I really wanted was climate *intra*-action?

Last year the potential of such relational materialist ontologies to trouble our existing ways of understanding climate change was exemplified for me. In early 2017, coal compelled Federal

Treasurer Scott Morrison to contravene parliamentary procedures by bringing a lump of it into Parliament to promote Australia's ongoing obsession with extracting and burning the world's dirtiest form of energy. Riding the wave of Tony Abbott's infamous 'coal is good for humanity' statement (Massola, Ker, and Cox 2014), Morrison declared 'this is coal: don't be scared, it can't hurt you' (Kenny 2017), passing it around the Liberal Party MPs who were visibly delighted with the stunt (see Hamilton, 2017 or Verlie & CCR15 2020 for the image of this political stunt). Because coal in its solid – and conveniently lacquered – form can be safely handled by a human body without causing immediate or obvious harm, Morrison contended, like the Minerals Council of Australia, that while coal is 'an amazing thing' which can 'provide endless possibilities', these possibilities are firmly limited to ones that humans desire and control (http://littleblackrock.com.au/). The subsequent public derision of this stunt drew on a significantly more animated approach to the more-than-human world: coal *can* harm us, and *does* harm us, even though we do not want it to. Therefore, we should act in ways that recognise and respond to this by keeping coal in the ground and closing coal mines (Regester 2017; Hamilton 2017).

However, this increased appreciation for non-human agency does retain some anthropocentric elements. Of course, keeping coal in the ground is a wonderful idea, one that recognises coal's ability to exceed human intentionality. But, in both the above approaches to climate change, human actors arrive fully formed and make rational decisions – albeit with different values – about what to do *to* the coal: burn it or keep it in the ground. Coal's agency is limited to that which humans can recognise and thus seek to control, manage, and use. That we might become affected by coal's agency in other, less tangible, less measurable, ways, and that we might become-with (Haraway 2008) coal is not fully considered. In short, the actor taking the climate action is presumed to be a pre-given, autonomous human who acts *upon* the world, rather than one that acts and emerges *with* it (Dürbeck, Schaumann, and Sullivan 2015; Sonu and Snaza 2015; Rooney 2018). Yet, to contend that climate change is carbon induced would appear to nullify human responsibility, potentially leaving us with an ineffective political relativism.

In this paper, I work with (Jackson and Mazzei 2012) Karen Barad's concept of intra-action (2007) to explore this tension, asking: what might happen to the goal of cultivating 'climate action' if we decentre the human from our pedagogies? I explore the notion of climate intra-action to conceptualise a climate change pedagogy focussed on 'intra-sectionally diffr-acting-with climate'. Climate intra-action begins with climate action, but acknowledges that because we are always acting-with the world, we cannot fully predetermine or limit what those actions should or will be and so we must be open to creating unanticipated, different climate actions (diffr-actions) *with* the world. This approach therefore understands climate and humans to co-emerge through practices: through intra-action, the boundaries between them change and they are mutually, although differently, effected. I explore the example of becoming climate killjoys as an example of this. In so doing, this paper contributes to the gap in feminist/relational materialist literature engaging specifically with climate change (Pearse 2017), and brings the politics of the relational materialisms to the fore (Clare 2016).

The issue of agency in climate change education

The end goal for climate change education (CCE) is 'climate action', behaviours or practices conducted by humans that reduce human impacts on the global climate, in order to reduce the impacts of the changing climate on humans and non-humans (Monroe et al. 2017; Henderson et al. 2017; Moser 2016). This focus on *doing* is shared with educational approaches building on new/feminist materialist, posthuman and relational (here termed 'relational materialist') theories (Edwards 2012; Puig de la Bellacasa 2010; Fenwick 2012; Fenwick and Edwards 2014; Taylor and Giugni 2012). However, such relational materialist theories critique the anthropocentrism they see in much education, including CCE. They argue that mainstream modern education often functions to establish 'compulsory humanity' (McKay, cited in Pedersen 2010b, 237) or 'anthroponormativity' (Taylor and Blaise 2014, 377) which 'continually reinscribes and "closes" categories of "human"' (Pedersen 2010b, 241–242). Such pedagogies are argued to reproduce human subjectivities where 'all nonhuman objects [become] potential "resources" to be extracted, refined, exchanged, and used' (Snaza 2015, 26), which are the very subjectivities that need to be challenged if we want to create more sustainable worlds (Pedersen 2010a).

In spite of its foundational concern for non-human nature, CCE can risk reiterating this anthropocentrism. If relying on dualistic pedagogies that depend heavily on the (very important) claim that climate change is human induced, and not natural, CCE can position humans as pre-existing, separated subjects who make meaning about and act upon the climate, which is thus treated as a passive object (Verlie 2017). Evidently, recognition that climate change is human induced serves to problematise the clear distinction between humans and nature: since we have 'impacted' it, climate is now no longer 'natural'. And of course, human-induced climate change in turn 'impacts' humans. Yet these are typically understood as parallel, discontinuous processes (Sasser 2016; Rooney 2018). That these processes of change might be indicative of a more mutual enmeshment, where neither 'entity' would exist in its current form without the other, often seems to elude our full appreciation.

If our pedagogies are underpinned by such epistemologies and ontologies that objectify climate and separate us from it, our ethical capacities to care for climate will be limited (Verlie 2017). As Sonu and Snaza put it, 'it would not be enough to have an ethical concern for the environment [if the environment] remains at its core the property of human sovereignty' (2015, 266). Sjögren (2014, 5) argues that the 'educable subject responsible for more sustainable futures needs to be rethought in light of theories of entangled humans and nonhumans' if we are to move beyond human exceptionalism. Relational materialist approaches to education respond to this call by rejecting *Anthropos* – the autonomous, rational, and outcome-oriented possessor of agency and knowledge (Braidotti 2013; Barad 2007; Haraway 2015) – as the assumed subject of education (Rautio 2012; Pedersen 2010b; Pedersen and Pini 2017; Fenwick and Edwards 2014). In contradiction, such approaches contend that subjectivities, knowledge and agency emerge in contingent, complex, more-than-human entanglements (Sjögren 2014; Snaza et al. 2016; Somerville 2017; Postma 2012; Leibowitz and Naidoo 2017).

And here we encounter the key tension this paper seeks to engage: a relational materialist ontology of climate–human relations, while perhaps theoretically less anthropocentric, may offer little guidance on how to proceed practically at cultivating conditions in which others – human and more-than-human – can flourish (Head and Gibson 2012; Sharp 2017; Duhn 2017). If understood simply as the blurring of borders or erasure of binaries, a relational account can lose the capacity to make political claims such as demanding climate action because such efforts typically depend on the articulation of boundaries in order to locate responsibility and attribute causality (Tuck and McKenzie 2014; Dürbeck, Schaumann, and Sullivan 2015; Tuana 2008; Clark 2010). Such an approach could even lead to us celebrating climate change as an example of our mutual, entangled, co-becoming and as evidence of humanity's lack of agency to control the climate.

Aims and methodology

In this paper, I aim to explore whether and how working with Barad's notion of intra-action can compose a practical, relational materialist, non-anthropocentric approach to CCE. By approach, I mean both the interpretation of educational occurrences and also the generation of pedagogical practices. I suggest that climate intra-action can potentially navigate the issue of accounting for non-human agency in climate changing practices while still demanding that humans take responsibility for these climate changing actions. Rather than critique one theoretical approach through the lens of another, I engage a 'diffractive' methodology (Barad 2007, 2014), where I take the insights and provocations of two or more matters (issues, disciplines and/or phenomena) and explore the tensions and insights that arise when they interfere with each other. In this case, the diffraction involves CCE, Barad's concept of intra-action, coal and other forms of carbon, and the experiences and insights that emerged in the 2015 Climate Change Responses course at RMIT University (CCR15).

RMIT University is situated in Naarm ('Melbourne'), on stolen and unceded Wurundjeri land. CCR15 was a new undergraduate social science course in the School of Global, Urban and Social Studies, which ran over a twelve-week semester from mid-July to the end of October. Following Lecturer Lauren Rickard's approach, the course focussed on mitigative and adaptive responses to climate change. This meant the course covered economic, political, social, cultural, technological and design-related issues. I was employed as a tutor, planning and facilitating one and a half hour workshops following the one and a half hour lecture. Between my two classes, there were about 45 students. The majority of my students were enrolled in the Bachelor of Society and Environment degree and were in their second or third year of study. My experience teaching, and the experience of my students learning with climate change forms the empirical part of my PhD thesis, and it is these empirical experiences that diffract throughout this paper.

I understand the written words produced here as stories of these empirical experiences (i.e. quotations from students, observations written by myself and the paper as a whole) as emerging from an entangled voice. There are two key ideas to this. Firstly, the production of experiences, stories, words and/or typed text are performative practices that co-enact agency, changing worlds, subjects and knowledges (Jackson 2008; McKenzie and Bieler 2016). Secondly, experiences, stories, words and/or text are not produced by, and do not belong to, stable, concretely bounded, individual human subjects; rather, they are produced by and contribute to the production of more-than-human entanglements (Mazzei 2013, 2016; Mazzei and Jackson 2016). This means that the words that are printed here are not solely mine, even when they appear to be so, and neither are they ever simply an individual human student's. In contrast, the words, ideas and experiences of both the voice narrating this paper and the voices of the 'data' have emerged through intimate, intense, sustained, dynamic, collaborative and complex eco-social relationships between my students, myself, the climate and many other more-than-human entities and forces. The paper is thus a form of 'weather writing', a transcorporeal (Alaimo 2008) production constituted through 'the participation of all kinds of bodies in collaboration with the human writer' (Neimanis 2015, 144). For this reason, CCR15 (understood as a more-than-human entanglement) is acknowledged as a co-author of the paper, recognising not only my students' but also climate change's material agency in producing the meaning in this paper.

Further to this, this paper works with the entangled voice's words as a form of insightful literature. As St Pierre (2008) argues, the distinction between data and literature is a false and problematic binary, and it is one that reiterates power structures between academics who produce knowledge as subjects and those they study who are treated as objects to be known. All our encounters with the world enable us to collectively produce knowledge and our research participants, including non-human participants, should be acknowledged for their contribution just as academic literature is, rather than being objectified as 'data'. As such, the entangled voice that narrates this paper refers to 'CCR15' and has '*quotations in italics*' when working with CCR15 as 'data' that is being storied. But it also draws on some of the statements and outcomes as insightful ideas, and in these instances CCR15 is cited as academic literature would be, i.e. as '(CCR 2015)'. Of course, in reality these modes of knowledge co-creation are messier than this clean framework suggests: all three modes overlap somewhat.

Climate intra-action: intra-sectionally diffr-acting-with climate

CCE and relational materialist approaches to education share appreciation for action, doing and responding. In humanist approaches to CCE, climate action is often understood as the final outcome occurring after a linear process of individual humans gaining knowledge and then becoming concerned about climate change (Verlie 2017). In relational materialist approaches, knowing, being and doing are not so neatly separated. Rather than conceptualising learning as the absorption and reproduction of a reservoir of facts, relational materialists often prefer to focus on 'response-ability'.

For Barad 'intelligibility is a matter of differential responsiveness' (2007, 149). Similarly, Latour (2004) argues that learning has occurred when a subject has become affected by the world and is more attentive and responsive to nuance, difference and diversity. Haraway (2016) contends that pedagogy is therefore about rendering each other capable of response, and in such accounts, 'learning is defined as expanded possibilities for action, or engaging in more sophisticated and flexible action' (Edwards and Fenwick 2013, 56). As Barad succinctly puts it, 'knowing is a matter of intra-acting' (2007, 149).

Barad's notion of intra-action has been widely used in relational materialist work to reconceptualise agency and subjectivity. Intra-action is similar to interaction, but Barad (2007) argues that 'interaction' sees agencies, subjects, objects and knowledges as closed, static entities that exist autonomously in the world prior to engaging with others. In contrast, 'intra-action' recognises that they are unpredictable, contingent, dynamic, open ended phenomena which emerge through relationships (Barad 2007). Therefore, nothing exists outside of or prior to its relations or entanglements with others. As an ontology of knowing climate, intra-action articulates how humans and climate do not pre-exist their relations, but are co-emergent, and how their relationship collaboratively enacts climate knowledges-and-responses (Neimanis 2015; Neimanis and Walker 2013; Rooney 2018; Walker 2013). As a pedagogy, climate intra-action thus asks us to attend to (van Dooren, Kirksey, and Münster 2016) – i.e. to notice and respond to – the ongoing, situated, dynamic reconfigurations of humans, climate, and human–climate relations that are enacted in everyday and extraordinary educational encounters.

There are three overlapping areas of focus that climate intra-action enables that I think are not (as well) facilitated by more humanist and anthropocentric approaches to CCE: a) 'intrasectionality' – the notion that identities and the boundaries between them are sociomaterially, relationally and performatively reconfigured; b) that all acting is always 'acting-with' a host of worldly others; and c) that we need to attend to the unpredictable, different actions that can emerge through this acting-with ('diffr-actions').

Intrasectionality

> Climate change is everything and it is going to affect everything, it is so immense and entwined via almost any issue you can think of in this world: ocean acidification, rising sea levels, biodiversity, loss of the ice sheets, extreme drought, fires and floods, mental illness, gender inequality, mass migration, mass extinction, conflict, hunger, desertification, destruction, loss: it is a live or die reality. (CCR 2015)

Intra-action understands worldly phenomena – such as students, teachers and the climate – as waves. Unlike particles which are closed entities that occupy positions in time and space, waves are material-energetic disturbances extended in and enfolding spacetime. Waves can overlap, combine, interfere and become entangled with each other, which is termed a 'superposition' (Barad 2007). Superpositions are characterised by indeterminacy, encompassing multiple potentialities for further differentiations as their constitutive waves amplify, nullify and/or reconfigure each other and interfere with the world.

Climate change epitomises this ontology that emphasises that we are all connected to everybody and everything else. However, through doing so, it teaches us that being connected to non-human nature is not a choice, nor should it be romanticised or idealised (Swanson et al. 2017) as is often common in relational materialist approaches (Clare 2016). Mass extinction and sea level rise do not happen in worlds composed of independent, autonomous entities, and this entanglement highlights how some bodies systematically – although perhaps indirectly or unintentionally – kill, poison, starve, displace and/or otherwise harm other bodies, or at least benefit from practices which do so. Climate intra-action is therefore not about dissolving boundaries but about accounting for which boundaries are produced and how, and who benefits from such articulations (van Dooren, Kirksey, and Münster 2016; Barad 2007). As Haraway reminds us, connections are never innocent; the details of the connections matter (2016).

Building on the important black feminist notion of intersectionality (Lorde 1984; Collins 2000; Crenshaw 1991) and Barad's posthuman relationality, Mellander and Wiszmeg (2016) developed the term 'intrasectionality' which usefully emphasises this. Intrasectionality foregrounds how important critical concepts relating to power and hierarchy like standpoints, social location and positionalities are not erased or flattened but dynamically reconfigured through the identity changing processes of intra-action. This approach of 'relational positionality' accounts for how identities come 'into being through relationships which are ever changing and constituted at multiple scales' (Neely and Nguse 2015, 141). In line with this, I find it useful to consider identities as 'superpositionalities'. Emerging through the ongoing interference of natural–cultural waves (such as gender and climate change), superpositionalities are momentarily articulable sociomaterial relational-locations which are both situated and dynamic. Understanding identities and/or subjectivities as intrasectionally composed superpositionalities helps us attune to how we are all waves of possibility (O'Brien 2016). This is not the exclusive domain of humans; more-than-human matter participates in these rippling relationalities (Huggan and Tiffin 2010; Kirksey and Helmreich 2010). Relational materialist approaches have been criticised for erasing or negating politics in their 'flat ontologies' (Tuck and McKenzie 2014; Quinn 2013; Pearse 2017; Neimanis 2017), but intrasectionality navigates this by accounting for the dynamic processes of connecting-excluding that compose situated-but-dynamic hierarchies.

Intrasectionality references processes we can understand as 'becoming-climate' (e.g. Haraway 2008): how all entities differ from themselves due to intra-action with/as climate; how climate is simultaneously changed; and also how the boundaries between entities and the climate are differentially enacted through these processes. That is, the moving, blurring and at times, dissolving, of particular boundaries is part of climate intra-action, as is the articulation of different or new boundaries.

Acting-with

Intra-action provides two notable challenges to anthropocentric understandings of agency: firstly, that agency is not attributable just to humans or even to individual non-human entities, as these entities 'do not pre-exist as such' (Barad 2007, 179), and secondly that agency is therefore not a correlate of intentionality. An intra-active approach contends that no being is ever alone and that therefore no being ever acts alone; rather, everything is always acting-with a 'confederation', 'assemblage' or 'entanglement' of human and more-than-human forces, intensities, entities and processes (Bennett 2010; Haraway 2016; Barad 2007). As Barad makes explicit, the problem is 'the framing of agency as a localisable attribution' (2007, 216). For Barad, matter does not possess agency, rather, matter is 'a congealing of agency' (2007, 151). That is, agency and entities co-emerge through their intra-actions. Climate change is such a 'congealing of agency' and all the actions that create, mitigate and/or adapt to climate change occur through acting-with the world.

Climate is a set of worldly relationships (Knox 2015), and climate *change* is a change in those relationships. Climate change is therefore not a thing created by a homogenised and autonomous 'humanity' that acts upon the world. Rather, climate change is a radically disrupted sequence of lively, more-than-human repertoires of responsiveness, emerging out of the intra-actions between some (not all!) humans, fossil fuels, industrialisation, animal metabolisms, technologies, extractivist cultures, and the climate's own patterned dynamics (Neimanis and Walker 2013; Sasser 2016; Walker 2013). A small minority of fossil-fuelled-humans benefit from climate change in the short term, but even that small minority never intended to make climate change happen. Climate change thus totally challenges the anthropocentric belief that human agency and intentionality are aligned; despite our best efforts, we cannot seem to control our 'own' effects on the world (Clark 2010; Dürbeck, Schaumann, and Sullivan 2015). Climate/change is acting-with.

Diffraction

Given we are always acting-with, the feasibility of consistently achieving standardised climate actions such as recycling better/driving less/eating less meat/writing to a politician is challenged. To predetermine pedagogical outcomes in such a way would be to presume the individual human is the sole and independent location of agency (Snaza 2015; Lenz Taguchi 2010; Rooney 2018) which can foreclose attention to previously unimagined means of responding to climate change. Thus, we need to reconsider the purpose or goals of CCE, as well as whether having 'goals' is itself feasible or desirable.

Relational materialists advocate exploratory, risky, open ended, processual, fluid, moving, curious and responsive pedagogies (Snaza et al. 2016; Sonu and Snaza 2015; Nxumalo 2017; Leibowitz and Naidoo 2017; Taylor and Ivinson 2013; Hickey-Moody, Palmer, and Sayers 2016) which 'meet the universe halfway' (Spector 2015, 449, paraphrasing Barad 2007). In tension with this, climate change is an incredibly urgent and massive problem and there are many obvious practices which can already effectively reduce greenhouse gas emissions. I believe Haraway and Barad's work on diffraction can help in developing relational materialist pedagogies in CCE that navigate this tension.

Diffraction is the effect of the patterns of movement that waves create as they disturb the world (Haraway 1997; Barad 2007). Diffractions are therefore generative, enabling new, novel, innovative, creative or different phenomena to emerge (Bozalek and Zembylas 2016; Mellander and Wiszmeg 2016; Neely and Nguse 2015). However, diffractions are eco-historically situated – they do not emerge out of nowhere or no-when (Barad 2007). Diffraction has been enacted widely as a research methodology (Kaiser and Thiele 2014; van der Tuin 2014), including within education research (Bozalek and Zembylas 2016; Lather 2016; Leibowitz and Naidoo 2017; Lenz Taguchi 2012, 2010; Mazzei, 2014; Phillips and Larson 2012; Taylor and Blaise 2014; Taylor 2013; Orlander and Ståhl 2018).

Diffraction has been less discussed as a pedagogy, although educational literature developing this is beginning to emerge (Spector 2015; Lenz Taguchi 2010). Within these accounts, a diffractive pedagogy is figured as one where learning emerges from the collisions of waves such as students and the wider material world. A diffractive pedagogy therefore cultivates creativity, reconfigures bodies and subjectivities, is dynamic, non-linear, transdisciplinary, multi-modal, disruptive, unchartered, transcorporeal, interwoven, and one that troubles established categories (Hickey-Moody, Palmer, and Sayers 2016; Postma 2012; Spector 2015; Lenz Taguchi 2010).

Taking the compositional parts of diffraction as diffr (different, difference, differentiation, differing) and action, we have an account that can understand that our actions that create, mitigate and/or adapt to climate change produce and are produced by difference: they contribute to, and are emergent enactments of, the ongoing differentiation of the climate-world. We can understand our climate actions as diffr-actions – as gatherings of confederations of human-and-more-than-human agencies which interfere and are interfered with as they surge, radiate into and become enfolded in the world's ongoing becoming. Thus, while we might start with intentions and directions, attention to climate diffractions allows us to be open to how the world might bump into, disrupt, reinforce, invert and/or displace our desired trajectories and inclinations, perhaps blocking or resisting our efforts, but also potentially generating novel, new, innovative, different forms of climate action and actors, and of course therefore, different climates.

It is through intra-sectionally diffr-acting-with climate change that our abilities to respond – our response-abilities – continuously emerge (Barad 2007). Taking a lead from Latour (2004), the diffractive climate educator therefore assembles, to the best of their intra-active abilities, entanglements in which students might become more response-able, able to create different futures with the world. In the section that follows, I explore this intra-active approach to CCE in partnership with empirical occurrences from CCR15.

Killing joy as a climate making project

In week 6 of CCR15 with the notion of intra-action in mind I asked students to discuss what they had learned so far, specifying that this might include how they had changed or what they were doing differently. Emerging as part of the entangled voice, one young woman contributed:

> I used to, like, not be so aggressive, to my housemates. But now, I'm just sort of, being a bit of a bitch [class laughs]. Like, going through the rubbish bin and telling them to recycle better, making them take green bags to the supermarket, just generally telling them to clean up their act.

I was, of course, glad that the student had engaged in the climate actions of recycling and using green bags. But an intra-active approach enabled me to appreciate that there was more going on than this individual student acquiring facts about carbon footprints and then acting *on* climate change.

The student's 'superpositionality' intrasectionally emerged as one aligned with what Sara Ahmed calls a 'feminist killjoy' (Ahmed 2017, 2013) subjectivity. Ahmed argues that a feminist killjoy is one who is angry 'because that's a sensible response to what is wrong' and who is 'willing to kill joy', for example by 'getting in the way of the happiness of others'. The feminist killjoy is willing to kill the joy and happiness that is generated in and enables violence, and she is 'prepared to be called a kill joy', or in this case, '*a bitch*', when doing so (Ahmed 2013). For this student, the feminist killjoy subjectivity emerged in relation with apparently 'mundane' (Taylor 2013) everyday material intra-actions between carbon emissions, green bags, housemates, recycling infrastructure and policies (or lack of), and (white-Australian) patriarchy – including the specific norms of positioning female leadership as bossy and aggressive (Paxton and Hughes 2010; Salmond and Fleshman 2010; Ahmed 2017) and of degrading women and animals using similar logics (Plumwood 1993).

Through acting-with this host of forces, the student's superpositionality changed: she became '*a bitch*'. As climate is a set of relationships (Knox 2015) these rearranged material relations constituted a change in the climate (even if this change is seemingly insignificant on a global scale). And, the boundaries between the student and the climate changed: climate change is no longer solely an external, global, geo-scientific phenomenon, but is now embodied in her affective and domestic practices.

Climate diffractions emerge through this intra-sectional acting-with. Following Ahmed's contention that 'killing joy is a world making project' (Ahmed 2016), we can see that killing joy can be a climate making project: it can constitute a form of climate action. That is, refusing to conform to norms of positive climate talk (Rickards, Wiseman, and Kashima 2014) can contribute to more liveable climates. But becoming climate killjoys was not one of the goals I had for my students – it was an unanticipated, novel, different kind of climate action than what I had intended. While it was surprising and divergent, it was intimately wrapped up in the performance of those more generic, standardised, predetermined actions like recycling.

Killing carbon joy, as a diffraction, emerged as a powerful and disruptive climate action in CCR15. Becoming aware of and revealing the violences that climate change creates, both students and I often became climate killjoys throughout the semester. For example, students disclosed that they were *'constantly butting heads with skeptics'* and that they *'nag housemates, workmates and friends who live less sustainably on a constant basis'*. Yet we did not just kill the joy of those outside class: sometimes we killed our own personal carbon joy, or that of other students in the class, as the following end of semester journal details:

> During a small group discussion in the week 10 tutorial, one of the members of the group I was sitting with made an assertion that they thought anyone who owned a cat could not "truly call themselves an environmentalist". This, even though said in passing, really shocked me. As I was sitting there in my coat covered in cat fur, I felt as though I had never really properly considered cat ownership through the lens of climate change mitigation. As someone who has always been environmentally conscious and concerned about climate change, especially at home, I started to question how legitimate my "environmentalism" could be if I had overlooked something like how cat ownership could be interpreted as an act of anti-environmentalism.

After this discussion of how her identity – as 'an environmentalist' – intrasectionally emerged (or dissipated, or was reconfigured) *with* her cat, her coat, concepts, ethics, and other students' practices of killing carbon joy, the student went on to list the forces and entities that she acted-with in this climate changing relationship:

- *'the exotic tinned food and the dry, plastic packaged, biscuit-like food I feed her'*
- *'her collection of cat toys, kitty litter, brushes, pillows, blankets and plants', and*
- *'indirect purchases like buying new curtains, blinds and upholstery sooner than would otherwise be needed due to rips and clawing'.*

Continuing to interrogate joy, the student then further discussed how cat ownership and killing the joy of it became diffractions, actions she had not previously considered to be climate-constituting, and how her cat's identity also changed intrasectionally through this diffraction:

> I wondered if being blinded by affection for my cat has caused a "cognitive dissonance" effect on my ability to seriously consider her contribution towards climate change … Without being entirely conscious of it, my actions meant that my cat was an active consumer within the Western neo-liberal capitalist economy at the same time as I was trying to reduce my own involvement within it.

Here the boundaries between climate and human are repositioned: climate change is no longer caused by humans who are independent of the world, rather, climate change is enfolded into the student's affectionate relationship with her cat.

When I had originally read this story, I had been pleased at the individual student's critical reflective interrogation of neoliberalism. Thinking with intra-action, I have been able to also appreciate how the convergence of the various forces – cat fur, class discussions, capitalism – affected this student, and how her increased sensitivity to these connections enabled her to generate a novel responsiveness (Latour, 2004), becoming not just a killjoy but also more empathetic to others:

> Before taking this course I had assumed a narrow view of climate change and saw it as separate from other social issues… However, I now realise that climate change is an issue that is embedded within all aspects of my life [and] understand in a more personal way possible challenges … and barriers to lifestyle changes that might prohibit individuals from taking steps towards reducing emissions.

Conclusion

If we return to 'that lump of coal' (Hamilton 2017), a less reported part of Morrison's stunt was that his party member Kelly O'Dwyer, Minister for Revenue and Financial Services, refused to handle the coal when passed to her, declining to share in the glee that the coal aroused in the other male MPs (See Hannam, 2017 or Verlie & CCR15 2020 for the image). O'Dwyer continues to support 'clean coal' and this lacquered coal is about as clean as coal can get. Thus while we can hardly promote her as a climate activist, the photo of her rejection of the coal is compelling, if elusive. I see her nonplussed expression as a momentary performance of killing carbon joy, one that disruptively emerged as the vibrancy of the coal intra-acted-with her (and potentially, the vital materiality of her heavily pregnant belly played an influential role too). We can only speculate about what O'Dwyer was thinking or intending at the time, or what stimulated her to reject the coal, but the vision of her doing so has been useful for me in coming to recognise the complexity, nuance and contingent nature of what can count as climate actions, how human and non-human materiality are always interrupting and colliding with our intentions, and how this contributes to the ongoing, dynamic, reconfiguration of our subjectivity.

While it was not in 2015, Scott Morrison's stunt with the lump of coal was indicative of the kind of 'carbon heavy masculinity' (Alaimo 2016), which, combined with white-settler-nationalism, led to the particularly Australian carbon joy we were trying to kill in CCR15. Insights from the entangled voice emerging in response to such coal joy provide useful guidance here for climate educators, insights that were generated in part due to coal emerging in class discussions and my pedagogy that encouraged students to articulate connections between humans and the more-than-human world:

> [I had originally thought] that science was the way to communicate about climate change ... But it is much more complicated than that. Australia's history is built off of coal. In most ways, we still rely on coal. The [climate] denial is almost cultural as it would not serve Australia well to believe in climate change. Our economic prosperity is tangled with it. Many people's identity is intertwined with it. When you think about it like that, like a change in identity, it is easier to see why people are in denial. (CCR 2015)

This student articulates how non-human materiality, social forces, identities, climate changing practices and climate knowledge are (en)tangled, and I think this is indicative of the more nuanced, relational, and less anthropocentric approach that intra-action can offer CCE. By diffracting intra-action, coal, CCR15 and CCE, I have come to think that 'goals' of 'climate action' too often determine in advance what is possible, limit what counts as mitigative or adaptive practice, and assume or aim for control over the world. If we are too militant and static in our education efforts, we can risk reinforcing anthropocentric subjectivities.

But climate intra-action does not preclude aiming for climate actions such as recycling, or keeping coal in the ground – we should still try! What an additional attunement to climate intra-action can do, however, is clarify the limited capacity we have to achieve such predetermined goals, through recognising that we (educators and students) are always acting-with myriad, sometimes known but often unknown and unknowable agential others, who can interfere with, disrupt, resist, reinforce and/or reorient our actions. With this attention to the potential for diffr-actions – different actions, actions that differ, actions that differentiate, difference that generates action - we can work *with the world*, rather than upon the world, to foster the capacity to respond to climate change in previously unthought ways, ways that may reconfigure, complement or amplify more standard climate actions. Rather than a disablement or negation of responsibility, decentring the human from CCE by attending to who and what we might be able to act-with and become-with can in fact grow our climate capacities (O'Brien 2016). Human response-abilities are not erased in this relational materialist climate change pedagogy, rather, they continuously emerge-with the climate's ongoing differentiation.

So, I still avidly want climate action. But, a focus on climate intra-action which understands climate action as a series of unpredictable, experimental, collaboratively enacted reconfigurations of worldly relationships has enabled me to better attune to the nuance and complexity of CCE and the need to be constantly navigating the dynamic, more-than-human field of partners that I must intra-sectionally diffr-act-with if I am to be an effective and response-able climate change educator.

Acknowledgments

This research was conducted on Wurundjeri and Dja Dja Wurrung country. I pay respect to Wurundjeri and Dja Dja Wurrung elders past, present and future, and recognise that colonisation is an ongoing process which requires active decolonisation.

I would like to thank my PhD supervisors, Iris Duhn and Lauren Rickards, as well as my Climate Change Responses class from 2015 for participating in this research project. I would also like to thank the three anonymous reviewers provided insightful feedback on earlier drafts of this paper.

Disclosure statement

No potential conflict of interest was reported by the authors.

References

Ahmed, Sara. 2013. "Hello Feminist Killjoys!" *Feminist Killjoys*. https://feministkilljoys.com/2013/08/26/hello-feminist-killjoys/.

Ahmed, Sara. 2016. "Dedication." *Feminist Killjoys*. https://feministkilljoys.com/2016/08/28/dedication/.

Ahmed, Sara. 2017. *Living a Feminist Life*. Durham: Duke University Press.

Alaimo, S. 2008. "Trans-corporeal Feminisms and the Ethical Space of Nature." In *Material Feminisms*, edited by S. Alaimo and S.J. Hekman, 237–264. Bloomington, IN: Indiana University Press.

Alaimo, S. 2016. *Climate Change, Carbon-Heavy Masculinity, and the Politics of Exposure*. Minneapolis, MN: University of Minnesota Press. http://www.uminnpressblog.com/2016/10/climate-change-carbon-heavy-masculinity.html.

Barad, Karen. 2007. *Meeting the Universe Halfway: Quantum Physics and the Entanglement of Matter and Meaning*. Durham: Duke University Press.

Barad, Karen. 2014. "Diffracting Diffraction: Cutting Together-apart." *Parallax* 20 (3): 168–187. doi:10.1080/13534645.2014.927623.

Bennett, Jane. 2010. *Vibrant Matter: A Political Ecology of Things*. Durham: Duke University Press.

Bozalek, Vivienne, and Michalinos Zembylas. 2016. "Diffraction or Reflection? Sketching the Contours of Two Methodologies in Educational Research." *International Journal of Qualitative Studies in Education* 30 (2): 111–127. doi:10.1080/09518398.2016.1201166.

Braidotti, Rosi. 2013. *The Posthuman*. Cambridge and Malden, MA: Polity Press.

CCR. 2015. *ENVI1212: Climate Change Responses class*. RMIT University.

Clare, Stephanie. 2016. "On the Politics of "New Feminist Materialisms"." In *Mattering: Feminism, Science, and Materialism*, edited by V. Pitts-Taylor, 58–72. New York, NY: New York University Press.

Clark, Nigel. 2010. "Volatile Worlds, Vulnerable Bodies." *Theory, Culture & Society* 27 (2–3): 31–53. doi:10.1177/0263276409356000.

Collins, Patricia Hill. 2000. *Black Feminist Thought: Knowledge, Consciousness, and the Politics of Empowerment*. Milton Park: Routledge.

Crenshaw, Kimberle. 1991. "Mapping the Margins: Intersectionality, Identity Politics, and Violence Against Women of Color." *Stanford Law Review* 43 (6): 1241–1299. doi:10.2307/1229039

Duhn, Iris. 2017. "Cosmopolitics of Place: Towards Urban Multispecies Living in Precarious Times." In *Reimagining Sustainability in Precarious Times*, edited by Karen Malone, Son Truong and Tonia Gray, 45–57. Singapore: Springer.

Dürbeck, Gabriele, Caroline Schaumann, and Heather Sullivan. 2015. "Human and Non-human Agencies in the Anthropocene." *Ecozon@* 6 (1): 118–136.

Edwards, Richard. 2012. "Theory Matters: Representation and Experimentation in Education." *Educational Philosophy and Theory* 44 (5): 522–534. doi:10.1111/j.1469-5812.2010.00719.x.

Edwards, Richard, and Tara Fenwick. 2013. "Performative Ontologies: Sociomaterial Approaches to Researching Adult Education and Lifelong Learning." *European Journal for Research on the Education and Learning of Adults* 4 (1): 49–63.

Fenwick, Tara. 2012. "Matterings of Knowing and Doing: Sociomaterial Approaches to Understanding Practice." In *Practice, Learning and Change: Practice-theory Perspectives on Professional Learning* edited by Paul Hager, Alison Lee, and Ann Reich, 67–83. Dordrecht: Springer.

Fenwick, Tara, and Richard Edwards. 2014. "Networks of Knowledge, Matters of Learning, and Criticality in Higher Education." *Higher Education* 67 (1): 35–50. doi:10.1007/s10734-013-9639-3.

Hamilton, Clive. 2017. "That Lump of Coal." *The Conversation*. https://theconversation.com/that-lump-of-coal-73046.

Hannam, P, 2017, Brandishing coal in a heatwave? Scott Morrison might as well fiddle through the energy crisis. The Sydney Morning Herald. https://www.smh.com.au/politics/federal/brandishing-coal-in-a-heatwave-scott-morrison-might-as-well-fiddle-as-australia-suffers-an-energy-crisis-20170213-gubasj.html

Haraway, Donna. 1997. *Modest – Witness@Second – Millennium.FemaleMan – Meets – OncoMouse: Feminism and Technoscience*. London: Routledge.

Haraway, Donna. 2008. *When Species Meet*. Minneapolis: University of Minnesota Press,.

Haraway, Donna. 2015. "Anthropocene, Capitalocene, Plantationocene, Chthulucene: Making Kin." *Environmental Humanities* 6: 159–165. doi:10.1215/22011919-3615934.

Haraway, Donna. 2016. *Staying with the Trouble: Making Kin in the Chthulucene*. Durham: Duke University Press.

Head, Lesley, and Chris Gibson. 2012. "Becoming Differently Modern: Geographic Contributions to a Generative Climate Politics." *Progress in Human Geography* 36 (6): 699–714. doi:10.1177/0309132512438162.

Henderson, Joseph, David Long, Paul Berger, Constance Russell, and Andrea Drewes. 2017. "Expanding the Foundation: Climate Change and Opportunities for Educational Research." *Educational Studies* 53 (4):412–425. doi:10.1080/00131946.2017.1335640.

Hickey-Moody, Anna, Helen Palmer, and Esther Sayers. 2016. "Diffractive Pedagogies: Dancing across New Materialist Imaginaries." *Gender and Education* 28 (2): 213–229. doi:10.1080/09540253.2016.1140723.

Huggan, Graham, and Helen Tiffin. 2010. *Postcolonial Ecocriticism: Literature, Animals, Environment*. Milton Park: Routledge.

Jackson, Alecia Y. 2008. "'What Am I Doing when I Speak of this Present?' Voice, Power and Desire in Truth-telling." In *Voice in Qualitative Inquiry: Challenging Conventional, Interpretive, and Critical Conceptions in Qualitative Research*, edited by Alecia Y Jackson and Lisa A Mazzei, 165–174. Milton Park: Routledge.

Jackson, Alecia Y., and L.A. Mazzei. 2012. *Thinking with Theory in Qualitative Research: Viewing Data across Multiple Perspectives*. Milton Park: Routledge.

Kaiser, Birgit Mara, and Kathrin Thiele. 2014. "Diffraction: Onto-epistemology, Quantum Physics and the Critical Humanities." *Parallax* 20 (3): 165–167. doi:10.1080/13534645.2014.927621.

Kenny, Mark. 2017. "Turnbull Government Sharpens Attack on Labor's Clean Energy." *The Sydney Morning Herald*. http://www.smh.com.au/federal-politics/political-news/turnbull-government-sharpens-attack-on-labors-clean-energy-20170209-gu9h5l.html

Kirksey, S. Eben, and Stefan Helmreich. 2010. "The Emergence of Multispecies Ethnography." *Cultural Anthropology* 25 (4): 545–576. doi:10.1111/j.1548-1360.2010.01069.x.

Knox, Hannah. 2015. "Thinking Like a Climate." *Distinktion: Journal of Social Theory* 16 (1): 91–109. doi:10.1080/1600910X.2015.1022565.

Lather, Patti. 2016. "Top Ten + List (re) thinking Ontology in (post) Qualitative Research." *Cultural Studies↔ Critical Methodologies* 16 (2): 125–131. doi:10.1177/1532708616634734.

Latour, B. 2004. "How to Talk About the Body? The Normative Dimension of Science Studies." *Body and Society* 10 (2–3): 205–229. doi:10.1177/1357034X04042943.

Leibowitz, Brenda, and Kibashini Naidoo. 2017. "The Potential for Posthuman Insights to Effect Socially Just Pedagogies." *Education as Change* 21 (2): 149–164. doi:10.17159/1947-9417/2017/2020

Lenz Taguchi, Hillevi. 2010. *Going Beyond the Theory/practice Divide in Early Childhood Education: Introducing an Intra-active Pedagogy*. Milton Park: Routledge.

Lenz Taguchi, Hillevi. 2012. "A Diffractive and Deleuzian Approach to Analysing Interview Data." *Feminist Theory* 13 (3): 265–281. doi:10.1177/1464700112456001.

Lorde, Audre. 1984. *Sister Outsider: Essays and Speeches*. Berkeley, CA: The Crossing Press.

Massola, James, Peter Ker, and Lisa Cox. 2014. "Coal is 'Good for Humanity', says Tony Abbott at Mine Opening." *The Sydney Morning Herald*. http://www.smh.com.au/federal-politics/political-news/coal-is-good-for-humanity-says-tony-abbott-at-mine-opening-20141013-115bgs.html.

Mazzei, Lisa A. 2013. "A Voice Without Organs: Interviewing in Posthumanist Research." *International Journal of Qualitative Studies in Education* 26 (6): 732–740. doi:10.1080/09518398.2013.788761.

Mazzei, Lisa A. 2014. "Beyond an Easy Sense: A Diffractive Analysis." *Qualitative Inquiry* 20 (6): 742–746. doi:10.1177/1077800414530257.

Mazzei, Lisa A. 2016. "Voice Without a Subject." *Cultural Studies ↔ Critical Methodologies* 16 (2): 151–161. doi:10.1177/1532708616636893.

Mazzei, Lisa A., and Alecia Y. Jackson. 2016. "Voice in the Agentic Assemblage." *Educational Philosophy and Theory* 49 (11): 1090–1098. doi:10.1080/00131857.2016.1159176.

McKenzie, M., and A. Bieler. 2016. "Critical Education and Sociomaterial Practice: Narration, Place, and the Social." In *[Re]thinking Environmental Education*, edited by Constance Russell and Justin Dillon. New York: Peter Lang.

Mellander, Elias, and Andréa Wiszmeg. 2016. "Interfering with Others – Re-configuring Ethnography as a Diffractive Practice." *Kulturstudier* 7 (1): 93–115. doi:10.7146/ks.v7i1.24056.

Monroe, Martha C., Richard R. Plate, Annie Oxarart, Alison Bowers, and Willandia A. Chaves. 2017. "Identifying Effective Climate Change Education Strategies: A Systematic Review of the Research." *Environmental Education Research* 1–22. doi:10.1080/13504622.2017.1360842.

Moser, Susanne C. 2016. "Reflections on Climate Change Communication Research and Practice in the Second Decade of the 21st Century: What More is There to Say?" *Wiley Interdisciplinary Reviews: Climate Change* 7 (3): 345–369. doi:10.1002/wcc.403.

Neely, Abigail H., and Thokozile Nguse. 2015. "Relationships and Research Methods: Entanglements, Intra-actions, and Diffraction." In *The Routledge Handbook of Political Ecology*, edited by T. Perreault, G. Bridge, and J. McCarthy, 140–149. Milton Park: Routledge.

Neimanis, Astrida. 2015. "Weather Writing: A Feminist Materialist Practice for (Getting Outside) the Classroom." In *Teaching with Feminist Materialisms*, edited by Peta Hinton and Pat Treusch, 141–157. Utrecht: Atgender.

Neimanis, Astrida. 2017. *Bodies of Water: Posthuman Feminist Phenomenology*. 1st ed. London: Bloomsbury Academic.

Neimanis, Astrida, and Rachel Loewen Walker. 2013. "Weathering: Climate Change and the "Thick Time" of Transcorporeality." *Hypatia* 29 (3): 558–575. doi:10.1111/hypa.12064.

Nxumalo, Fikile. 2017. "Stories for living on a damaged planet: Environmental education in a preschool classroom." *Journal of Early Childhood Research* 00 (0): 1–12. doi:10.1177/1476718X17715499.

O'Brien, Karen L. 2016. "Climate Change and Social Transformations: is it Time for a Quantum Leap?" *Wiley Interdisciplinary Reviews: Climate Change* 7: 618–626. doi:10.1002/wcc.413.

Orlander, Auli Arvola, and Marie Ståhl. 2018. "Towards an Understanding of Diffractive Readings of Narratives in the Field of Science Education." In *Cultural, Social, and Political Perspectives in Science Education: A Nordic View*, edited by Kathrin Otrel-Cass, Martin Krabbe Sillasen and Auli Arvola Orlander, 139–152. Cham: Springer International Publishing.

Paxton, P., and M. M. Hughes. 2010. "Women as Presidents, Prime Ministers, and Government Ministers." In *Gender and women's leadership: A reference handbook*, edited by K. O'Connor, 321–330. Thousand Oaks, CA: SAGE.

Pearse, Rebecca. 2017. "Gender and Climate Change." *Wiley Interdisciplinary Reviews: Climate Change* 8 (2): 1–16. doi:10.1002/wcc.451.

Pedersen, Helena. 2010a. "Education Policymaking for Social Change: A Post-humanist Intervention." *Policy Futures in Education* 8 (6):683–696. doi:10.2304/pfie.2010.8.6.682.

Pedersen, Helena. 2010b. "Is 'The Posthuman' Educable? On the Convergence of Educational Philosophy, Animal Studies, and Posthumanist Theory." *Discourse: Studies in the Cultural Politics of Education* 31 (2): 237–250. doi:10.1080/01596301003679750.

Pedersen, Helena, and Barbara Pini. 2017. "Educational Epistemologies and Methods in a More-than-human World." *Educational Philosophy and Theory* 49 (11): 1051–1054. doi:10.1080/00131857.2016.1199925.

Phillips, Donna Kalmbach, and Mindy Legard Larson. 2012. "The Teacher–student Writing Conference Entangled: Thinking Data with Material Feminisms." *Cultural Studies ↔ Critical Methodologies* 12 (3): 225–234. doi:10.1177/1532708612440256.

Plumwood, Val. 1993. *Feminism and the Mastery of Nature, Opening Out*. London: Routledge.

Postma, Dirk. 2012. "Education as Sociomaterial Critique." *Pedagogy, Culture & Society* 20 (1): 137–156. doi:10.1080/14681366.2012.649419.

Puig de la Bellacasa, Maria. 2010. "Ethical Doings in Naturecultures." *Ethics, Place and Environment* 13 (2): 151–169. doi:10.1080/13668791003778834.

Quinn, Jocey. 2013. "Theorising Learning and Nature: Post-human Possibilities and Problems." *Gender and Education* 25 (6): 738–753. doi:10.1080/09540253.2013.831811.

Rautio, Pauliina. 2012. "Being Nature: Interspecies Articulation as a Species-specific Practice of Relating to Environment." *Environmental Education Research* 19 (4): 445–457. doi:10.1080/13504622.2012.700698.

Regester, Sam. 2017. "Please Stop Ruining the Planet for us: An Open Letter to the Government, from Young People." *Junkee* (10/02/2017). http://junkee.com/please-stop-ruining-planet-us-open-letter-government-young-people/95712

Rickards, Lauren, John Wiseman, and Yoshi Kashima. 2014. "Barriers to Effective Climate Change Mitigation: The Case of Senior Government and Business Decision Makers." *Wiley Interdisciplinary Reviews: Climate Change* 5 (6): 753–773. doi:10.1002/wcc.305.

Rooney, Tonya. 2018. "Weather Worlding: Learning with the Elements in Early Childhood." *Environmental Education Research* 24 (1): 1–12. doi:10.1080/13504622.2016.1217398.

Salmond, K., and P. Fleshman. 2010. "The Development of Children's Perception of Leadership." In *Gender and Women's Leadership: A Reference Handbook*, edited by K. O'Connor, 11–19. Thousand Oaks, CA: SAGE.

Sasser, Jade S. 2016. "Population, Climate Change, and the Embodiment of Environmental Crisis." In *Systemic Crises of Global Climate Change: Intersections of Race, Class and Gender*, edited by P. Godfrey and D. Torres. Milton Park: Taylor & Francis.

Sharp, Hasana. 2017. "Spinoza and the Possibilities for Radical Climate Ethics." *Dialogues in Human Geography* 7 (2): 156–160. doi:10.1177/2043820617720063.

Sjögren, Hanna. 2014. "Educable Futures?: Managing Epistemological Uncertainties in Sustainability Education." *Resilience: A Journal of the Environmental Humanities* 1 (2): 1–14. doi:10.5250/resilience.1.2.005.

Snaza, Nathan. 2015. "Toward a Genealogy of Educational Humanism." In *Posthumanism and Educational Research*, edited by N. Snaza and J. Weaver, 17–29. Milton Park: Routledge.

Snaza, Nathan, Debbie Sonu, Sarah E Truman, and Zofia Zaliwska. 2016. "Introduction: Re-attuning to the Materiality of Education." In *Pedagogical Matters: New Materialisms and Curriculum Studies*, edited by Nathan Snaza, Debbie Sonu, Sarah E Truman and Zofia Zaliwska, XV–XXXI. New York, NY: Peter Lang Publishing, Inc.

Somerville, Margaret. 2017. "The Anthropocene's Call to Educational Research." In *Reimagining Sustainability in Precarious Times*, edited by Karen Malone, Son Truong and Tonia Gray, 17–28. Singapore: Springer.

Sonu, Debbie, and Nathan Snaza. 2015. "The Fragility of Ecological Pedagogy: Elementary Social Studies Standards and Possibilities of New Materialism." *Journal of Curriculum and Pedagogy* 12 (3): 258–277. doi:10.1080/15505170.2015.1103671.

Spector, Karen. 2015. "Meeting Pedagogical Encounters Halfway." *Journal of Adolescent & Adult Literacy* 58 (6): 447–450. doi:10.1002/jaal.382.

St. Pierre, Elizabeth Adams. 2008. "Decentering Voice in Qualitative Enquiry." In *Voice in Qualitative Inquiry: Challenging Conventional, Interpretive, and Critical Conceptions in Qualitative Research*, edited by Alecia Y. Jackson and Lisa A. Mazzei, 221–236. New York, NY: Routledge.

Swanson, Heather, Anna Tsing, Nils Bubandt, and Elaine Gan. 2017. "Introduction: Bodies Tumbled into Bodies." In *Arts of Living on a Damaged Planet: Ghosts and Monsters of the Anthropocene*, edited by A.L. Tsing, H.A. Swanson, E. Gan and N. Bubandt, 1–13. Minneapolis, MN: University of Minnesota Press.

Taylor, Affrica, and Mindy Blaise. 2014. "Queer Worlding Childhood." *Discourse: Studies in the Cultural Politics of Education* 35 (3): 377–392. doi:10.1080/01596306.2014.888842.

Taylor, Affrica, and Miriam Giugni. 2012. "Common Worlds: Reconceptualising Inclusion in Early Childhood Communities." *Contemporary Issues in Early Childhood* 13 (2): 108–119. doi:10.2304/ciec.2012.13.2.108.

Taylor, Carol A. 2013. "Objects, Bodies and Space: Gender and Embodied Practices of Mattering in the Classroom." *Gender and Education* 25 (6): 688–703. doi:10.1080/09540253.2013.834864.

Taylor, Carol A., and Gabrielle Ivinson. 2013. "Material Feminisms: New Directions for Education." *Gender and Education* 25 (6): 665–670. doi:10.1080/09540253.2013.834617.

Tuana, Nancy. 2008. "Viscous porosity: Witnessing Katrina." In *Material Feminisms*, edited by S. Alaimo and S.J. Hekman, 188–213. Bloomington, IN: Indiana University Press.

Tuck, E., and M. McKenzie. 2014. *Place in Research: Theory, Methodology, and Methods*. London: Taylor & Francis.

van der Tuin, Iris. 2014. "Diffraction as a methodology for feminist onto-epistemology: On encountering Chantal Chawaf and posthuman interpellation." *Parallax* 20 (3):231–244. doi:10.1080/13534645.2014.927631.

van Dooren, Thom, Eben Kirksey, and Ursula Münster. 2016. "Multispecies Studies Cultivating Arts of Attentiveness." *Environmental Humanities* 8 (1): 1–23. doi:10.1215/22011919-3527695.

Verlie, Blanche. 2017. "Rethinking Climate Education: Climate as Entanglement." *Educational Studies* 53 (6): 560–572.

Verlie, B & CCR15 (2020) From action to intra-action? Agency, identity and 'goals' in a relational approach to climate change education, Environmental Education Research, 26:9–10, 1266–1280, doi: 10.1080/13504622.2018.1497147.

Walker, Rachel Loewen. 2013. "Environment Imagining Otherwise." *Journal of Curriculum and Pedagogy* 10 (1): 34–37. doi:10.1080/15505170.2013.790001.

OPEN ACCESS

Entangled threads and crafted meanings: students' learning for sustainability in remake activities

Hanna Hofverberg

ABSTRACT

This article explores the significance of students' encounters with materiality in general and with crafting materials in particular when learning for sustainability. The aim of the explorative study is to illustrate a research approach that can show what students and the material do in correspondence and what stories emerge from this activity. An explorative analysis is conducted via video recordings of a remake project in a Grade 8 handicrafts class in Sweden. The stories that the students recognise are the material's texture, shape and construction, which emerge from the materiality intrinsic to the crafting process and the intentions of the students, as these are visible in action. These stories provide possibilities, as well as set limits for, what is possible to remake. The stories are elaborated on by threading back to materiality concerns found in historical remake practice to recognise the educational possibilities for remaking pedagogy.

Introduction

As a reaction against the neglecting of materials and how matter *matters* (Barad 2003), the turn to materiality in research (Coole and Frost 2010) is concerned with decentralising the human subject among materials (Fenwick 2015; Sørensen 2009). In environmental and sustainability education research (ESER), questions about how materials have come to matter have intensified in the last decade due not only to the materiality turn but also to the call for lifestyle re-orientations in the 'expansion of mankind' age – the Anthropocene (Crutzen and Stoermer 2000; Steffen et al. 2015). At the core of these ESER discussions about how materials matter is a concern to find other ways of knowing and understanding the entanglements of how materials and humans live and learn together to engage with environmental and sustainability issues (see for example Clarke and Mcphie 2016; Lloro-Bidart 2017; Mannion 2007; Payne 2016; McKenzie and Bieler 2016; Pyyry 2017; Russell 2005; Rautio 2013; Rautio et al. 2017; Somerville 2016, Taylor 2017; Van Poeck and Lysgaard 2016). In this article, I seek to contribute to this scholarly discussion with an explorative study about the significance of students encountering materiality in general and crafting material in particular in conjunction with the students' learning for sustainability. The case in focus for this study is a school project in Sweden in which students were given the task to remake old clothes into new products.

The reason for exploring a remake project with textiles is that such projects are common recycling activities in environmental and sustainability education (ESE) and seen by policy makers as setting an example for how one can save resources (Swedish National Agency for Education 2011, 203). In fact, in Sweden today, eight kilograms of textiles per person contribute to household waste each year (SMED 2011, 6); thus, remaking projects are often seen as relevant for ESE. Despite that remaking activity is common in education, usually as part of the Swedish handicrafts subject, 'educational sloyd', research into remaking activities is limited. However, related research focusing on remaking activities in ESE does exist. For example, Danish scholars studied waste activities in ESE and found that teachers are using 'artistic activities as an entry point for dealing with waste' (Jørgensen, Madsen, and Læssøe 2018, 810). Some of the interviewed Danish teachers expressed that working with 'reusable materials support children's fantasy, ingenuity and creativity' (811). Another example is Odegard (2012), who studied how preschool children encounter 'junk' materials in remaking activities by analysing focus group conversations with preschool teachers. In the group conversation, the teachers brought with them pedagogical documentation of the activities, such as photos and texts. Odegard focused on how materials are recognised and emerge in what is referred to as 'intra-activity' (Barad 2007). Based on her findings, Odegard argues that when the children encounter the materials, the materials are 'undefined', which she further argues provides 'an articulation that emphasises their [the material's] properties rather than their uses' (387). However, she continues to explain that how the children worked with the materials was, to a large extent, 'dependant on the teachers' expectations of an upcoming product' (398), thus showing the importance of remaking pedagogy. One final example of a remaking project is from South Africa, where McGarry (2013) co-created artwork together with five young men who were engaged with informal waste collection from local landfills. In the project, the waste was used to make puppets and the puppets were used to explore the participants 'perceptions of personal and social progress, essentially how they could improve their situations with their existing means, and look at what way these means/abilities could be further enriched or developed' (112). For the participants, working with waste in new ways resulted in changes, even though, as McGarry concludes, it was hard to see what specific practices, instruments or processes led to the changes or 'where learning was taking place' (p. 113). These examples from Denmark, Norway and South Africa show the potential of remake projects in ESE. However, observations of students' encounters with the material in remake projects has not been explored, and this area is in need of empirical research. Of relevance to this article is that there is limited knowledge about students' encounters with materiality, and further, about how these human–material encounters can contribute to a remake pedagogy, which could further motivate empirical research of student–material encounters in remake activities.

To contribute to the scholarly discussion of the significance of students encountering materiality in general and crafting material in particular, this explorative study aims to illustrate a research approach inspired by Ingold's theory of making. This aim is addressed by asking the following empirical questions:

1. What functions does the material have for students when learning to remake?
2. What stories about human–material relations are created when students encounter the material in the remake activity?
3. In the students' remaking activity, does a conjunction of the present and the past emerge between the stories created and the historical practice of remaking? And if so, how could this be used to further develop students' remaking skills in relation to sustainability?

The outline of the paper is as follows: First, Ingold's (2011, 2013) concepts of 'practice of correspondence' and 'storying' are presented. These concepts helped me identify student–material encounters in crafting activities and the embodied stories that emerged from these correspondences. Thereafter, the case of the remake project is presented along with the selection of the empirical data. Then four illustrative examples of the correspondence between students and the material and what stories that emerge from these correspondences are presented. This third section gives an account of how the first and the second research questions are answered. Fourthly, the illustrative examples are discussed as materiality concerns found in historical remake practice, thus answering the third research question. Lastly, how the research approach can create new knowledge fruitful for remake activities and remake pedagogy are discussed.

A theory of crafting: human–material correspondence

Crafting is all about making things – transforming material into new forms and functions – and one can easily get the idea that the craft activity starts with an idea about what one wants to achieve, followed by imposing a form on the material 'out there' (Ingold 2013, 20). In this article, I follow Ingold's (2011, 2013) understanding of making, which suggests something different, namely, that humans do not act upon the world but rather act from within in what Ingold describes as 'a process of growth' (2013, 21). He continues,

> This is to place the maker from the outset as a participant in amongst a world of active materials. These materials are what he has to work with, and in the process of making he 'joins forces' with them, bringing them together or splitting them apart, synthesising and distilling, in anticipation of what might emerge. (ibid.)

Hence, Ingold argues that, in crafting activities, a craftperson joins forces with the material, and this process is what Ingold (2013) refers to as a 'practice of correspondence' (108). This contrasts with a process of interaction, which Ingold describes as two closed parties that connect 'through some kind of bridging operation' (107). In a practice of correspondence, people are already set in relation to the materiality of the world. To correspond with the world, Ingold argues, 'is not to describe it, or to represent it, but to *answer to it*' (108). From a crafting perspective, this means that learning occurs when we answer to the world and join forces with the material in a new way, and in doing so, engage in new correspondences with the material. In the article, I explore these student–material encounters in the crafting activity. In the exploration, I do not consider that students act upon passive materials, but rather, in the crafting process, students join forces with the materials as they answer to them in a practice of correspondence.

As students learn to answer to the material in correspondence, the material or the tools are not separated from the crafting activity. Instead, as Ingold (2011) argues, 'The entire system of forces and relations set up by the intimate engagement of the saw, the trestle, the workpiece and my own body' (56), where the system and the forces are perceived as the activity. This means that a material or tool has no intrinsic attributes. But, as Ingold continues, 'To describe a thing as a tool is to place it in *relation* to other things within a field of activity in which it can exert a certain effect' (ibid.). Hence, a tool or a material is always known in relation to other things within an activity. To understand the meaning-making process that emerges from a specific practice (i.e. a system of forces), Ingold uses the concept of 'stories'. He explains that a tool 'must be endowed with a story, which the practitioner should know and understand in order to recognise it as such and use it appropriately' (ibid.). Ingold goes even further when he states that 'considered as tools, things *are* their stories' (ibid). These stories should be understood as creating a continuation between the past practice (history), the current practice (the present) and the purpose or task in terms of the excepted outcome (future). Ingold (2011) further explains,

> Just as stories do not carry their meanings ready-made into the world so, likewise, the ways in which the tools are to be used do not come pre-packaged with the tools themselves. But neither are the uses of tools simply invented on the spot, without regard to any history of the past practice. Rather, they are revealed to practitioners when, faced with a recurrent task in which the same devices were known previously to have been employed, they are perceived to afford the wherewithal for its accomplishment. Thus, the functions of tools, like the meaning of stories, are recognised through the alignment of present circumstances with the conjunctions of the past. Once recognised, these functions provide the practitioner with the means to keep on going. (57)

In other words, Ingold argues that a specific meaning is neither fixed in, nor imposed on, a tool but rather emerges as a co-creating process recognised through the circumstances of the present (which has a specific purpose) in conjunction with the past within this specific practice. However, the story that emerges in conjunction with the past should not be understood as embedded in a static and closed tradition because what is being recontextualised and recognised is not pre-determined, but rather it emerges in the practice of correspondence. Arguably, the contextualisation process of continuity and change (in terms of time in the past, time in the future, and the present/place) is key to pedagogy, as it affects what stories will emerge.

In the article, I explore a remake project that is carried out in the Swedish craft subject, educational sloyd. The crafting materials that the students answer to and join forces with in the remaking activity are set in a 'system of forces' with the aim to be remade. However, remaking clothes are by no means a new activity; throughout history, clothes were seldom thrown away. More commonly, clothes are inherited, reused, mended and resewn (Palmsköld 2013, 76); for example, larger garments were unstitched, washed and remade into new products and smaller garments were made into clothes for children (Resare 2002, 273). Even the smallest bit of fabric was often saved and reused as decoration such as the fringe on pillow corners (Palmsköld 2013, 82). Further, if the garments were not remade into other clothes, they were cut and used for rag rugs or as patchwork (Franow 2002, 274). These examples illustrate that clothes were remade into other objects, and this activity has both functional and aesthetic values in the remaking practice, and accordingly, the stories about the material were recognised as such. In addition, to be able to recognise worn-out clothes as something that can be remade, crafting skills (for example, how to darn socks or patch clothes) are needed. It could be argued that a remaking activity carried out in Sweden today could potentially have historical 'threads' to a remaking practice. The connection occurs if the story that is told has an alignment or an affinity to the stories that were produced in older practices, in other words, when the material is recognised through the alignment of present circumstances in conjunction with the past (Ingold 2011, 57). But, importantly, as mentioned, these stories are never pre-determined but rather emerge as a practice of correspondence.

In the analysis of the remake project, I first use Ingold's concept of the practice of correspondence to explore the student–material encounters and how they evolve in the remaking activity. When I identify the correspondences in the analysis, I explore what functions the material have as the students answer to the material in correspondence (question 1). Second, I explore what stories emerge from this activity. Analytically, how the student tells these stories is considered not only as a verbal story but also an embodied activity. Here, I follow Ingold as he argues that 'to tell' refers to, on the one hand, being able to recount stories of the world, and on the other hand, being able to recognise subtle cues in one's environment and respond to them with judgement and precision (Ingold 2013, 109–110). By using stories as an analytical concept in the remake activity, I contend that we need to have a story for learning to emerge – in this case, a story about and with the remake material. Hence, the question that becomes relevant to explore asks what stories emerge (question 2). As the remake activities have historical threads to a remake practice (i.e. a conjunction with the past), in the third step of the analysis, I also explore if any conjunctions exist between the stories that have been created in students' remaking activities and historical remaking activities. And if so, I explore how these conjunctions could be used to further develop students' remaking skills in relation to sustainability (question 3). To highlight this connection between the past and the future in the remake activity, I will define this connection with history as 'threading back' and its connection with the future as 'threading forward'. I do this to discuss pedagogical opportunities as students learn for sustainability.

The remake project, data selection and data analysis

The project was carried out in the Swedish handicrafts subject called educational Sloyd, which, in Sweden, is a mandatory school subject in Grades 3–9. In the curriculum, it is stated that students should be given 'opportunities to develop knowledge of how to choose and handle materials in order to promote sustainable development' (Swedish National Agency for Education 2011, 203), and as a respond to this statement, remaking projects are common. This particular remake project was observed by using video recordings of a Grade 8 class. The task for the students was to make a new product from used garments by using a sewing machine. The class was composed of 15 students, and the raw data amounted to about 20 h of film distributed over a period of 10 weeks (i.e. 80 minutes per week). Two cameras were used: one was attached to the teacher's body and the other was held and operated by the researcher. In an earlier study of the same empirical material in the remake project, I identified how the students had to 'transact' with ideas for future products, with the crafting materials' capabilities, and with the crafting techniques (reference to author). In the present article, I focus on the human–material relations and what sustainability stories emerge from the correspondence.

I found two activities particularly relevant for my analysis, as they made the human–material correspondence visible. The first activity was when the students were deciding on the design of the remaking project (i.e. what to do, which material to use, or the character of the imagined end-product), and the second activity was when the students were trying to realise the design through crafting. Taken together, it was possible to see from the empirical video data how six of the students (three boys and three girls) were in a decision-making process of what to remake. Given this visibility, these six students' decision-making processes were targeted for further analysis in the first activity. In the second activity, in which the students realise their imagined product, there were many human–material correspondences. To scope the human–material correspondences, I decided to follow four students who were all remaking jeans. This means that I did not consider any lecturing by the teacher (for example, about how to make a paper pattern) or when the students write descriptions of their imagined process if the remake material was not part of the process.

The analysis was conducted in three steps. In the first step, I watched the selected passages and identified the correspondence between the student and the remake material. For example, if a student remade a pair of jeans into a pillow, I noted how the correspondence developed as the design of a pillow emerged. In the second step, the stories that the student recognised in the design process are identified, which emerge from the materiality intrinsic to the crafting process and the intentions of the students, as these are visible in action. For example, if the form of the jeans is recognised by the students, and the form is what students use when making a pillow, the form is identified as a story. In the third step, the constructed data from the correspondences and stories are discussed in relation to historical remake practice (i.e. in conjunction with the past). By taking a materiality focus on the remake practice, I 'thread back' and 'thread forward' to discuss pedagogical opportunities as students learn for sustainability.

To present the findings, I first give four examples from the correspondences by presenting excerpts and descriptions from the video data. I have also added sketches from screenshots to provide a better understanding for the situation. The reason for choosing both excerpts and descriptions is due to the character of the activity. Often, if the activity did not contain much verbal dialogue, then descriptions are used. In one example with Jonas, the verbal dialogue between the teacher and the student was quite long, and therefore, the excerpts made the dialogue more transparent. The excerpts and descriptions were translated from the video recordings into written text, focusing on approximating a literal translation (Linell 1994, 11), which means that what happens in the activity needs to be visible to the reader, and the transcript must be legible. In these presentations, I pay special attention to both the student and the recycled material. On one hand, translating an embodied activity into a text has limitations, as it will likely lose some complexity. On the other hand, the simplification of an activity also provides an opportunity to highlight something more clearly, in this case, the encounter between the student and the remake material.

Two illustrations of creating a design in correspondence with the material

In the first example, we meet two students, Clair and Oliver, as they choose what to remake; and in the process of choosing, they create a design in correspondence with the material. In the following text, I first describe the two activities and then discuss the correspondences and the stories that emerge from the correspondences.

Clair starts by saying, 'I think I will make a pillow' (151110, 001MTS, 4:28). She holds up a pair of jeans, looks at them and then puts them aside. She then starts to touch and pat a piece of fake fur and a dress made of lace. After a while, she makes a heart shape with her fingers on the fake fur (6:36). She then starts to sketch a pillow on her working instruction paper to illustrate her design. After a while, she speaks up loudly, 'I do not know how to draw a pillow' (7:28), but no one answers and she continues to draw on the paper with her pencil. Then she picks up the lace dress, makes it lie flat on the table, and with her arms, measures a square to see what it looks like (8:14). Then she takes the fake fur and puts it on the dress and makes a square again with her underarms (8:23) as she continues to draw and write down her ideas (Figure 1).

The student next to Clair, Oliver, sits with his pair of jeans in front of him and decides to make a pillow using the jeans. Clair asks how he is going to design them (9.26). Oliver responds, 'Yeah, I could make it a big pillow where I will have the upper part here', pointing to the back of his shoulder, and the two legs could come down on each side, which he shows by pointing with his hands (9:31). Clair laughs. Everyone at the table starts to give suggestions on how he could do it. However, Oliver concludes that he will make a pair of shorts instead because that seems much easier to do (10:10) (Figure 2).

Figure 1. A sequence illustrating Clair's design process.

Figure 2. A sequence illustrating Oliver's design process.

If we look at the correspondences in the described passage from the beginning of the lesson, Clair says that she wants to make a pillow. Why Clair wants to make a pillow is not possible to know from the excerpt, but she clearly has not yet decided how this pillow will look. Here, the correspondence emerges from the point where Clair feels the garments with her hands as she pats them and her fingers make a shape of a heart at the fabric. She is also corresponding with the lace dress by using her arms to imagine how big the pillow should be. These correspondences help her design her remaking product. The story about the material that emerges from these correspondences is the texture of the fabric because she continuously uses her hand to feel the fur and the lace as she is designing the pillow.

From the second example, we see Oliver starting to imagine how he can make a pillow. He describes to his fellow students how the jeans could be put around his shoulders with the jeans' legs coming down on each side. The correspondence here starts with the shape of the jeans, which Oliver corresponds with and, arguably, it is the shape and how the jeans are constructed that prompt Oliver to imagine a possible design. In other words, it is the shape and the construction of the jeans that Oliver recognises and thus emerge as part of the story when Oliver imagines what to remake, although, in the end, he gives up this design and chooses another idea instead.

Two illustrations of how to realise the design in correspondence with the material

When students have decided what to do and try to come up with a solution for how to make the design, resistance often arises. What is interesting here is that the resistance does not necessarily emerge from interpersonal student–student or student–teacher relations but rather from the entanglements intrinsic to the material in the crafting process, where the intentions of individuals are sometimes blocked or resisted by the chosen material. In the following, I give two examples of when students handle the material differently when corresponding with it. I first describe the two activities and then discuss the correspondences and the stories that emerge from these correspondences.

First, we meet Jonas. When the teacher arrives at his desk, Jonas explains that he wants to keep the original shape of the jeans and make a pillow that is 35 × 40 cm, but he does not know how to realise his design (151110: GPO10344:0.29). The teacher and the students begin discussing, and the conversation ends with the teacher asking Jonas to solve how to realise his design and make a pattern according to his design (1:50). For a long time, Jonas sits at his desk with the jeans in front of him, measuring the jeans over and over again by using a measuring tape in his effort to make the design. When the teacher arrives the next time, Jonas has decided to simplify the product, and the design has now changed into a rectangular pillowcase. He has decided to cut off one of the legs and make a pillowcase out of it. However, referring to the assessment, the teacher thinks that this idea is too simple and asks Jonas to come up with another solution (15:53). So Jonas continues to measure the jeans. The teacher approaches him a third time, and they have the following exchange (Figure 3):

1. Jonas: There is something wrong with the jeans. They are wider here than there [points to the jeans].
2. They are wider here than there [points to the paper pattern] (15:41–15: 50).

The teacher explains that this is exactly what Jonas needs to deal with; that is, he needs to come up with a solution for how to reuse the jeans. When the teacher approaches him a fourth time, Jonas has decided to return to the initial idea of making a square pillow, and the conversation proceeds as follows:

3. Teacher: Are you still using these [points to the jeans]?

4. Jonas: Yes, I am, and I'll use both legs.
5. Teacher: There'll be a seam somewhere, won't there?
6. Jonas: Yes, but it is diagonal, like that [points to the paper pattern].
7. Teacher: Is that your intention?
8. Jonas: Yes, like that [points to the paper pattern again].
9. Teacher: Yes, okay. Is it the same at the back?
10. Jonas: Hmm. Should I measure it? ... when I make that one, should I measure it like a square?
11. like that... so that it's forty, forty, forty, forty [points to each side of the paper] ...
12. so that it is [Jonas draws a square in the air with his finger] ... and then cut it.
13. Teacher: Yes, and if you do that, you'll be able to get a higher grade, which will be good (151110: GP020344: 5.30–6.14).

In a second example of how a student handles the material, we meet Anna, who is about to make a small toilet bag out of pair of jeans. When making the design, she measures the jeans, makes a pattern, and then cuts the jeans accordingly. When we meet her, Anna has the jeans in front of her and asks the teacher if she should zigzag the edges first, as she normally would do. (GP020351.MP4 151201 3:58). Now, resistance emerges. According to Anna's plan, she wants to use a zipper, but to use a zipper, a seam allowance is needed, otherwise it is not possible to attach the zipper. Anna has the pieces of jeans in front of her, where the sides are sewn together. Here she realises that it will not be possible to attach the zipper, as there is not enough fabric for the seam allowance. The teacher, who stands next to her, responds that she

Figure 3. A sequence illustrating Jonas realising his design.

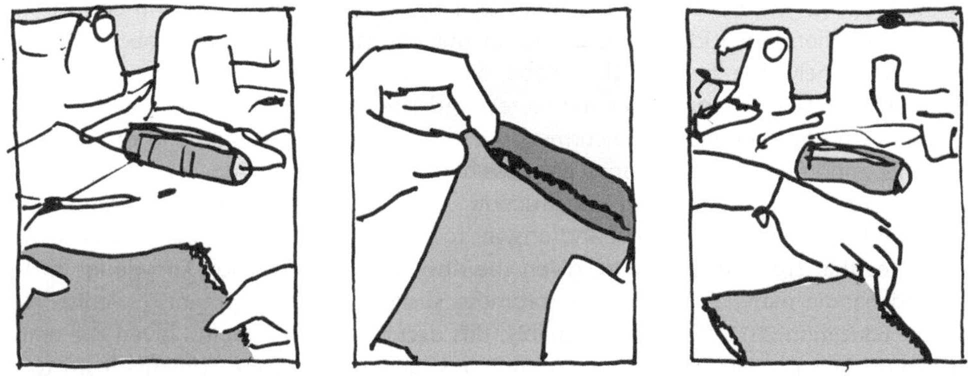

Figure 4. A sequence illustrating Anna realising her design.

had not realised that Anna wants to use a zipper. Anna is silent for quite some time. The jeans are lying there in front of her. They are already cut, and she does not know what to do. After a while, Anna says that she will not use a zipper because the zipper, if sewn with the required seam allowance, will make the toilet bag too small. She decides to simplify the construction and use a button and a loop of yarn to make it possible to close the bag (6:27) (Figure 4).

I have described two passages where the students have decided what to do and then continued to work. In both examples, the students are joining forces or trying to join forces with the material in a practice of correspondence. These correspondences will now be in focus (Figure 4).

In the first example, Jonas had decided to make a pillow, but realising that idea proves to be quite hard. If we look at the correspondences here, the jeans are in front of him and Jonas measures the jeans several times. In this correspondence, the jeans are clearly troubling Jonas, as he cannot come up with a solution. Instead, he suggests simplifying the imagined product by making a rectangular pillow, but the teacher asks him to come up with another solution. As Jonas is corresponding with the material, he utters that there is something wrong with the jeans (line 1). The shape of the jeans is wider on the upper side compared to the lower side, which is something that Jonas recognises as he measures the jeans, and this correspondence creates specific limitations, as Jonas is about to come up with a solution. From the excerpt, it is also possible to see that the correspondence is not solely a cognitive process. When the teacher asks Jonas for his new solution, Jonas replies by first pointing to the paper where he has drawn the sketch of the pillowcase (line 8) and then continues to explain his idea by drawing a square in the air with his finger (lines 10–12). Thus, he uses his body to show (and perhaps also to develop) his ideas. The story about the material that emerges in this activity is the shape of the jeans. The shape provides limitations and makes Jonas doubt his first idea to make a square pillow. However, by continuing to correspond with the shape and the construction of the jeans (using both legs), these entanglements make it possible for Jonas to realise the final design, thus allowing him to continue with the remake activity.

We now turn to the correspondences in the second example, with Anna working on her toilet bag. Anna has measured the jeans and cut them, but unfortunately, they are cut too small to include a zipper. Here, Anna is forced to adjust her design to the available fabric. Instead of a zipper, Anna imagines using a button to realise the product. Here, the story that emerges from this example is a story of construction and the miscalculation of the jeans' form, which causes Anna to simplify her imagined product.

Stories of sustainability in remaking practices as a pedagogical opportunity

In the previous section, I illustrate the student–material encounters and how the correspondences evolve in the remaking activity, and by doing that, I explore what functions the material has as the students join forces with the material and answer to the material in correspondence (question 1). Furthermore, I identify three stories that the students recognise about the material (question 2), namely, the texture, the shape or form, and the construction. By providing a research approach that acknowledges the correspondence between the human and the material, I have sought to show the reciprocal correspondence between the human and the material, and this is highly important for what stories are possible to learn in remaking activities. These stories – the material's texture, shape and construction – are what the students recognised as they learnt to join forces with the material and answer to the material in the remake project. The curriculum states that students should be given the opportunity 'to develop knowledge of how to choose and handle materials in order to promote sustainable development' (Swedish National Agency for Education 2011, 203), and arguably, this occurs in the examples, given the embodied experience of the material's texture, shape and construction, which the students learn to recognise as they promote sustainable development.

The question that remains is whether there are any conjunctions between these three stories and historical remaking activities (question 3). If one is to 'thread back' in conjunction with the past (Ingold 2011), I would argue that these examples of how the material is recognised have a bearing on the Swedish historical remaking practice. For example, historically, when remaking clothes, the craftsperson would have had to answer to the material and encounter the texture, as well as determine the product's shape and construction, as these are inevitable when remaking. However, how these stories were told, for example, with economic, social or ecological means, are likely to differ, as the activities are set in another time and context. Nevertheless, there is a link here, which I argue can be used in a remake pedagogy. My point is that by acknowledging historical threads in a remake activity, the activity can be regarded as having pedagogical opportunities that take departure from the students' own correspondences with the material. In addition to these three stories that the students recognise, I want to mention three complementary materiality concerns that threading back provides.

The first materiality concern is knowing the *source*. In the past, recycling fabrics and making a virtue of necessity was something you had to do when you could not buy everything you needed (Hallén 2002, 274). When making a virtue of necessity, knowing about the source was helpful in this remaking practice. By source, I mean what the garments are made of, for example, cotton, linen, wool or leather (and today it also concerns types of plastic material such as polyester or acrylic). What the material is made out of is particularly important if the garments are meant to be remade into new products, as that will affect what is possible to remake. Not only that, but by knowing the source, one will also, in correspondence with the material, know what to expect from the material and how to take care of the material. That is not to say that the correspondence is predetermined but rather, based on previous knowledge of human–material relations, one can know what is likely to happen. For example, knowing how wool fibres or viscose fibres perform when they are wet (wool fibres are cleansing, and viscous fibres break easily) could help students expand their possible actions in remaking activities. Knowing about the source is also about recognising and developing an understanding of how fibres feel when squeezing them (for example, what natural fibres feel like compared to plastic fibres). The second materiality concern is knowing about the *fabric*. When mending garments, knowing about the fabric – that is, whether it is woven, knitted or felted – is helpful, as the techniques of these human–material correspondences are not the same. That is, a pair of jeans woven in twill are mended with specific techniques and a specific needle so that the needle or the thread does not break. When mending a knitted fabric (for example, a t-shirt or a sweater), the correspondence of the human–material differs in that the fabric is made from loops and needs to be repaired accordingly. The third materiality concern is to use every piece of the garment, which historically would have been common practice (Palmsköld 2013, 74). What this materiality concern gives us is a story of *zero waste*, which provides attention to all parts of the garments, as the smallest pieces of fabric were saved and used for decoration (Palmsköld 2013, 82). These three materiality concerns – the source, the fabric and zero waste – create potentially relevant additions to the stories that the students make as they learn for sustainability in a remake activity that is understood as part of a larger remaking practice. In other words, the conjunctions with the past provide pedagogical opportunities that the teacher can consider as the students are remaking for sustainability.

Discussion

We know from previous research that arts and crafts activities involving recycling are promising, as they can create new ways of supporting children's imagination, ingenuity and creativity (Jørgensen, Madsen, and Læssøe 2018, 811) as well as be a way to recognise a material's properties (Odegard 2012, 387) and act as a springboard for working with personal and social change (McGarry 2013, 110–112). My study adds to this research, as it explores and illustrates students' encounters with crafting material. Therefore, I expand on what it might mean to be creative in a remake activity

(Jørgensen, Madsen, and Læssøe 2018) or what it means to recognise and experience a material's properties (Odegard 2012). For example, in the remake activity, the students needed to decide what to remake and come up with solutions for how to remake the products, which resulted in a creative process where the students created a design in correspondence with the material. Thus, by applying a specific research approach, I have been able to create knowledge of (1) what a creative process with material can look like and (2) how the correspondences with the material provided both possibilities for, and a hindrance to, the students' remaking activity. What then becomes visible in the analyses is that what students learn as they encounter the material is dependent on how the human–material relation (correspondence) develops in the crafting process and on the purposes that emerge in the activity. According to McGarry (2013), remaking activities could be a springboard for working with personal and social change; however, my research does not address issues of personal and social change. Nevertheless, what I have sought to show in this paper is that the materiality acts back upon those who engage with it. From the illustration of the research approach, it is possible to conclude that – as it is impossible to know how materiality or materials act on people in advance or on itself – the research approach also highlights the importance of human agency in the correspondence. One could argue that this leaves us with the knowledge that personal change can be made by working with remake projects, but knowing precisely how this happens is another matter.

The research approach also adds to the specific practice of educational sloyd, in that it can be used to create knowledge about how, in a remake project, students join forces with and answer to the material. Such knowledge opens up for pedagogical actions that can be used by teachers to facilitate possibilities in a sustainability context and to recontextualise old practices of remaking. If teachers want to educate for, in and about sustainability, I would argue that a key question is how humans learn to correspond to materials in socio-material activities. What I have illustrated with this research approach is that the specific materiality in these human–material relations both provide possibilities and present challenges, which are important to acknowledge when engaging with environmental and sustainability issues.

Learning to join forces and correspond with materials can also open up for experiences that humans want to have in relation to environmental and sustainability issues. For example, we as humans are entangled in human–material relations in our consumption patterns, with the waste we produce, and by our efforts to recycle, reduce, or reuse materials. By taking the correspondence approach to learning for sustainability, an opportunity arises for ESE pedagogy to contextualise human–material relations within our environments and with and through the socio-material contexts in which we live our lives. By doing so, educators can open up for embodied correspondences and experiences, and through this, the students can hopefully become prepared to correspond in new unforeseen ways and to be flexible enough to correspond wisely to unexpected changes in the environment. From this perspective, it becomes important to continue to create more empirically grounded knowledge concerning students' encounters with materiality, that is, what materiality the students recognise and how they – with their hands, skin, eyes, ears, bodies and minds – learn to correspond accordingly.

Funding

This work was supported by the 'Teaching and Learning Practical Embodied Knowledge' project funded by The Swedish Research Council [grant number 2013-2200].

References

Barad, K. 2003. "Posthumanist Performativity: Towards and Understanding of How Matter Comes to Matter." *Signs: Journal of Women in Culture and Society* 28(3): 801–831. doi:10.1086/345321.
Barad, K. 2007. *Meeting the Universe Halfway: Quantum Physics and the Entanglement of Matter and Meaning*. Durham: Duke University Press.
Clarke, D. A. G., and J. Mcphie. 2016. "From Places to Paths: Learning for Sustainability, Teacher Education and a Philosophy of Becoming." *Environmental Education Research* 22(7): 1002–1024. doi:10.1080/13504622.2015.1057554.
Coole, D., and S. Frost, eds. 2010. *New Materialisms*. Durham: Duke University Press Books.
Crutzen, P. J., and E. F. Stoermer. 2000. "The Anthropocene." *IGBP Newsletter* 41(17): 17–18.
Fenwick, T. 2015. "Sociomateriality and Learning: A Critical Approach." In *The SAGE Handbook of Learning*, edited by D. Scott and E. Hargreaves, 83–93. London: SAGE Publications Ltd. doi:10.4135/9781473915213.n8.
Franow, I. 2002. "Rag-Rugs [Trasmattor]." In *Fabrics Everywhere [tyg överallt], nordiska musset*, 232–247. Stockholm: Nordiska Mussets Förlag.
Hallén, M. 2002. "Rags Too Small to Use but Very Handy to Have [För Små Att Använda Men Bra Att ha.]." In *Fabrics Everywhere [tyg överallt], nordiska musset*, 220–231. Stockholm: Nordiska Museets Förlag.
Ingold, T. 2011. *Being Alive – Essays on Movement, Knowledge and Description*. London: Routledge Taylor and Francis Group.
Ingold, T. 2013. *Making – Anthropology, Archaeology, Art and Architecture*. London: Routledge Taylor and Francis Group.
Jørgensen, N., K. D. Madsen, and J. Læssøe. 2018. "Waste in Education: The Potential of Materiality and Practice." *Environmental Education Research* 24(6): 807–817. doi:10.1080/13504622.2017.1357801.
Linell, P. 1994. *Transcript of Speech and Conservation: Theory and Practice. [Transkription av tal och samtal: Teori och praktik]*. Linköping: Linköpings universitet Tema Kommunikation.
Lloro-Bidart, T. 2017. "A Feminist Posthumanist Political Ecology of Education for Theorizing Human-Animal Relations/Relationships." *Environmental Education Research* 23(1): 111–130. doi:10.1080/13504622.2015.1135419.
Mannion, G. 2007. "Going Spatial, Going Relational: Why 'Listening to Children' and Children's Participation Needs Reframing." *Discourse: Studies in the Cultural Politics of Education* 28(3): 405–420. doi:10.1080/01596300701458970.
McGarry, D. 2013. "Empathy in the Time of Ecological Apartheid a Social Sculpture Practice-Led Inquiry into Developing Pedagogies for Ecological Citizenship." PhD diss., Rhodes University.
McKenzie, M., and A. Bieler. 2016. *Critical Education and Sociomaterial Practice: Narration, Place, and the Social*. New York: Peter Lang Publishing.
Odegard, N. 2012. "When Matter Comes to Matter – Working Pedagogically with Junk Materials." *Education Inquiry* 3(3): 387–400. doi:10.3402/edui.v3i3.22042.
Palmsköld, A. 2013. *Textile Remake – Material and Cultural Usage. [textilt återbruk – om materiellt och kulturellt slitage]*. Halmstad: Gidlunds Förlag.
Payne, P. G. 2016. "What Next? Post – Critical Materialisms in Environmental Education." *The Journal of Environmental Education* 47(2): 169–178. doi:10.1080/00958964.2015.1127201.
Pyyry, N. 2017. "Thinking with Broken Glass: Making Pedagogical Spaces of Enchantment in the City." *Environmental Education Research* 23(10): 1391–1401. doi:10.1080/13504622.2017.1325448.
Rautio, P. 2013. "Being Nature: Interspecies Articulation as a Species - Specific Practice of Relating to Environment." *Environmental Education Research* 19(4): 445–457. doi:10.1080/13504622.2012.700698.
Rautio, P., R. Hohti, R.-M. Leinonen, and T. Tammi. 2017. "Reconfiguring Urban Environmental Education with 'Shitgull' and a 'Shop." *Environmental Education Research* 23(10): 1379–1390. doi:10.1080/13504622.2017.1325446.
Resare, A. 2002. "Worth Preserving [Värt Att Vårda]." In *Fabrics Everywhere [tyg överallt], nordiska musset*, 207–219. Stockholm: Nordiska Museets Förlag.
Russell, C. L. 2005. "Whoever Does Not Write Is Written: The Role of 'Nature' in Post-Post Approaches to Environmental Education Research." *Environmental Education Research* 11(4): 433–443. doi:10.1080/13504620500169569.
Somerville, M. 2016. "Environmental and Sustainability Education: A Fragile History of the Present." In *The Sage Handbook of Curriculum, Pedagogy and Assessment*, edited by D. W. L. Hayward and J. Pandya, 506–522. London: Sage.
Sørensen, E. 2009. *The Materiality of Learning – Technology and Knowledge in Educational Practice*. New York: Cambridge University Press.
Steffen, W., W. Broadgate, L. Deutsch, O. Gaffney, and C. Ludwig. 2015. "The Trajectory of the Anthropocene: The Great Acceleration." *The Anthropocene Review* 2(1): 81–98. doi:10.1177/2053019614564785.

Svenska MiljöEmissionsData (SMED). 2011. *Mapping the amount and flows of textile waste* [Kartläggning av mängder och flöden av textilavfall]. Accessed 28 October 2018. http://www.smed.se/wp-content/uploads/2011/05/SMED_Rapport_2011_46.pdf

Swedish National Agency for Education. 2011. *Curriculum for the Compulsory School, Preschool Class and the Recreation Centre 2011*. Stockholm: Ordförrådet AB.

Taylor, A. 2017. "Beyond Stewardship: Common World Pedagogies for the Anthropocene." *Environmental Education Research* 23(10): 1448–1461. doi:10.1080/13504622.2017.1325452.

Van Poeck, K., and J. A. Lysgaard. 2016. "The Roots and Routes of Environmental and Sustainability Education Policy Research." *Environmental Education Research* 22(3): 305–318. doi:10.1080/13504622.2015.1108393.

More-than-human stories: experimental co-productions in outdoor environmental education pedagogy

Scott Jukes and Ya Reeves

ABSTRACT

This research draws upon pedagogical experimentation on a ski-touring journey in the Australian Alps, building upon place-responsive pedagogies in outdoor environmental education with insight from new materialist and posthuman theory. In particular, this research focusses on the generative potential of considering co-productions and assemblages that include the materiality of a more-than-human world. Combining place-responsive and new materialist ideas, the concept of more-than-human stories is offered as a pedagogical strategy that may challenge anthropocentrism and develop different ways of thinking about and with more-than-human places. An example of these ideas in practice is offered, where Ya, an undergraduate student, presents her more-than-human story—*One Single Moment*—a picture-story book created on the ski-touring journey. Following her picture-story book, she maps the conditions of possibility, the various pedagogical influences that led to the creation of *One Single Moment*. Our aim through this research is to offer methodological and pedagogical insight into using new materialist theories in creative and productive ways.

Introduction

This paper embraces a picture storybook, a student as co-author and a 'multi-modal methodology' (Coole and Frost, 2010, 32) of the new materialisms. The paper is a co-production between Scott, Ya and the Bogong High Plains in the Australian Alps. Our conversations that led to this paper began on a ski touring trip for the subject 'Winter Alpine Environments'[1] in the Bogong High Plains.[2] The pairing for this project emerges from a sharing and grappling of ideas, where Scott as an educator working in collaboration with Ya as a student at the time offers some unique perspectives of educator-student-place relations. In part, this research partnership is inspired by Honan's (2014) post-qualitative stance urging researchers to disrupt habits of (re)presenting research and knowledge production along with rethinking the relations between the researcher and researched. Verlie and CCR15 (2018) also provides a prompt for such an approach, with her new materialist research 'co-authored' with the university course Climate Change Responses (CCR 15). What such a statement signifies is the 'entangled voice' or co-created knowledge that her research emerged from. For Scott as a researcher interested in outdoor environmental education (OEE) pedagogy, the topic of students and their developing understandings and responses to teachings are crucial. From here on we use 'we' to signify the collaborative co-production that this research emerges from.

Within this paper, we build upon place-responsive pedagogies[3] in OEE by drawing insight from new materialist and posthuman theory. Reid (2019), in a recent editorial in this journal, explains there is a 'spike in interest raised by the curious work of "new materialisms"' (165). We are also intrigued by the emerging new materialisms and how we might add to this 'bubble of activity' (Reid, 2019, 165) for OEE research and pedagogy. Somewhat alongside or enmeshed with new materialisms is posthuman critical theory, which Braidotti (2013, 2018a, 2019) describes as a convergence of post-humanism and post-anthropocentrism in the development of ethically accountable, critical and creative new ways of approaching thinking. Without wanting to conflate new materialist and posthuman approaches, Braidotti (2013, 2018a, 2019) and Coole and Frost (2010) describe the possibilities as transdisciplinary, and in this paper, we introduce particular aspects employed in our thinking, learning, research and pedagogies. For now, we offer a brief review of place-responsive pedagogies as a significant body of work, which is used as a springboard for this project.

Place-responsive pedagogies

OEE is often associated with experiential approaches to education, with adventurous activities being a significant area of practice. Throughout the last few decades OEE has turned towards theories of place with aims to foster relationships between people and places (for example, Raffan, 1993; Baker, 2005; Stewart, 2004, 2008; Wattchow and Brown, 2011). Authors such as Lugg (2004) have called for more attention to the environment with a reappraisal of activities used in OEE. Other authors, such as Stewart (2004, 2008) and Wattchow and Brown (2011), have called for pedagogies of place-responsiveness, to ensure places don't become 'featureless backdrops' to outdoor learning (Stewart, 2008, 94). Somerville (2010) explains the potential for the development of place relations and place studies in broader educational contexts:

> While research in the physical sciences has typically been seen as the solution for complex environmental problems, emerging research in the eco-humanities suggests that changing our relationship to our places is as important as techno-scientific solutions. In this way place can offer an important framework for an integrated educational curriculum which seeks to address environmental issues from a range of perspectives that could cross traditional subject boundaries ... (331)

We follow other outdoor environmental educators in seeing the educational potential of place studies in both our field and other fields of education. However, ideas surrounding whose place is prominent and how place is defined are still concerns for place-based educators (Lynch and Mannion, 2016). What places and whose places are highlighted? We agree with Gough's (2008) suggestion that places are not profoundly pedagogical but may become so via the ways they can be 'envisioned, named, transformed and traversed' (72). Part of our intention in this paper is to explore how we may reconceptualise place from new materialist and posthuman perspectives, challenging anthropocentric and static notions of place.

The use or creation of stories or storylines as a way to envision place is a strategy explored and advocated by several researchers (for example, Wattchow and Brown, 2011; Mikaels and Asfeldt, 2017; Stewart, 2008; Somerville, 2010, 2013; Somerville et al., 2009). Wattchow and Brown (2011) explain in their signposts to a place-responsive pedagogy the power of place-based stories and the importance of representing experiences. Studying a place's histories and ecologies is part of place-responsiveness (Wattchow and Brown, 2011), and we add, requires attentiveness to the various more-than-human and material elements of places. As Wattchow and Brown explain, 'places are brimful of personal, geological, ecological, historical, economic and political stories' and place-responsive educators would be well placed to cultivate worthwhile stories (193). In Somerville's (2010) framework for a pedagogy of place, she also advocates for stories, expressing *'our relationship to place is constituted in stories and other representations* [author's emphasis]' (342). Narrowing our focus for this paper, we are curious about storied understandings of places and we wish to see if they can work in conjunction with the new materialist and posthuman thinking we introduce later.

Elsewhere, Somerville (2013) further advocates for storylines and learning through storied understandings of places. Drawing upon the research of Pretty et al. (2009), Somerville explains there is 'substantial evidence supporting the significance of local ecological knowledge expressed as *stories, ceremonies, and discourses* [author's emphasis]' (5). She argues through her work with indigenous people of the Murray-Darling Basin that such storied knowledge is constantly in process and relation to local places. The significance is that stories can help guide a society's actions. A theme that we will ponder and return to throughout this paper is the role of story, whilst considering whose stories are told? If we read and come to know a place, what stories can be seen in the landscape (Stewart, 2008)? Where do stories come from and how are they told?

Prins and Wattchow (2019) also support storytelling practices, stating 'it is important to acknowledge that stories are not just a representation or an account of what has happened but rather, a continual communication that is also a way humans make sense of their world' (7). Yet we also echo the interests of Mannion and Lynch (2016), who raise important concerns about practices focussing on human centred stories. In addition to stories making sense of a human world, we argue throughout this paper that stories can help draw us into and part of the more-than-human world when we consider different perspectives, meanings and complex entanglements with matter. Hence, we move towards new materialist motivated thinking.

From places to (new materialist inspired) assemblages

New materialist and posthuman critical theories have been narrowly explored in OEE research, with Gough (2016) and Pleasants and Stewart (2019) offering appraisals for their unique potential. Following the emergence in broader educational research, some authors (for example, Clarke and Mcphie, 2014, 2016; Clarke, 2017; Mcphie and Clarke, 2015, 2018; Mikaels and Asfeldt, 2017; Ruck and Mannion, 2019; Payne, 2016) in OEE are utilising new materialist and posthuman approaches in place of humanistic methodologies. In particular, certain research involving place-responsive pedagogy incorporates these posthuman and new materialist methodologies, with a particular interest in the concept *assemblages* (see Stewart, 2018; Mannion, Fenwick and Lynch, 2013; Mannion, 2019; Jukes, Stewart and Morse, 2019). As Mannion (2019) explains, the concept assemblage was created by Deleuze and Guattari (1987), translated from the French verb agéncement. The verbing connotation is purposeful, indicating an active coming together of things. Consequently, any assemblage, be that a place, or a research project such as this, is an assemblage of various human, more-than-human, and material elements that impact, affect and implicate each other in their coming together (Mannion, 2019). We find the concept useful for considering places as made by dynamic heterogeneous relationships of human and more-than-human actants (Jukes et al., 2019). Mannion (2019) proposes an assemblage pedagogy and assemblage research, offering advice on how outdoor, environmental and sustainability education may incorporate such theories into practices.

But why is it important or valuable to approach research and pedagogy in such a way? As Mannion (2019) justifies, humanistic methodologies employed in educational research don't as yet seem successful in adequately addressing issues of our time, such as climate change and global precarity; critiques of humanism are permeating and building forceful arguments. Mannion (2019) refers to Malone, Truong and Gray (2017) in stating some of these arguments:

> The need for a de-centring of the human in our ontologies and epistemologies (for example, in solely looking to the needs of humans in definitions of sustainability or in child-centred approaches), a countering of human exceptionalism in general towards a relational orientation (in terms of, for example, human-plant, human-animal, human-place), a revision of 'stewardship' views of the environment (with its paternalistic associations of sympathy, mastery and control), a greater acceptance of environmental crisis of the Anthropocene (particularly climate change, but in connected ways food insecurity, migration, and so on), and a foregrounding of the importance of alternative ways of knowing. (2)

In agreeance with Stewart (2018), we are in a pressing time where we need to think differently about 'how humans conceptualize and relate to the more-than-human world' (139). We find these emerging ideas in our field, and from broader educational research, particularly thought-provoking. Hence, this paper investigates experiments with pedagogy from such a philosophical-methodological disposition.

Philosophical-Methodological disposition: A post-anthropocentric material turn

Numerous authors in new materialist/posthuman critical theory explore ways of paying attention to the agency and materiality of bodies, including those of non-humans (for example, Coole and Frost, 2010; Fox and Alldred, 2015; Bennett, 2010; Barad, 2007; Braidotti, 2013). As van der Tuin (2018) states, 'the new materialisms are mainly a research methodology for the non-dualistic study of the world within, beside and among us, the world that precedes, includes and exceeds us' (277). Noting the plurality of new materialisms and posthuman critical theory, Gough (2016) explains they are a collection of concepts and approaches in formation[4] (see Gough and Whitehouse, 2018 for an exploration of various viewpoints including ecofeminism and new material feminism).

At this stage, it is important to explain what we mean by concepts. Many new materialist and posthuman researchers follow Deleuze and Guattari (1987, 1994) and their pluralistic and process-oriented ontology. Deleuze and Guattari (1994) describe the function of philosophy as the creation of concepts; 'there are no simple concepts. Every concept has components and is defined by them. It therefore has a combination. It is a multiplicity…' (15). Colebrook (2002) adds that a 'concept is not a word; it is a creation of a way of thinking' (20). What a concept does is 'provoke us, [it] dislodges us from our ways of thinking and opens experience up to new "intensities": a way of seeing differently' (Colebrook, 2002, 20). That is not all that concepts can do, as Mcphie and Clarke (2018) explain, concepts work in real and material ways; 'they perform ecologically' (2). Concepts are not 'fixed entities', explain Ruck and Mannion (2019, 5), but *becomings* that work in practice to facilitate understandings. Lenz Taguchi and St. Pierre (2017) argue for conceptual-based practices in educational research as a way of enacting different thinking. Hence, many in new materialist and posthuman critical theory consider concepts as ways of thinking, and thinking differently, in order to investigate our contexts.

As it is beyond the scope of this paper to lay out all the concepts and possibilities of new materialist/posthuman critical theory, instead we highlight a combination of ideas we have found generative to think with and have organised them into three overlapping categories: *material agencies* and *entangled relationships* as assemblages of *a more-than-human world*. We unpack these below with reference to how we use them or how they have been used in some (outdoor) environmental education research.

Material agencies

Matter can often be depicted by dominant Western thought as dead, inert, brute or passive, prioritising superior human agency over static objects (Coole, 2010; Coole and Frost, 2010; Barad, 2007; Bennett, 2010). In a countermovement, new materialisms examine the agentic capacity of matter. New materialists see 'materiality as always something more than "mere" matter: an excess, force, vitality, relationality, or difference that renders matter active, self-creative, productive, unpredictable' (Coole and Frost, 2010, 9). The active capacity of matter to make a difference in the world is an aspect examined by new materialists, moving away from the subjugation of non-human material agencies. People are not the only material bodies that form the world, and it is the various scales of materiality that come together forming and shaping the world—other material actants include, for example, bacteria, omega-3 fatty acids, electricity grids, rubbish, snow, weather, climate, trees, stones, green barrels and the list can go on: 'Materiality is a rubric that tends to horizontalize the relations between humans, biota, and abiota' (Bennett, 2010, 112). The work of Bennet explores the vibrancy of matter and its affective agencies, highlighting the aesthetic or enchanting aspects of materiality. Furthermore, she urges for attention to the vital materiality of the world, human habits of consuming matter and the capacity of non-human things. What such attention to material agencies does is acknowledge the political ecology of worldly matter in non-dualistic ways.

When considering material agencies, Barad (2007, 151) clarifies agency is not something matter possesses; agency is enactment, a performance of different activity. Mattering, for Barad (2007), is a process: 'Matter isn't situated in the world; matter is worlding in its materiality' (180-181). The expansive process ontology of Barad offers insight for understanding phenomena differently. Thinking is not separate from the body, the body is not separate from the world, culture is not outside of nature. Materiality from this sense is not governed by an outside order, but its agency performs in conjunction and relation to other things, with its own immanent desire and will for positive pluralistic creation (Bennett, 2010; Coole, 2010). Simply put, in thinking with materialist theory we eschew adages such as mind over matter, and don't think of materialism as ownership of material things—instead we pay attention to the materiality of the world, it's iterative becoming and our inherent human entanglement with it. A focus on materialism offers educators a lens to consider the world, how matter acts with (or without) us and where the physical stuff in our learning contexts can even assist our pedagogical aims (Jukes, et al., 2019). In this sense, matter may also be considered as 'storied,' having a narrative agency (Oppermann, 2018). Which is to say that stories of the world may be expressed in and through matter.

Mutual entangled relationships—assemblages

Coole and Frost (2010) discern 'an overriding characteristic of the new materialisms [is] their insistence on describing active processes of materialisation of which embodied humans are an integral part, rather than monotonous repetitions of dead matter from which humans are apart' (8). New materialisms help to highlight the importance of thinking through our relationships with the material world in non-dualistic ways, and that thinking with different ideas helps create new understandings.

Building on ideas of agency being performative, agency is also relational. For exploring material relationships, it is again helpful to return to Barad (2007, 2008). Her conception of intra-action re-draws the boundaries of relations—relations are within and part of the world and its performance. Intra-active relationships are the entanglement of bodies that co-constitute the world and what it is becoming. Phenomena, for Barad, are specific intra-actions, relationships or entanglements. An example phenomenon may be a rapid in a river: the intra-active relationships of water, stones and gravity (among other things) merge to become a rapid. This entangled relationship is an assemblage of heterogeneous actors performing in particular ways creating the phenomena.

Furthermore, matter and discourse are entangled in what Barad describes material-discursive practices. She describes the world as ongoing material-discursive phenomena, where human and non-human matter performs intra-actively—that is mutually entangled in process. Simply, matter and discourse co-constitute meanings. Matter and discourse coexist implicating each other in processes of change. Again, the concept assemblage helps account for mutual entangled relationships of mattering and meaning making.

For our project, it is important to bring such ideas into educational discourse as it offers inclusive ways of examining learning contexts. Hultman and Lenz Taguchi (2010) and Gough (2016) build on such mattering; firstly by encouraging educators to challenge notions of anthropocentrism and humans as completely autonomous agents. Secondly, by acknowledging learners are emergent in relational fields where more-than-human matter influences becomings. Such thinking enables pluralistic explorations of human and more-than-human as co-existing. Mannion et al., (2013) support these notions in OEE, explaining 'relations and interactions of teacher, pupils and places as examples of how culture and nature interact coextensively and contingently' (804). The aim is to be inclusive, allowing consideration of multiple and numerous influences to a learning context.

What implications might this have for (outdoor) environmental education research? Malone and Truong (2017) suggest that 'reconsidering how a relationship with the planet that doesn't focus on the human, but on the relationship between the human and the more-than-human may provide a new imagining—a new space for changing how human and nature are being considered' (5). Elsewhere, Malone (2016, 2017) put such a suggestion to work exploring the assemblage of 'child-dog-bodies' in La Paz, Bolivia. Her investigation of coexisting child-dog-bodies describes the co-merging of children and dogs that live in the streets of La Paz. Such an exploration challenges some conventional notions of children-in-nature such as idyllic romantic childhoods, disrupting some dominant ways of knowing. There are other ways of knowing and relating, and other ways to research becoming-with the more-than-human world. Another example could be the student-snail relations at a campsite on the Snowy River one of us examines elsewhere (Jukes, 2018). There Scott uses the concept becoming (Deleuze and Guattari, 1987), to consider 'becoming-snail', which isn't a literal becoming, but a reciprocal exchange or relational encounter. The slimy multitude of snails of the campsite had an aesthetic and affective influence on students, changing the relationship and understanding of non-human others. Bennett (2010) suggests that thinking assemblages of people with other material bodies horizontally is stepping toward an improved ecological sensibility (10).

Becoming part of a more-than-human world

> We still *need* that which is other than ourselves and our own creations [author's emphasis]. (Abram, 1996, ix)

The phrase more-than-human world is highlighted by Abram (1996), where he believes in a certain sense modern humans participate 'almost exclusively with other humans and our own human-made technologies' (ix)—this has led to a precarious relationship with the more-than-human world. Drawing upon Malone (2016) and Clarke and Mcphie (2014), we argue from a posthuman standpoint it's impossible to not participate with things other than humans, that we are always already participating with a more-than-human world. Abram's (1996) thesis aims to attract attention to sensuous engagement and embodied reciprocity within a more-than-human world, taking insight from the phenomenological work of Merleau-Ponty. Although many aspects of phenomenological work are not commensurate with new materialist approaches, the new materialist work of Coole and Frost (2010), Coole (2010) and Connolly (2010) are amenable to Merleau-Ponty's focus on embodiment, corporeality and ideas of perception. As Coole and Frost (2010) suggest, they 'give materiality its due' (20). However, where new materialist perspectives diverge considerably from phenomenology is the foundation in difference and pluralism and extension of agencies beyond the human in the former rather than essentialism in the latter (among other reasons). We still find resonance with the work of

Abram (1996, 2010) due to our interest in the ethical perspective of humans belonging to a more-than-human world. Importantly, what the concept of a more-than-human world does is highlight how humans are part of a world beyond ourselves. Human beings are decentred, horizontalized and brought back to earth among other non-human things. Such a movement is ethical, as it offers a challenge to ownership over the world and a (re)consideration of relationships with more-than-human others. As St. Pierre, Jackson and Mazzei (2016) advocate:

> If humans have no separate existence, if we are completely entangled with the world, if we are no longer masters of the universe, then we are completely responsible to and for the world and all our relations of becoming with it. We cannot ignore matter (e.g., our planet) as if it is inert, passive, and dead. It is completely alive, becoming with us, whether we destroy or protect it. (101)

St. Pierre et al.'s (2016) statement instils in us the need for affirmative actions in education where students may learn of the issues of our time whilst also not falling into despair.

To return to Abram's (2010) work, he describes the reciprocity people continually have with the material world—seeing is also being seen, touching is also being touched—as an unfolding of our human existence in a more-than-human world. Our minds inhabit our material bodies which inhabit the world: no separation exists just an extending relationality. Such relational reciprocity is part of everyday existence if we choose to perceive it. But to take this thought into the posthuman realm, what if when perceiving our material reciprocity, we (conceptually) step into the other we are entangled with? This idea is explored in a picture storybook introduced later. It is also an idea explored by Stewart (2018) when he conducts a pedagogical thought experiment with his students to think like a Murray Cod. Although, as Stewart states 'how can I ever know what a Cod might think?' (136), such a pedagogical practice allows him and his students to conceptually and experientially inhabit the life and circumstances of the fish and its habitat of the river system. This allows the various relational assemblages of what influences this more-than-human other to be explored.

Similarly, Mcphie and Clarke (2015) consider a walk in the park, exploring animistic qualities of things they encounter, deconstructing the idea of what something *is* by instead considering what it *does*. Their walk in the park challenges many notions of Western thought and considers the materiality of things encountered. Furthermore, they express how humans are an active part *of* the world, '(re)placing both moralistic and connection narratives with a narrative of a relationally underpinned metaphysics of immanence' (246). The walk in the park is performative, experiential and experimental, opening different possibilities for relational encounters in (outdoor) environmental education.

In efforts to think differently about our (present-changing) relationships with a more-than-human world, and build the theoretical terrain of this project, we highlight aspects of posthuman critical theory (Braidotti, 2013, 2018a, 2018b, 2019). Posthuman knowledge production considers humans in conjunction with others (bodies, things, actants, technologies), whilst challenging dominant notions of what it means to be human. Braidotti's work in developing a posthuman theoretical framework offers ways forward for such a task, urging for critical thinking which acknowledges the co-creation of the world whilst also imagining creative and positively different futures. A material and more-than-human world full of life and vitality envelops 'us' (us being the diverse heterogeneous multiplicity called human). Material forces shape, shift and move us—from inside and outside our bodies. If OEE is to foster relationships with the natural world, we would argue considering the material bodies of the more-than-human world as a vital task. Considering a posthuman world challenges notions of anthropocentrism, human exceptionalism and dominant modes of thought (Braidotti, 2013, 2018a, 2018b, 2019). However, as Gough (2016) describes:

> Challenging *hierarchical anthropocentrism* (i.e. challenging the assumption of human superiority) does not prevent us from acknowledging an *irreducible anthropocentrism*, that is, accepting that we necessarily experience the world with species-specific biophysical limitations *and possibilities*. However, we must also

Figure 1. Becoming skiers: A student-snow-ski assemblage. (Photo by Scott Jukes).

> consider how an understanding of irreducible anthropocentrism might be changed by accepting that we increasingly experience the world as *posthumans*, with perhaps (eventually) *fewer* species-specific biophysical limitations and with further possibilities provided by biophysical extensions, enhancements and assemblages. [author's emphasis] (61-62)

What might posthuman theorising allow us to see? And in what ways are we already posthuman? A way we may become posthuman in an outdoor educational context is how students and educators become skiers when attached to skis, on snow, in alpine areas (Figure 1). Such an assemblage extends our capabilities, allowing us to travel in alpine areas in particular ways, that allow us to participate and intra-act with that environmental assemblage: this assemblage has particular affordances. Furthermore, our focus in such thinking moves towards places as belonging to more than just us humans, towards material bodies and pedagogies of responsiveness. We belong to the assemblages; they don't belong to us. Thinking and (re)thinking about our place and other non-human's place in the world is an ethical act. This is not a transcendent or theological movement, but an acknowledgment of the vital materiality of things, including their agency, vibrancy and power (Bennett, 2010).

Assembling a concept: *More-than-human stories*

We have been encouraged by Malone and Truong's (2017) call to engage 'with alternative discourses that consider other ways of knowing the human and more-than-human world… by exploring creative pedagogical approaches' (7). As outdoor environmental educators and researchers we are faced with the problem: how might we prompt student understandings of human and more-than-human relationships that have shaped, currently shape and might shape the future of different places? And more specifically, how can we make these ideas into a pedagogy that is useful and digestible for our students? St. Pierre et al. (2016) describe working with the new materialisms:

> Put simply, we can't tell someone how to do this new work, *how to think* [author's emphasis], how to experiment, how to tip an assemblage toward the plane of immanence. Our best advice is to read and read and read and attend to the encounters in our experiences that demand our attention. If we have something different to think with, we might be able to live on a different plane of thought, a different plane of experience. (106)

Braidotti and Hlavajova (2018) explain the posthuman era, this geological moment of the Anthropocene, needs critical and creative scholarship to address the precarity of our time—this time of mass species extinction and climate change induced by human actions. Developing ethical understanding and different perspectives of the more-than-human world require different ways of thinking. This includes conceptual creativity and experimentation, where we must 'trust in the powers of the imagination' (Braidotti and Hlavajova, 2018, 10). Hence we return to concepts:

> Concepts are intensive and create orientations for thinking. It is in this respect we might begin to think of concepts as methods, precisely because concepts are at once prehuman (emerging from the problems or plane of thinking in which we find ourselves), but that also reconfigure or reorient the plane precisely by being prompted by a problem. (Colebrook, 2017, 654)

As Colebrook (2017) proposes, there is an educational power in concepts.

In assembling ideas we have thus far explored, we wish to arrange them into a concept, a pedagogical way for thinking: we call it more-than-human stories. What more-than-human stories as a concept brings together is storylines, aspects of place responsive-pedagogy and new materialist/posthuman theories. The concept aims to foster the imaginative potential of student's formation of knowledge about the more-than-human world, both factually and creatively, envisioning storied matter (Oppermann, 2018). The idea of more-than-human stories is also influenced by Haraway's (2016) speculative fabulations; her creative storying of the worlds of species. Haraway's work compliments that of van Dooren (2014), who advocates for storying ways of life to create new understandings and relationships with those other-than-human. Van Dooren explains that places can be storied by humans and more-than-humans, and should be understood as material-discursive phenomena (67). For van Dooren, stories of those other than us are worth telling in this time of mass extinction, loss and precarity: 'stories are part of the world, and so they participate in its becoming. As a result, telling stories has consequences: one of which is that we will inevitably be drawn in new connections, and with them new accountabilities and obligations' (van Dooren, 2014, 10). Our aim through more-than-human stories is developing different perspectives that enable ethical consideration of the entangled web of various species on earth. We acknowledge there are different genres of stories, from factual to folk, and different storied practices, from reading, telling and making/authoring (Payne, 2010). Although aiming to be factual, more-than-human stories may shift what storylines are made relevant, displacing the human as the central actant in a story. They may be made or told. The art of the story may help us enter, explore and consider the lives of more-than-human others.

More-than-human stories as a pedagogical strategy may prompt students to think humans as part of assemblages of the more-than-human in different and creative ways. As Colebrook (2017) suggests, education should not be a mass normalizing project, but instead an open-ended project of creative difference: 'education is at once necessary to bring forth a future distinct from what we already are, and yet that orientation toward a world of relations that is not oneself comes with the essential risk of stupidity' (651). Giving students the freedom to think, and think differently, to see where the agentic capacity of the more-than-human world guides them has risks. Yet the relinquishing of control is something worth trying, having political consequences (Morse, Jickling and Quay, 2018). Also, it must be noted that such ideas are not the only aspect of the educational milieu. For example, the ski touring journey this research emerges from included teaching about natural histories, cultural influences and the elements of Australian alpine environments in winter. Such an OEE experience may unfold material-discursive phenomena (Barad, 2007, 2008)—the entanglement of matter and meaning—where more-than-human stories may play a part.

Empirical materials: An emergent more-than-human story

> Method in the new, then, might be characterized as emergent in the act of creation… (St. Pierre, et al., 2016, 105)

The ski touring journey, as part of OEE subject Winter Alpine Environments, included pedagogical experimentation where a more-than-human story emerged. Informed by Denzin (2013), we present empirical materials, rather than using the descriptor of data. These empirical materials include: i) an example of a more-than-human story, and ii) a postscript. Using 'empirical materials' implies that they are not a universal truth or brute fact gathered, instead they are a performative insight into this particular educational context and its thinking with concepts. The empirical materials were something created—a proliferation emerging from an immanent assemblage of events.

Mannion (2019) highlights that thus far it is less common for new materialist researchers to suggest exemplars for educators and other researchers. Although the empirical materials below may not be exemplary, they are examples of experiments with new materialist ideas:

i. *One Single Moment*—a picture storybook made by Ya—is an example of a more-than-human story. The picture storybook was drawn and written by Ya on the ski touring journey. Whilst partly a thought experiment, *One Single Moment* can be considered a 'narrative experiment' (Gough, 2014), that uses watercolour pictures with words. The picture storybook is more than a representation; it is a figuration, an 'expression of alternative representation' (Braidotti, 2013, 164). It involves making and performance, emerging from contextual situatedness (subjectivity). The re-reading by others is continued performativity and an echo of affects that allows thinking with a more-than-human other.
ii. Following *One Single Moment*, is what we call a postscript—a mapping by Ya of the conditions of creation/possibility that led to her making the picture storybook (Braidotti, 2013, 2018a). In other words, the mapping in the postscript is a rhizome[5] that led to the proliferation (line of flight) of the picture storybook (Deleuze and Guattari, 1987).

These empirical materials blur the boundaries of conventional qualitative inquiry; we envision this research as postparadigmatic (Gough, 2016), posthuman (Braidotti, 2013, 2018a) and new materialist (St. Pierre, et al. 2016). Mcphie and Clarke (2018) explain 'new materialist approaches aim for creation and production' (3). The non-conventional efforts in this research aim to be different. As St. Pierre (2018) encourages: 'inquiry should *begin* with the too strange and the too much' (607) as it allows differences to emerge. The things that do not seem strange are 'what everyone knows, what everyone does, the ordinary, repetition. Post qualitative inquiry asks … us to trust that something unimaginable might come out that might change the world bit by bit, word by word, sentence by sentence' [and drawing by drawing] (St. Pierre, 2018, 607). Co-producing this paper offers opportunities for new co-created ways of knowing to emerge, enabling discussion of both educator's pedagogical intention and student's learning and response—which both acknowledge the more-than-human influences in the educational milieu. Furthermore, as Braidotti (2018) points out, 'we need sharper focus on the complex singularities that constitute our respective locations' (23). Hence we offer this paper as a focus on the singular and productive aspects of our location.

'One single moment'—by Ya Reeves

SNOW DUMP ON MY ARMS FORCED THEM DOWN AND DOWN. OUT AND DOWN, SPLAYED AND RUNNING LOW.
BECAME A MARKER, A LEANING PLACE, A LISTENING POST, A GUARDIAN.
5
WATCHED AS FIRE WALLS WERE WHIPPED FROM THE BACKSIDE UP AND UP, LEAVES ROCKETED SKYWARD IN A VORTEX INFERNO.
I WAS BLACK FACED, THEN FURRY. I WAS NEW NECKED, SPREADING AND TWISTING OVER AGAIN.
6

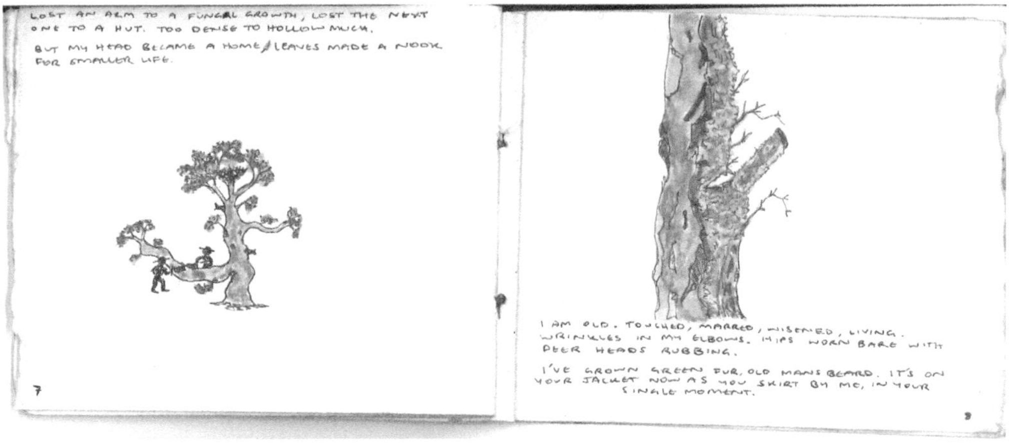
LOST AN ARM TO A FUNGAL GROWTH, LOST THE NEXT ONE TO A HUT. TOO DENSE TO HOLLOW MUCH.
BUT MY HEAD BECAME A HOME, LEAVES MADE A NOOK FOR SMALLER LIFE.
7
I AM OLD. TOUCHED, MARRED, WISENED, LIVING. WRINKLES IN MY ELBOWS. HIPS WORN BARE WITH DEER HEADS RUBBING.
I'VE GROWN GREEN FUR, OLD MANS BEARD. IT'S ON YOUR JACKET NOW AS YOU SKIRT BY ME, IN YOUR SINGLE MOMENT.

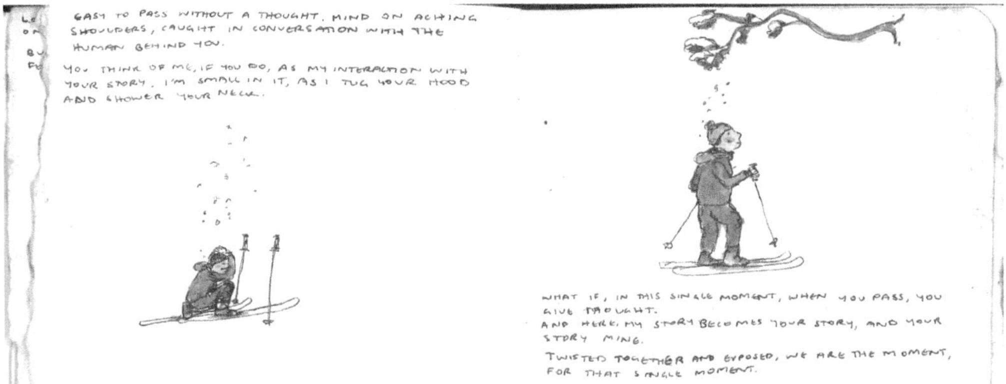
EASY TO PASS WITHOUT A THOUGHT. MIND ON ACHING SHOULDERS, CAUGHT IN CONVERSATION WITH THE HUMAN BEHIND YOU.
YOU THINK OF ME, IF YOU DO, AS MY INTERACTION WITH YOUR STORY. I'M SMALL IN IT, AS I TUG YOUR HOOD AND SHOWER YOUR NECK.
9
WHAT IF, IN THIS SINGLE MOMENT, WHEN YOU PASS, YOU GIVE THOUGHT.
AND HERE, MY STORY BECOMES YOUR STORY, AND YOUR STORY MINE.
TWISTED TOGETHER AND EXPOSED, WE ARE THE MOMENT, FOR THAT SINGLE MOMENT.
10

A postscript: Conditions of creation/possibility—mapping a becoming

This section, written by Ya nine-months after making the storybook, involves a mapping of the various influences and circumstances that led to its creation as a way of 'apprehending points of growth' (Mannion, 2019, 8). Mapping the conditions of creation is an effort to avoid essentialising the production of *One Single Moment* to an abstract concept because, truthfully, there is a multiplicity of heterogeneous effects immanent to the educational context. Below Ya presents some for her:

This book was a culmination of things; things I both understood and didn't, a process of producing and grappling with new knowledge.

- The first part of the trip, three days with an educator that appeared genuinely interested in the way my brain worked, and who encouraged me to keep thinking and articulating, was electrifying. Scott was open about his pedagogical experimentation, challenging us to think differently, trying to shift our focus away from the human and towards the material agency of the place.
- *Difference* was something that Scott searched for, urging difference to be sought out and paid attention to. This was an educator's invitation to take off running with any and all ideas. A hint that no productive response could be incorrect.
- Scott facilitated an activity that encouraged us to explore a fascination.[6] His fascination, which he shared later on, was to do with Snow Gums and their way of existing comfortably in this place that we find so challenging. This reignited an old fascination of mine. Could I read the history of a tree in the tree itself?
- As a child between 1999–2008, I watched one Snow Gum grow in my backyard in Dinner Plain (a small town in the Australian Alps). The tree grew slower than anything, every year those branches would touch the ground with the weight of snow, catapult back up in the Spring. Every year, the catapult became less, branches turned downward more. By the time we left, the tree had changed shape, not just grown taller, grown out. More than that, I could see those heavy years written in the creases of the tree elbows.
- Whilst I always wrote, finding intrigue in words and telling stories even as a kid, I had a complex relationship with drawing. I drew as a way of journaling when I got tired of words. I always thought paintings needed to be representations, accurate, capturing some exact Truth. I was never so good at that. But drawing and painting were becoming extensions of word expression.

(*continued*)

- Whilst working for another university, Scott had facilitated an assessment task that involved writing a story from the perspective of something other than a human. He told me about this task, and it stuck in my mind. I was conscious of not anthropomorphising the subject of the story. I felt like it needed to be grounded still in its own existence, and there was a danger of imposing a human character onto, say, a tree, in order to give it value or to empathise with it.
- Scott explained to me aspects of his research and the theory/philosophies that intrigued him. He explained the concepts of the rhizome, intra-action, and material agency, his attempts to challenge anthropocentrism in research and practice. Whilst I found these concepts difficult to grasp, I started to actively try and view the place through some of these conceptual lenses.
- During the second part of the ski trip (5 days), Scott and I were in different groups, and mine skied North towards Bogong (Victoria's highest mountain). Without having Scott there to query, I was left alone to make sense of our conversations without his guidance. Drawing and writing helped me to unpack the ideas.
- On day three of the trip, I turned on my phone to discover that a former student/friend of mine, Tom, had committed suicide a day earlier. He was 18. I associated Bogong with Tom, as it was a place I had spent important time with him, a place that he had been exploring since he was a small kid. The areas we were skiing presented clear views over to Bogong, that big horseshoe of a mountain. This news was all-encompassing, and I decided that I needed a productive focus for distraction.
- I started writing *One Single Moment* that morning, day 3. Started channelling all of my energy into that assessment task that Scott had told me about.
- I started drawing the book as we travelled from place to place. Drew from what I saw at the time, the trees in front of me, the colours, tufts of lichen, angles in branches. The pictures follow the trail of our tour yet are punctuated with my own memories (of fearfully watching the 2003 fires tear into the Alps, of smudgy musk sunsets). The book started to be an embodiment of the place, more than just a story of one snow gum, and that place was also becoming an embodiment of my memories of Tom. Among the other aspects of the trip, it became a three-day mission of creating through grief and emotion, wrestling with understanding, and adding more layers to my own understanding of the High Plains.
- Made the book in the most uncomfortable of places. Ass numb at 10pm in the hut, cramped in my sleeping bag in the morning, adding scribbles halfway through windy lunches, pinning the scraps of paper down with carrots and Sriracha sauce. I utilised whatever I had, including string from my repair kit to bind it, and mat repair glue to paste the front and back cover.
- I finished it on the last morning before we left the mountain. Handed it over to Scott unceremoniously in a carpark as a means of thanking him. Didn't make any copies. Nine months later, Scott sent me a copy.
- Just as the above factors contributed to the production of the book, the book itself produced a proliferation of thoughts in me when I eventually re-read it all those months later. Here are just a few:

  - EMBRACING THE IN-BETWEEN: This book was caught somewhere in between understanding, theory, emotion, place, stability, new friendships, new perspectives. The book itself is a process, as I was learning whilst creating. It fails to (neither does it attempt to) capture some essential truth, represent a singular theory, or strictly adhere to rules—for me, it inhabits this in-between quite comfortably.
  - PROVOCATION: It wouldn't have come about (or perhaps come about in the same way) without provocation from Scott, by way of his conversation, interest, sharing of concepts, and the obvious way he sought difference and encouraged divergent thinking. Nor without his easy ability to expose his own pedagogical ideas and approaches to his students.
  - LAYERS OF UNDERSTANDING: My history in the place has informed much of the story's content. When I say history, I am referring to my experiences as well as my understanding of cultural and natural history. They are essential to the ability to read (or even consider) the history of the tree. The book comments on:

    - Natural History: the weathered and nutrient-poor condition of Australia's very old mountain, the predominant winds, subniviun spaces as homes to marsupials, the recent and devastating fire history, and the Snow Gum's ability to adapt to its conditions, bending rather than breaking.
    - Cultural History: the extensive history and importance of this Alpine Environment to Indigenous people, the many sacred sites, as well as the European/cattle grazing tradition of building huts across the mountains.
    - Current Political Debates: surrounding feral animals, their impact, and their subsequent existence in the mountains (the storybook mentions the 'deer heads rubbing' against the tree).

    Neither through concepts, nor emotion alone could I have produced *One Single Moment*.

  - VOICE OF THE PLACE: *One Single Moment* wouldn't exist without the landscape being what it was, when it was, where it was. The quality and richness of that trip, in all of its complexity and messiness, could never be entirely replicated, as the conditions will and can never be the same. But alpine environments in winter in Australia have a rare, isolated, ephemeral, shifting, and productive quality. This book is located in and a product of that place, and in many ways, I believe it seeps a similar character.

Complexity, productivity and offerings

Investigating the multiple threads of this research shows that matter and place are complex and multi-storied phenomena. As Somerville (2013) describes, 'writing about place is an ontological act, producing the self at the same time as writing the words. It is predicated on unwinding the spiral of "material form and interpretative understandings or experiences" to enable new possibilities to come into being' (19). *One Single Moment* is a story about the history, material form and vitality of Snow Gums on the Bogong High Plains. The picture storybook, combined with the mapping in the postscript, is also an enactment of thinking with and through concepts, place, pedagogy and educational experiences. Gough (1999) asserts that there is a contested nature to epistemological concerns in educational and environmental research. Thus we acknowledge that environmental education research as an epistemological endeavor is entangled with ontology, with materiality, our student's pasts, the environments history. The presentation of *One Single Moment* and the Postscript are a non-reductive mapping of onto-epistemological formations. It shows part of the complexity of education, pedagogy and the student experience. However, we note here the generative capacity of education, that although things may be complex, even messy, it is not something to avoid—messiness is truthful to the reality of educational contexts. Ya as a student, grappling with multiple complex subjectivities was able to formulate a story that imaginatively reaches for new ways of understanding. The story writing/drawing process allowed expression and communication of complex ideas. The material and social milieu of OEE contexts could do well to allow time and space for creatively thinking through the environment and the more-than-human in open-ended ways. Enabling students to create their own discourses, as Gough (1999) suggests, has valuable educational potential. We add that this creation is material-discursive, always in conjunction with material agencies. In this way, such a pedagogy may be envisioned as 'wild' (Morse et al, 2018), as it acknowledges the self-will and agency of both students and the more-than-human world in the formation of knowledge and subjectivity.

Hence, more-than-human stories offer a conceptual heuristic for considering our place in a more-than-human world, more-than-human inhabitants and different spatio-temporal perspectives. In other words, we offer more-than-human stories as a strategy or tool for the educator in the assemblage theory toolbox (Mannion, 2019) where place becomes the co-storyteller (Payne, 2010). In a similar vein to Mannion (2019), we imagine this as a step towards more-than-human pedagogies. What it might allow is a (re)envisioning of current and future relationships, recognising the entangled and mutually constitutive relationships between the humans and that which is more-than-human. Yet, the risk with an open-ended pedagogy is that it may not reach such a lofty aim. Nevertheless, we must guide, provoke and catalyse students and look for the *assemblage convertors* (Bennett, 2010, 42; Deleuze and Guattari, 1987, 378) that prompt ethical rethinking in this time of the Anthropocene. For as Braidotti and Hlavajova (2018) state, '"We" are in *this* anthropocentric crisis together' (12). 'We' are not all the same (Braidotti, 2019) yet we environmental educators might enact our pedagogy in creatively different and context specific ways. Environmental educators should be encouraged to experiment with new ideas, play with pedagogy and not baulk at complexity. Embracing relational openness and affirmative action's when working with our students and more-than-human others may be a step to co-producing a positive future.

Notes

1. Winter Alpine Environments, is an Outdoor Environmental Education subject at La Trobe University. The subject description, written by Mangalsdorf (2018), explains: 'Students examine aspects of natural history, land management, snow deposition and ecology directly relevant to selected Alpine environments of South Eastern Australia. Students develop technical expertise of cross-country ski touring to enable safe exploration and promote experiential and interpretative knowledge of winter alpine environments.'
2. Australia's alpine areas (including surrounding slopes and valleys) make up 0.3% of Australia's landmass (Slattery, 2015). The Bogong High Plains make up a small part of these alpine areas. As a sight of study, they offer a unique landscape with endemic species and contested natural history. The high plains are part of the Victorian Alpine National Park, established in 1989, after a long struggle for 'protection' (Slattery, 2015). The alpine areas have been (and remain in some aspects) a contested space between interest groups such as grazers, hydroelectric power, ski resorts, timber industry, conservationists and recreational users. Current issues

facing the high plains are climate change and invasive species that are threatening endemic species. These topics are discussed at various times through the Winter Alpine Environments subject.

3. Importantly, we do not imagine place in a static or anthropocentric way. As we develop throughout this paper, places don't belong to humans—we are part of them. Places are inherently more-than-human, unbounded and inconsistent. Our new materialist and posthuman perspectives envision places as open-ended assemblages.
4. We acknowledge debate over whether new materialisms are actually new, considering they draw upon a range of philosophical traditions. Nonetheless, the neologism 'new materialism' works and it is the newer work from this disposition we predominantly cite.
5. A rhizome is a concept created by Deleuze and Guattari (1987) as a way of thinking differently to linear images of thought, allowing heterogeneous connections and movements. Gough and Price (2004) explain that 'abandoning arborescent [tree like/linear] thinking means becoming nomadic, allowing thoughts to wander beyond familiar territories and to produce new texts/terrains' (31).
6. 'Fascinations' is an activity/thought experiment Scott has been playing with, aiming to acknowledge the agency of the more-than-human world, prompted by previous research (Jukes et al., 2019; Jukes, 2018). 'Fascinations' aims to develop what Bennett (2010) describes as an aesthetic-affective openness. The activity/thought experiment is also influenced by 'wild pedagogies' in allowing the 'self-will' of the student and the more-than-human world to influence the educational context (Morse et al., 2018). Intriguingly, the etymology of fascination derives from Middle French and Latin words with meanings like bewitch/er, enchant, spell and speak.

Acknowledgements

We would like to acknowledge and thank Alistair Stewart and Marcus Morse who provided advice and comment on earlier versions of this paper. We would also like to thank the anonymous reviewers who offered insightful feedback on the manuscript.

Disclosure statement

No potential conflict of interest was reported by the authors.

Funding

Scott's contribution to this research was funded by an Australian Government Research Training Program Scholarship.

ORCID

Scott Jukes http://orcid.org/0000-0003-3758-9413
Ya Reeves http://orcid.org/0000-0001-6257-9132

References

Abram, David. 1996. *The Spell of the Sensuous: Perception and Language in a More-than-Human World*. New York: Vintage Books.

Abram, David. 2010. *Becoming Animal: An Earthly Cosmology*. New York: Vintage Books.
Baker, Molly. 2005. "Landfullness in Adventure-Based Programming: Promoting Reconnection to the Land." *Journal of Experiential Education* 27 (3): 267–276. doi:10.1177/105382590502700306.
Barad, Karen. 2007. *Meeting the Universe Half Way: Quantum Physics and the Entanglement of Matter and Meaning*. Durham: Duke University Press.
Barad, Karen. 2008. "Posthumanist Performativity: Toward an Understanding of How Matter Comes to Matter." In *Material Feminisms*, edited by Stacey, Alaimo, and Hekman, Susan, 135–169. Bloomington, IN: Indiana University Press.
Bennett, Jane. 2010. *Vibrant Matter: A Political Ecology of Things*. USA: Duke University Press.
Braidotti, Rosi. 2013. *The Posthuman*. Cambridge: Polity Press.
Braidotti, Rosi. 2018a. "A Theoretical Framework for the Critical Posthumanities." *Theory, Culture & Society* 36 (6): 31–61. doi:10.1177/0263276418771486.
Braidotti, Rosi. 2018b. "Posthuman Critical Theory." In *Posthuman Glossary*, edited by Rosi, Braidotti, and Hlavajova, Maria. New York: Bloomsbury Academic.
Braidotti, Rosi. 2019. *Posthuman Knowledge*. Cambridge: Polity Press.
Braidotti, Rosi, and Maria Hlavajova. 2018. "Introduction." In *Posthuman Glossary*, edited by Rosi, Braidotti, and Hlavajova, Maria. New York: Bloomsbury Academic.
Clarke, David. 2017. "Educating beyond the Cultural and the Natural: (Re)Framing the Limits of the Possible in Environmental Education." In *Reimagining Sustainability in Precarious Times*, edited by Karen, Malone, Truong, Son, and Gray, Tonia, 305–319. Singapore: Springer.
Clarke, David, and Jamie Mcphie. 2014. "Becoming Animate in Education: Immanent Materiality and Outdoor Learning for Sustainability." *Journal of Adventure Education and Outdoor Learning* 14 (3): 198–216. doi:10.1080/14729679.2014.919866.
Clarke, David, and Jamie Mcphie. 2016. "From Places to Paths: Learning for Sustainability, Teacher Education and a Philosophy of Becoming." *Environmental Education Research* 22 (7): 1002–1024. doi:10.1080/13504622.2015.1057554.
Colebrook, Claire. 2002. *Gilles Deleuze, Routledge Critical Thinkers*. New York: Routledge.
Colebrook, Claire. 2017. "What is This Thing Called Education?" *Qualitative Inquiry* 23 (9): 649–655. doi:10.1177/1077800417725357.
Connolly, William. 2010. "Materialities of Experience." In *New Materialisms: Ontology, Agency, and Politics*, edited by Diana Coole and Samantha Frost, 178–200. Durham: Duke University Press.
Coole, Diana. 2010. "The Inertia of Matter and the Generativity of Flesh." In *New Materialisms: Ontology, Agency, and Politics*, edited by Diana Coole and Samantha Frost, 92–115. Durham: Duke University Press.
Coole, Diana, and Samantha Frost. 2010. "Introducing the New Materialisms." In *New Materialisms: Ontology, Agency, and Politics*, edited by Diana, Coole and Samantha, Frost, 1–43. Durham, NC: Duke University Press.
Deleuze, Gilles, and Felix Guattari. 1987. *A Thousand Plateaus: Capitalism and Schizophrenia*. Translated by B. Massumi. London, UK: Bloomsbury.
Deleuze, Gilles, and Felix Guattari. 1994. *What is Philosophy?* Translated by H. Tomlinson, and Burchell, G. New York: Columbia University Press.
Denzin, Norman. 2013. "The Death of Data?" *Cultural Studies ↔ Critical Methodologies* 13 (4): 353–356. doi:10.1177/1532708613487882.
Fox, Nick, and Pam Alldred. 2015. "New Materialist Social Inquiry: Designs, Methods and the Research-Assemblage." *International Journal of Social Research Methodology* 18 (4): 399–414. doi:10.1080/13645579.2014.921458.
Gough, Noel. 1999. "Rethinking the Subject: (de)constructing human agency in environmental education research." *Journal of Environmental Education Research* 5 (1):35–48. doi:10.1080/1350462990050102
Gough, Annette, and Hilary Whitehouse. 2018. "New Vintages and New Bottles: The "Nature" of Environmental Education from New Material Feminist and Ecofeminist Viewpoints." *The Journal of Environmental Education* 49 (4): 336–349. doi:10.1080/00958964.2017.1409186.
Gough, Noel. 2008. "Ecology, Ecocritiscm and Learning: How Do Places Become 'Pedagogical'?" *Transnational Curriculum Inquiry* 5 (1): 71–86.
Gough, Noel. 2014. "Undoing Anthropocentrism in Educational Inquiry: A Phildickian Space Odyssey?" In *Posthumanism and Educational Research*, edited by Nathan Snaza and John Weaver. New York: Taylor and Francis.
Gough, Noel. 2016. "Postparadigmatic Materialisms: A 'New Movement of Thought' for Outdoor Environmental Education Research?" *Journal of Outdoor and Environmental Education* 19 (2): 51–65. doi:10.1007/BF03400994.
Gough, Noel, and Leigh Price. 2004. "Rewording the World: Poststructuralism, Deconstruction and the 'Real' in Environmental Education Research." *Southern African Journal of Environmental Education* 21: 23–36.
Haraway, Donna. 2016. *Staying with the Trouble: Making Kin in the Chthulucene*. Durham: Duke University Press.
Honan, Eileen. 2014. "Disrupting the Habit of Interviewing." *Reconceptualizing Educational Research Methodology* 5 (1): 1–17. doi:10.7577/rerm.929.

Hultman, Karen, and Hillevi Lenz Taguchi. 2010. "Challenging Anthropocentric Analysis of Visual Data: A Relational Materialist Methodological Approach to Educational Research." *International Journal of Qualitative Studies in Education* 23 (5): 525–542. doi:10.1080/09518398.2010.500628.

Jukes, Scott. 2018. "Thinking Outdoor Environmental Education Differently: Material Relations on a Snowy River Journey." Masters Diss., La Trobe University.

Jukes, Scott, Alistair Stewart, and Marcus Morse. 2019. "Acknowledging the Agency of a More-than-Human World: Material Relations on a Snowy River Journey." *Journal of Outdoor and Environmental Education* 22 (2): 93–19. doi: 10.1007/s42322-019-00032-8.

Lenz Taguchi, Hillevi, and Elizabeth St. Pierre. 2017. "Using Concept as Method in Educational and Social Science Inquiry." *Qualitative Inquiry* 23 (9): 643–648. doi:10.1177/1077800417732634.

Lugg, Alison. 2004. "Outdoor Adventure in Australian Outdoor Education: Is It a Case of Roast for Christmas Dinner?" Journal of Outdoor and Environmental Education 8 (1): 4–11. doi:10.1007/BF03400790.

Lynch, Jonathan, and Greg Mannion. 2016. "Enacting a Place-Responsive Research Methodology: Walking Interviews with Educators." *Journal of Adventure Education and Outdoor Learning* 16 (4): 330–345. doi:10.1080/14729679.2016.1163271.

Malone, Karen. 2016. "Reconsidering Children's Encounters with Nature and Place Using Posthumanism." *Australian Journal of Environmental Education* 32 (1): 42–56. doi:10.1017/aee.2015.48.

Malone, Karen. 2017. "Ecological Posthumanist Theorising: Grappling with Child-Dog-Bodies." In *Reimagining Sustainability in Precarious Times*, edited by Karen, Malone, Truong, Son, and Gray, Tonia, 161–172. Singapore: Springer.

Malone, Karen, and Son Truong. 2017. "Sustainability, Education, and Anthropocentric Precarity." In *Reimagining Sustainability in Precarious Times*, edited by Karen, Malone, Truong, Son, and Gray, Tonia, 3–16. Singapore: Springer.

Malone, Karen, Son, Truong, and Tonia Gray. 2017. *Reimagining Sustainability in Precarious Times*. Singapore: Springer Nature.

Mangalsdorf, Anthony. 2018. *Winter Alpine Environments OED2WAE: Subject Learning Guide*. Unpublished document. Bundoora, Australia: La Trobe University.

Mannion, Greg. 2019. "Re-Assembling Environmental and Sustainability Education: Orientations from New Materialism." *Environmental Education Research*: 1–20. doi:10.1080/13504622.2018.1536926.

Mannion, Greg, and Jonathan Lynch. 2016. "The Primacy of Place in Education in Outdoor Settings." In *Routledge International Handbook of Outdoor Studies*, edited by Barbara. Humberstone, Heather, Prince, and Karla, Henderson, 85–94. Abingdon, Oxon: Taylor and Francis.

Mannion, Greg, Ashley Fenwick, and Jonathan Lynch. 2013. "Place-Responsive Pedagogy: Learning from Teachers' Experiences of Excursions in Nature." *Environmental Education Research* 19 (6): 792–809. doi:10.1080/13504622.2012.749980.

McPhie, Jamie, and David Clarke. 2015. "A Walk in the Park: Considering Practice for Outdoor Environmental Education through an Immanent Take on the Material Turn." *The Journal of Environmental Education* 46 (4): 230–50. doi:10.1080/00958964.2015.1069250.

Mcphie, Jamie, and David Clarke. 2018. "Nature Matters: Diffracting a Keystone Concept of Environmental Education Research—Just for Kicks." *Environmental Education Research*: 1–18. doi:10.1080/13504622.2018.1531387.

Mikaels, Jonas, and Morten Asfeldt. 2017. "Becoming-Crocus, Becoming-River, Becoming-Bear: A Relational Materialist Exploration of Place(s)." *Journal of Outdoor and Environmental Education* 20 (2): 2–13. doi:10.1007/BF03401009.

Morse, Marcus, Bob Jickling, and John Quay. 2018. "Rethinking Relationships through Education: Wild Pedagogies in Practice." *Journal of Outdoor and Environmental Education* 21 (3): 241–254. doi:10.1007/s42322-018-0023-8.

Oppermann, Serpil. 2018. "Storied Matter." In *Posthuman Glossary*, edited by Rosi, Braidotti, and Hlavajova, Maria, 411–414. New York: Bloomsbury Academic.

Payne, Phillip. 2010. "Remarkable-Tracking, Experiential Education of the Ecological Imagination." *Environmental Education Research* 16 (3–4): 295–310. doi:10.1080/13504621003613160.

Payne, Phillip. 2016. "What Next? Post-Critical Materialisms in Environmental Education." *The Journal of Environmental Education* 47 (2): 169–178. doi:10.1080/00958964.2015.1127201.

Pleasants, Kathleen, and Alistair Stewart. 2019. "Entangled Philosophical and Methodological Dimensions of Research in Outdoor Studies? Living with(in) Messy Theorisation." In *Research Methods in Outdoor Studies*, edited by Barbara Humberstone and Heather Prince, 9–20. New York: Routledge.

Pretty, Jules, Bill Adams, Fikret Berkes, Simone Ferreira de Athayde, Nigel Dudley, Eugene Hunn, Luisa Maffi, et al. 2009. "The Intersections of Biological Diversity and Cultural Diversity: Towards Integration." *Conservation and Society* 7 (2): 100–112. doi:10.4103/0972-4923.58642.

Prins, Alex, and Brian Wattchow. 2019. "The Pedagogic Moment: Enskilment as Another Way of Being in Outdoor Education." *Journal of Adventure Education and Outdoor Learning*: 1–11. doi:10.1080/14729679.2019.1599295.

Raffan, James. 1993. "The Experience of Place: Exploring the Land as Teacher." *Journal of Experiential Education* 16 (1): 39–45. doi:10.1177/105382599301600109.

Reid, Alan. 2019. "Blank, Blind, Bald and Bright Spots in Environmental Education Research." *Environmental Education Research* 25 (2): 157–171. doi:10.1080/13504622.2019.1615735.

Ruck, Andy, and Greg Mannion. 2019. "Fieldnotes and Situational Analysis in Environmental Education Research: Experiments in New Materialism." *Environmental Education Research*: 1–18. doi:10.1080/13504622.2019.1594172.

Slattery, Deirdre. 2015. *Australian Alps: Kosciuszko, Alpine and Namadgi National Parks*. 2nd ed. Clayton South, Australia: CSIRO Publishing.

Somerville, Margaret. 2010. "A Place Pedagogy for 'Global Contemporaneity'." *Educational Philosophy and Theory* 42 (3): 326–344. doi:10.1111/j.1469-5812.2008.00423.x.

Somerville, Margaret. 2013. "Water in a Dry Land: Place-Learning through Art and Story." In *Innovative Ethnographies*, edited by Phillip Vannini. New York: Routledge.

Somerville, Margaret, Kerith Power, and Phoenix de Carteret. 2009. "Introduction: Place Studies for a Global World." In *Landscapes and Learning: Place Studies for a Global World*, edited by Margaret Somerville, Kerith Power and Phoenix de Carteret, 1–20. Rotterdam, The Netherlands: Sense Publishers.

St. Pierre, Elizabeth. 2018. "Writing Post Qualitative Inquiry." *Qualitative Inquiry* 24 (9): 603–608. doi:10.1177/1077800417734567.

St. Pierre, Elizabeth, Alecia Jackson, and Lisa Mazzei. 2016. "New Empiricisms and New Materialisms: Conditions for New Inquiry." *Cultural Studies ↔ Critical Methodologies* 16 (2): 99–110. doi:10.1177/1532708616638694.

Stewart, Alistair. 2004. "Decolonising Encounters with the Murray River: Building Place Responsive Outdoor Education." Journal of Outdoor and Environmental Education 8 (2): 46–55. doi:10.1007/BF03400803.

Stewart, Alistair. 2008. "Whose Place, Whose History? Outdoor Environmental Education Pedagogy as 'Reading' the Landscape." *Journal of Adventure Education & Outdoor Learning* 8 (2): 79–98. doi:10.1080/14729670801906125.

Stewart, Alistair. 2018. "A Murray Cod Assemblage: Re/considering Riverscape Pedagogy." *The Journal of Environmental Education* 49 (2): 130–141. doi:10.1080/00958964.2017.1417224.

van der Tuin, Iris. 2018. "Neo/New Materialism." In *Posthuman Glossary*, edited by Rosi, Braidotti and Maria, Hlavajova, 277–279. New York: Bloomsbury Academic.

van Dooren, Thom. 2014. *Flight Ways: Life and Loss at the Edge of Extinction*. New York: Columbia University Press.

Verlie, Blanche, and CCR15. 2018. "From Action to Intra Action? Agency Identity and Goals in a Relational Approach to Climate Change Education." *Environmental Education Research*: 1–15. doi:10.1080/13504622.2018.1497147.

Wattchow, Brian, and Mike Brown. 2011. *A Pedagogy of Place: Outdoor Education for a Changing World*. Clayton, Vic: Monash University Publishing.

Informal environmental learning: the sustaining nature of daily child/water/dirt relations

Sarah Crinall and Margaret Somerville

ABSTRACT

This post-inquiry paper looks to the intimate matter of young children and their worlds outside of school. With new materialism, water and art as philosophical muse, a new kind of 'sustaining nature' for environmental education emerges, problematizing 'sustainability' as an aim. We gathered and exchanged short videos and field notes (iPhone artifacts) during morning and afternoon walks with young children in our everyday lives. In a playful moment, these are engaged with in relation to theoretical material on 'art and sensation' from philosophers, Deleuze (2003), Grosz (2008) and ontologies of water that *do* post-qualitative research playfully (Crinall 2017; Somerville 2017). We mutually emerge *with* relational pedagogies, conjuring the grace and gift of daily life with the matter of *clay, dirt, deadbird, water, and home* into an environmental learning that attunes bodies/minds to the significance, grace and gift of sustaining, artful everyday life.

Introduction

We take walks frequently with young children, following them in their spontaneous play down roads, into ditches, puddles and swales where children, water and dirt un/entangle. In this special edition on *New materialisms in environmental education research* we engage with these meanderings and a recent problematizing of 'sustainability' by environmental learning scholars, who are urging us to understand humans as matter mutually emerging *with* the world (Taylor 2017). With the micro, intimate matter of children and their worlds outside of school this paper joins relational pedagogies at the time of the post-qualitative turn, when ontologies are being engaged with as research practice to disrupt traditional linear modes of rationalism that have long dominated education inquiry (St. Pierre 2014).

Walking with new materialism child scholars

New material/child scholars of sustainability/environmental education are now looking to the co-emerging intimate everyday lives of matter, young children and the world, asking difficult questions about power relationships between species (e.g. Nxumalo and Pacini-Ketchabaw 2017—stick insects and human; Taylor 2017—kangaroos; Taylor and Pacini-Ketchabaw 2015—ants), and species and non-living matter (e.g. Rautio 2013—children and stones; Pacini-Ketchabaw and Clark 2016—water; Horton and Kraftl 2017—water; Rooney 2018—the weather). These scholars are problematizing education programs within dominant understandings of 'sustainability,' where children are framed as separate to nature, with power over matter (living and non-living), and able to both destroy and save the planet (Nxumalo and Pacini-Ketchabaw 2017).

For example, Pacini-Ketchabaw and Nxumalo unsettle the colonial nature/culture divide of Western epistemology 'through a series of situated, small, everyday stories from childcare centres [relating] raccoon-child-educator encounters in order to consider how raccoons repeated boundary-crossing, potentially [revealing the] impossibility of the nature/culture divide' (Pacini-Ketchabaw and Nxumalo 2015, 151).

Messying the nature/culture divide

As the nature/culture divide is messied, new material theoretical frames are finding new relations and emergences, re-defining what it means to live with other matter and be in the world. We notice that what we are 'sustaining' here shifts during the post-qualitative turn. Energising the wonder and glow of other matter, the data has agency (MacLure 2013a; see Koro-Ljungberg, Löytönen and Tesar 2017), they stay with the trouble (via Haraway 2016) of the ethics of life, death and human responsibility for planetary sustenance (Nxumalo and Pacini-Ketchabaw 2017). Relational philosophers are blurring bounds of nature/culture (e.g. Rautio, 2013; Pacini-Ketchabaw and Nxumalo 2015) and new modes of writing are sought to be with dirty, messy place/bodies coming to know (MacLure 2013b; Crinall 2017).

Relational pedagogies such as child/kangaroo, child/ant, child/stick insect, child/bird and child/eel, and entanglements with the non-living agential entities of child/mud and child/water, ask how human and other bodies are emerging with the world locally, amongst tumultuous change globally. These pedagogical enquiries are often in institutional settings and structured excursions that intend to co-mingle children and other nature bodies. These studies disrupt institutional 'business as usual' learning programs (e.g. Taylor 2017), asking practitioners to step back from setting agendas (Green and Somerville 2015) to look to children, 'full of small achievements' with whom we can learn new ways of world collaboration through the kind of 'relations that many young children have with the world already' (Taylor 2017, 12). We are encouraged here to 'learn new ways' to be with children following the threads of ethics of ethico-onto-epistemologies by way of seeking answers to the world's becoming (see section below and Barad 2010), where humans are accepted as curious animals of the planet (see Barad 2010; Rautio 2013; Crinall 2017).

Exploring post-inquiry as emergence

In the post-qualitative turn, St. Pierre (2014) notes a research proceeding beyond (not replacing) the epistemologically-oriented works of the qualitative, humanist frame of rational, logical and cognitive framing and affirmations of what 'is'—provided by Cartesianism—to offer something different. Qualitative research is no longer the only way of not/knowing in education. A post-inquiry is 'a practice of freedom' (St. Pierre 2014, 3) where we are unconstrained by planned methods and conclusive findings. Promiscuous feminisms work outward toward issues for the globe from issues for women and are comfortably uncomfortable in their 'disloyal fidelity' (MacLure 2013b, 625). Un/comfortably, in post-inquiry, we are drawn to Barad's ethico-onto-epistemology of intra-action[1] (2012), becoming in place of knowing, outside of the qualitative realm. Where Cartesian ontology knows matter that 'interacts' and combines two separate entities into one through connection, 'intra-action' understands matter as emerging out of an assemblage as agency emerges within particular spaces, times and bodies (human and other). Intraactive (spacetime)matter is always, already enfolded back into the assemblage, diffractively, and potentially 'other' simultaneously in contradiction (see Barad 2007, 2010, 2012). This is research as 'doing', doing post-qualitative research.

> So it is not that there are no separations or differentiations, but that they only exist within relations. Putting the point another way, phenomena are differential patterns of 'mattering'—diffaction patterns dispersed across differently entangled spaces and times, or rather *spacetimematterings*. (Barad 2012, 77)

In immersing ourselves in the material collected from our everyday walks that we engage with below, we think of everyday life as a form of art, or as artful everyday life, drawing on Deleuze's understanding of art as sensation and knowing (Deleuze 2003; Deleuze in Grosz 2008).

Art as sensation and knowing

For Deleuze, art is an alternative form of knowing to rational forms of philosophy. Grosz, in *Chaos Territory Art* (Grosz 2008) reiterates Deleuze's art as the communicable form of the rationally unspeakable; the representational form of the cognitively unrepresentable. And so the world is knowable directly through the nervous system via art, bypassing the cognitive brain and experienced by the skin, muscles, and senses, felt as a sensation experienced in another body part to the eyes that see it.

Where art is excess for Grosz, and art is nature, nature is shifted through provocations of lack so common in sustainability aims, which seek to reduce, reuse, recycle, into rhythm (Crinall 2017). Where art, according to Deleuze, moves through catastrophe into rhythm, as nature, Crinall (2017) agentially cuts a world-naming beyond lack with/in/as everyday living:

> Grosz argues that so much of nature is art (2008, 8), and art, sensation and nothing else according to Deleuze (2003), art is excess… I slip into another thought – In *Logic of Sensation*, Deleuze proposes that art, which is its own form of becoming, not creator nor viewer, attends provocations and problems moving through catastrophe into rhythm. What I find on the other side of the blogged assemblage then is potentially a world-naming that enables planetary sustenance in a daily rhythm of being-with. (Crinall 2017, 230)

A sustainability education theory that engages art to think differently at a time when logic and reasoning have dominated, does not intend to redefine art to include the everyday. It engages art to be with lived matter in a new way.

> I am not interested in providing qualitative evaluations of works of art, in distinguishing 'good' from 'bad' or high from low art, in setting up some sort of heirarchy of the various arts, but in exploring the peculiar relations that art establishes between the living body, the forces of the universe and the creation of the future. The most abstract of questions, which if they are abstract enough, may provide us with a new way of understanding the concrete and the lived. (Grosz 2008, 3)

With the artful everyday as our base, the writings of Karen Barad, Gilles Deleuze and Elizabeth Grosz potentially materialise a new kind of life: 'Art is where the becomings of the earth couple with the becomings of life to produce intensities and sensations that in themselves summon up a new kind of life'. (Grosz 2008, 79)

Walking with children and dirt/clay/water

Children take the lead and we follow, being-with the children/water/dirt/clay bodies and finding walks from home deviating into unpredicted asides (see also Crinall 2017, 206). We reach for our iPhones and take photos or films, and make field notes. Seeking to understand 'sustainability' in relation to the alternative knowledges attained and relationships negotiated, the material of our exchanges is energised by two inquiries:

> How are children naming life-sustaining relationships across species and matter in daily life?
>
> What new imaginings become possible when sustaining relationships are named?

Everyday spacetimematterings gather 'in our pockets' as we, children, *the dirt ditches, deadbird, water, home,* venture out together. These 'autotelic practices of humans such as picking up and carrying stones [is] potentially relevant in further understanding and conceptualizing the ways in which humans are nature in relation to and constituted by all other animate or non-animate co-existing entities' (Rautio 2013, 1).

The entities of an autotelic practice, engaged as spacetimematterings, the discrete objects and subjects are now co-existing entities. From there we scribe our sustaining creative process in a curious and playful way from our exchanges with each other across time and space. The peculiar relations between bodies, forces and the creation of the future play out a new way of being (with and as) our lived materialities. A reader may engage with this paper as an experience of sensation. We invite you into the space of responsive thought, where tangential threads may dangle into your own everyday lives.

Water ontologies

Being and coming to know water is explored using writing with/as water through our body/place writing which continuously moves (as we are moved) along a life through various states of matter across a topology of ecosystems and spheres in a space other to rational knowing. Our words are moving water, *pausing to mark the stillness of rock and stone* (Somerville 2013, 74).[2]

Water makes its mark

on massive round rocks

line upon line of white

and dark grey striations

tracing long gone deep water

across rows of curved rock surface

timeless and still yet

overlaid by ever-moving shadows cast

from creek's fallen timber

shimmering water reflections

change moment by moment

with shifting sun cloud and

light movement of air

against stillness of rock and stone (Somerville 2017, 43, field notes)

Water marks our daily walks and performs her dances, timeless and still. Water offers us watery ontologies as banks for an onto-epistemological river.

An ethico-onto-epistemology with water*

I step down into healthland now

Walking home

Now slipping along the swales

Berry searches for nipple

I smoke breath

Puff

draw one after the other

My addiction

These walls *walks* and fresh ocean air

If bachelard said water was milk

Then I am Berry's water and she laps at my stream

I dress up as water

Water is my muse

I drink her milk

I act like her

I am becoming her

Do I mimick water on this daily walk and perform her dances here?

While walking the swales to the sea

like the particles of water that drizzle from the storm

let me know a sustaining life?

I ask water about an ontology of water

Toward a justice for … (Crinall 2017 133, interpersonal blog post[3]) (*Barad 2010)

Water takes us up in its swimming, soaking, cycling,[4] sustaining nature. When we 'draw the *corpus/corpse* (MacLure 2013a) of water into view and the human corpus/corpse into water' (Crinall 2017, 131–132) a new mode of becoming-human floods from in-between place, space and body with an awareness of sustenance.

Water ontologies shift our listening. In rational epistemologies water is parochially tended by humans, deemed necessary for water's health. Becoming *with* water disupts linear, predictable modes of knowing and being. Water dissipates life into flows of not/knowing where we *listen to children*, and *listen to water* (Pacini-Ketchabaw and Clark 2016, 100).

An ethico-onto-epistemology with-water that is ontologically prior emerges:

> Water sounds, feels, smells, moves, floods, offers habitat, has power, gathers up all in its force. Water as data reconfigures what we understand as matter, it is always in relation, gathering up everything in its onto-epistemological force, including the human subject. It may be recorded as data by humans in the form of embodied sensory responses, memory, affect, curiosity, small videos, or still photographs but it is always in excess of these, always already there and continuing on long after any of us, a powerful reminder of the transience of human life. (Somerville 2017, 47)

With water ontologically prior, and always in excess, it was always already here, in/between our bodies. Here humans are emerging (out of) water, water that is 'always in relation, gathering up everything in its onto-epistemological force, including the human subject' (Somerville 2017, 47). An ethico-onto-epistemology with-water understands a life becoming-with as beginning with water, and as water. Thinking with this original water data fluidly, we can *befriend* water (Rautio and Vladimirova 2017), and attend her *curiously* with etiquette (Haraway 2016). Queries here precipitate on the surface, a kinship with our watery self/other. A sustaining, with water disrupts, moves, seeps, is lifeful, deadly, diffusing and cyclical. This sustaining was and is always moving between bodies that are simultaneously one and apart and is not solely lifeful, nor only threatening. Pacini-Ketchabaw and Clark (2016) articulate, when thinking with water in the early childhood classroom, that with water 'our hearing and listening are always partial and imperfect because we are still learning to live with water' (100).

Children, water, dirt, roads as data

Specific, embodied artifacts of child/water/dirt/road spacetimematterings are gathered in our pockets, shared as short videos and field notes of prose and photos between iPhones. With the swipe of a finger down the screen of iPhone in the messages App[5] we ask ourselves 'where are we in all this'? A conversation trickles across Australia's East Coast from south on Phillip Island, Victoria up to Emu Plains, West of Sydney.

How do everyday exchanges between us, between the children and the everyday places of road, water, child and swales sustain?

Artifacts

Autumn, 2018

Dear Margaret,
The sky is purple grey this morning. I am held by worn, brown armchair looking to lawson street's swales. I thought of you yesterday when children emerged in their puddlesuits in boots with two bread bags twisted over the top (*we need to go to the gumboot shop, mum*). Child, so eager to meet with water and dirt depression, pulled at my arm.

We are home

we walk along the swamp paperbark tree-lines
dirt ditches (swales)
we are by our house –
frog croak,
deadbird,
dirt,
'weedy' grass,
litter,
water
I say: *'Let's go home, Bean'*
Bean replies: *We are home*
(Sarah Crinall, Autumn, 2016, field notes[6])

It rained again today

on and off light starry rain
I'm dry now by the fire
looking in
looking back
Bean's pink overalls hang over the white chairs
And Berry's too
I want to leave them at the door ready for tomorrows rain walk mum
Bean pleads
they will dry here i reply
the rain touched us
softly
softly
on foot
in boots
lets go!
babies in hand
Berry with me siena doll with bean
rain on heads
light tickle
damp dropping hair
warm warm torsos
walking
stomp here mum!
and here!
there are puddles everywhere
big oblong

rotund
round
oval opaque brown water pools
Berry drops from hands to toes
leather shoes dancing in the surface
she can't feel it so it must be something else that has her
gleening
the splash and splash
the unpredictable nature of the water moving
in all directions
everywhere
faces
legs
hands
the two are there in the puddle and I am watching
dry
clean
photographing and filming
and I skip to Bean's common query
where are the parents in this book?
why is the little girls alone?
I drop too
down
i flinch
water in through jeans immediately
puddle sucks me in
and down
Girls screech with joy
and we walk home
can we play in the bad weather again tomorrow mum?
of course I sing.

Love, Sarah

Autumn, 2017

Dear Sarah,

It's turned suddenly cold and wet late summer and into early autumn and Lulu and Charmaine want to go for a rain walk to the fairy festival. We set off two girls under pink frilly umbrellas for neaBerryy scrappy wasteland beside railway track until we come to a big pile of yellow clay where they stop briefly, frilly pink umbrellas discarded on ground, mainly to examine the hole dug into the side of the fairy festival. Charmaine says it has got bigger as she kneels and peers in, Lulu joins – maybe it's a lizard – can lizards dig? Yes, remember when they dug a hole in the dirt near the river to lay their eggs?

We move on, fairy festival does not have much traction today. Walking by wire fence frilly umbrellas get stuck in wire and long grass is wet, until captured by a tiny flow of water running down the track and making little gullies. But they want to look for tadpoles further up the road/track. At a large pothole in what was originally a bitumen road, we peer into small pool full of black spotted jelly tadpole eggs but this doesn't capture either, it's the flow of the water into and out of the tiny pool in this water formed erosion channel cutting deeper into the old bitumen road.

Eventually, I take a tiny video.

Captivated myself at how this can be so engaging, this minute water flow in a tiny clay erosion gully in an old bitumen road running through scrappy bush by railway line. Pinky clay coloured erosion hollow running along edge of decaying bitumen. Leaf and stone debris on sides of hollow. Narrow channel of clay water with red ochre colour bleeding into one spot. Abandoned bag and umbrella on bitumen edge. Two girls squatting in channel arms outstretched holding sticks towards water.

Press the 'triangle' on the video and a small scene emerges:

Lulu arm-with-stick makes wide movement following flow of water talking to the water

'so it goes down' towards where Charmaine squats says you said it down right hand on stick left hand lifts other small fat stick from waterway

plops back

Charmaine stands

throws fat stick 'Poppy stop it'

squats again

Lulu bends further towards water talking to it softly

some inaudible

'it all goes straight down'

with small splashing plop sounds

fine fingers lift small fragments of stick from water channel and drop them to the side, allowing water to flow.

Lulu simultaneously low growly voice

mud near the? ?

bouncing down to water 'oh oh oh the rock' lifts and throws small rock from water channel to bank chanting in low guttural voice towards water

'red rock red rock' as she scrapes out channel with stick

Lulu following water channel

'there's lot of water flow'

moving back and forward in the tiny space of the water flow

'there's a lot of water flow'

'I'm letting lot of water flow down and down'

sing song voice to water

Adult voice: What will happen to the tadpoles if we let all the water flow down there, do you think they'll live without water?

Lulu: oh they will, that's a dam

Love, Margaret

Meandering (after)thoughts

These artifacts are of the world, already being named by children's bodies in relation to watery place bodies, and or own, meandering daily lives with them, with water and as water.

With water

There may be more spectacular places to immerse bodies than roads, but the children choose them and they choose the children. With water, the roads and children hold each other's attention and so sustain each other, spacetimemattered from the assemblage,

stomp here!

splash, splash.

I'm letting the water down and down.

It's going down and down.

As water

As the children are letting the water flow, down and down, we, as water, flow with children/water/dirt/road. We are not the choreographers of this moment having chosen to follow their flow. We take to our flow, *captivated at how this can be so engaging, this minute water flow in a tiny clay erosion gully in an old bitumen road running through scrappy bush by railway line.*

When we lifted our phones to capture these ordinary, everyday moments we were moved, like the water the children have released in flow and splash. A sense of sustenance transpires from the precipitated child/water/dirt/road assemblages, soaking into our nervous systems as though they are 'art' itself: and so we flow. Lift arm, finger to red circle, press record. This brings us to wonder: Are the child/water/dirt/road assemblages, a sensing form of sustenance, as art is the sensing form of knowing accessed via sensation? Are we experiencing everyday life as art?

As art, the road/water/child assemblage from which we emerge and enfold becomes a collaboration of various spacetimematter given and giving agency with our bodyplace awareness and always, though not only, critical attention to worldly health.

This kind of everyday life, as art, or artful everyday life, names various specific bodies, children/water/dirt/roads/us, as already in an on-going practice of sustaining, relational everyday living. This is not so much a world-learning for sustainability as a world-making. The on-going flow of these relations, sustains by inclination beyond expectation, and via sensation beyond fact. This world-making sustains relations that hydrate the children, and ourselves, and arguably continues on via this reflection for you today. These pieces may or may not affect via sensation in the breath and raze of hairs on skin, ripple of cells, aversion of eyes, or breath lengthened.

All responses are the flow of how learning as living moves, with and as water, with and as art, through and around us in various, multiple modes, including the academic writing process.

the splash and splash

the unpredictable nature of the water moving

in all directions.

Specific encounters soak into our skin, mark us and wash us in particular ways. This is a different kind of sustainability to the humanist kind. It is one which emerges mutually with the world in a continuous becoming-with water, dirt and other matter into and out of the assemblage. At a time when movements of lack and ration often frame sustainability in everyday lives - reducing, refusing, recycling etc. for planetary health, we are held by the sense of sustenance and abundance dripping, soaking in through jeans, from the worldly assemblage of child/water/dirt/road assembling in our phones as artifacts of the world, artworks. This sense of sustenance and abundance in the sensible form of child/water/dirt/road assemblage leaves this thread dangling:

How can we be '*touched by the rain, softly, softly,*' of children and water already always flowing and splashing sustenance hither and thither in our ordinary, everyday lives together/apart?

We offer the practice of body/place writing made with smart phone snapshots and short videos collaged as dirt/water/child artifacts, as evidence and practice of the world's sustaining assemblage. Through this multimodal practice, we make and exhibit, not only images/words/sounds, but become aware of and enter into these soaking, flowing, sustaining relations that move. Taking but one thread from the many possibilities for sustaining relationships with specific body/place pedagogies, we observe, entangle and send sustenance outward with the worldly resonance of the artifacts we have attuned to. While we may not save the planet, we can sustain, hold a space for continuing, specific, fleeting, relational encounters for and with the abundance of the various other bodies we are and are with daily.

Conclusion

Creating this paper has been sustaining (in) nature for us, an act of sustaining the world. With others re-imagining sustainability in Environmental Education Research with new materiality (see e.g. Rautio 2013; Green and Somerville 2015; Crinall 2016; Gannon 2017; Malone, Truong and Gray 2016; Taylor 2017) we have shifted our focus from *future-orienting* goals for sustainability into a post-inquiry with artful everyday life, water and children. The sustaining nature of research in early childhood, in the experience of everyday lives with young children, emerges as, if it must be named, an artful sustainability learning.

We have attuned our body/minds to the significance, grace and gift of the sustaining, artful everyday with children, oxygenated by water and art as our muse. An artful sustainability learning is open to the exchanges between bodies and places in everyday lives—*with* as a sustaining practice in itself. Being playful with the dirty, messy (dis)comforts of not-knowing is a crucial aspect. Where art precedes and summons up 'sensations to come … where intensities proliferate, where forces are expressed for their own sake, where sensation lives and experiments, where the future is affectively and perceptually anticipated' (Grosz 2008, 78), art can shift daily becomings-with beyond Cartesian, rationalist modes of framing the world. Our mutual emergence with the world can be enacted playfully through the *intensities* and *sensations* experienced.

It has been our pleasure to swim in the abundant excess of everyday life's roadside matter, *as* the matter of our sustainability learning, and we offer these entanglements—of child, water, dirt, road and scholar—where the intimate matter of daily life meets theoretical material on art and sensation, and water as ontology—for others to experience.

We conclude with questions that maintain the watery flows rippling here for education theorists and practitioners interested in pedagogies that swim in the sustaining nature of everyday life's matter with children:

What possibilities are there for learning centres and practitioners to follow threads, listening and responding in everyday ways to children?

How might practitioners continue to draw in and engage with the ordinary, mundane everyday lives of children and the world?

How can educators take up this challenge to tap into sensation and their own sustenance? And finally, How can a practice of mutually emerging *-with* disrupt power relationships not only between species, and species and matter, but human to human?

Made with children and everyday matter we are all already always becoming-with, in our immediate surrounds. How artful sustainability learning, emerging from within the flow of everyday life, translates into a formal education setting is a potential challenge for practitioners and future environmental education researchers.

Notes

1. 'the notion of 'intra-action' queers the familiar sense of causality (where one or more causal agents precede and produce an effect), and more generally unsettles the metaphysics of individualism (the belief that there are individually constituted agents or entities, as well as times and places). According to my agential realist ontology, or rather ethico-onto-epis-temology (an entanglement of what is usually taken to be the separate considerations of ethics, ontology, and epistemology), 'individuals' do not preexist as such but rather materialize in intra-action.' (Barad 2012)
2. Water ontologies used here have been established in our previous research pursuits (Crinall 2016, 2017, 2019; Somerville 2013, 2017; Somerville and Perkins 2010).
3. The blog is Sarah's private writing/making space and not accessible to the public. Blog posts are interpersonal, sent specifically between Sarah and Margaret, at times.
4. We are taken specifically with the swimming, soaking, cycling nature of water in this paper and intentionally stay with this pleasurable sense water provides us at this time. For those readers keen to engage with water's complex multifaceted, dangerous, devastating nature see Pacini-Ketchabaw and Clark (2016) and Horton and Kraftl (2017). Both offer insights into early childhood education and water.

5. Consent has been given by the guardians of these young children to be involved in this article. One of the authors is the mother to two of the children. Three of the four children have been involved in conversation directly between authors and child about the coming together of this article, and have willingly included themselves. It is understood that the children may not comprehend, to the level of an adult, what they are consenting to. We feel it important to raise the issue of consent as unsolveable, or at least unresolved at the moment. What are the ehtics of researching with one's own children? We also wonder, what are the consequences for academia of not speaking with the children in our lives?
6. This blog post above is from a recent research project that sought to understand 'sustainability' in relation to the alternative knowledge attained and relationships negotiated between art/ists, children and water in everyday life (Crinall 2017).

Acknowledgements

We would like to acknowledge the ARC collective *Naming the World: enhancing early years sustainability learning* with whom we share the emergence of new ideas within new materialist approaches; The Space Place Body research group for scholarly and everyday support and feedback; and Dr. Karen Crinall for editing work. Sarah would like to thank her daughters, for without who this work and thinking would not be.

Disclosure statement

No potential conflict of interest was reported by the authors.

ORCID

Sarah Crinall http://orcid.org/0000-0003-4066-1324
Margaret Somerville http://orcid.org/0000-0001-8804-5825

References

Barad, K. M. 2007. *Meeting the Universe Halfway: quantum Physics and the Entanglement of Matter and Meaning*. Durham, NC: Duke University Press.

Barad, K. M. 2010. "Quantum Entanglements and Hauntological Relations." *Derrida Today* 3 (2): 240–268.

Barad, K. M. 2012. *Summer. Intra-actions: Interview of Karen Barad by Adam Kleinman*. Milan, Italy. *Mousse Magazine* 13. https://www.academia.edu/1857617/_Intra-actions_Interview_of_Karen_Barad_by_Adam_Kleinmann_

Crinall, S. M. 2016. "Bodyplacetime: Painting and Blogging 'Dirty Messy' Humannatured Becomings." In *Reimagining Sustainaibility in Precarious Times*, edited by K. Malone, S. Truong, and T. Gray, 95–114. Singapore: Springer Press.

Crinall, S. M. 2017. "Blogging Art and Sustenance: Artful Everyday Life (Making) with Water." PhD diss., Western Sydney University, Sydney, Australia.

Crinall, S. M. 2019. *Sustaining Childhood Natures: The Art of Becoming with Water*. Singapore: Springer Press.

Deleuze, G. 2003. *Francis Bacon: The Logic of Sensation*. (D. W. Smith, Trans.). Minneapolis, MN: University of Minnesota.

Gannon, S. 2017. "Saving Squawk? Animal and Human Entanglement at the Edge of the Lagoon." *Environmental Education Research* 23 (1): 91–110. doi:10.1080/13504622.2015.1101752.

Green, M., and M. Somerville. 2015. "Sustainability Education: Research Practice in Primary Schools." *Environmental Education Research* 21 (6): 832–845. doi:10.1080/13504622.2014.923382.

Grosz, E. 2008. *Chaos, Territory, Art*. New York, NY: Columbia University Press.

Haraway, D. 2016. *Staying with the Trouble*. Durham, NC: Duke University Press.

Horton, J., and P. Kraftl. 2017. "Rats, assorted Shit and 'racist Groundwater': Towards Extra-sectional Understandings of Childhoods and Social Material Processes." *Environment and Planning D: Society and Space* 36 (5): 926–948.

Koro-Ljungberg, M., L. Löytönen, and M. Tesar. 2017. *Disrupting Data in Qualitative Inquiry: Entanglements with the Post-Critical and Post-Anthropocentric*. New York, NY: Peter Lang Publishers.

MacLure, M. 2013a. "Promiscuous Feminists Postscript." *Journal of Qualitative Studies in Education* 26 (5): 625–628. doi:10.1080/09518398.2013.786848.

MacLure, M. 2013b. "The Wonder of Data." *Cultural Studies - Critical Methodologies* 13 (4): 228–232. doi:10.1177/1532708613487863.

Malone, K., S. Truong, and T. Gray. 2016. *Reimagining Sustainability in Precarious Times*. Singapore: Springer Press.

Nxumalo, F., and V. Pacini-Ketchabaw. 2017. "Staying with the Trouble' in Child-Insect-Educator Common Worlds." *Environmental Education Research* 23 (10): 1414–1426. doi:10.1080/13504622.2017.1325447.

Pacini-Ketchabaw, V., and V. Clark. 2016. "Following Watery Relations in Early Childhood Pedagogies." *Journal of Early Childhood Research* 14 (1): 98–111. doi:10.1177/1476718X14529281.

Pacini-Ketchabaw, V., and F. Nxumalo. 2015. "Unruly Raccoons and Troubled Educators: Nature/Culture Divides in a Childcare Centre." *Environmental Humanities* 7 (1): 151–168. doi:10.1215/22011919-3616380.

Rautio, P. 2013. "Children Who Carry Stones in Their Pockets: Autotelic Material Practices in Everyday Life." *Children's Geographies* 11 (4): 394–408. doi:10.1080/14733285.2013.812278.

Rautio, P., and A. Vladimirova. 2017. Befriending Snow: On Data as an Ontologically Significant Research Companion. In *Disrupting Data in Qualitative Inquiry: Entanglements with the Post-Critical and Post-Anthropocentric*, edited by Mirka Koro-Ljungberg, Teija Löytönen and Marek Tesar, 23–35. New York, NY: Peter Lang Publishers.

Rooney, T. 2018. "Weather Worlding: Learning with the Elements in Early Childhood." *Environmental Education Research* 24 (1): 1–12. doi:10.1080/13504622.2016.1217398.

Somerville, M. 2013. *Water in a Dry Land: Place-learning Through Art and Story*. New York, NY: Routledge Publishing.

Somerville, M. 2017. "(Becoming-with) Water as Data." In *Disrupting Data in Qualitative Inquiry: Entanglements with the Post-Critical and Post-Anthropocentric*, edited by M. Koro-Ljungberg, T. Löytönen and M. Tesar, 35–49. New York, NY: Peter Lang Publishers.

Somerville, M., and T. Perkins. 2010. *Singing the Coast*. Canberra, Australia: Aboriginal Studies Press.

St. Pierre, E. 2014. "A Brief and Personal History of Post Qualitative Research: Toward 'Post Inquiry." *Journal of Curriculum Theorising* 30 (2): 1–19.

Taylor, A. 2017. "Beyond Stewardship: Common World Pedagogies for the Anthropocene." *Environmental Education Research* 23 (10): 1448–1461. doi:10.1080/13504622.2017.1325452.

Taylor, A., and V. Pacini-Ketchabaw. 2015. "Learning with Children, ants, and Worms in the Anthropocene: towards a Common World Pedagogy of Multispecies Vulnerability." *Pedagogy, Culture & Society* 23 (4): 507–529. doi: 10.1080/14681366.2015.1039050.

What if schools were lively more-than-human agencements all along? Troubling environmental education with moldschools

Tuure Tammi

ABSTRACT
The recent more-than-human turn has increased interest in writing about relations between humans and other animals. In addition, scholars have called for a need to complement the animal turn with a turn to microbes. Microbes entangle all life in relations and participate in processes of living and dying, but thus far, they have been largely absent in environmental education research. This article attempts to think with microbes in environmental education by zooming into the phenomenon of 'moldschools' in Finland. Employing the concept of agencement, the article first explores how school buildings, toxic molds, and humans have effectuated each other and introduces the idea of indoor climate crisis. Then the article explores the complexity of everyday life in one moldschool, asking, how was the school becoming felt and practiced differently by children and teachers in relation to material-discursive mold. Finally, the article asks what moldschools might teach us or ask from us in terms of environmental education and ethics. The notion of schools as always already more-than-human agencements is offered to make space for hesitation and curiosity on various scales of connection.

More-than-human turn and environmental education

We are living an era when humans' impact on the planet has attained geological magnitude. There is a growing awareness that we are responsible for and are mortally vulnerable to these changes (Gibson, Bird Rose, and Fincher 2015) and that our 'own delusional, self-centered, self-aggrandizing thinking created this predicament' (Nxumalo and Pacini-Ketchabaw 2017, 2). As a response, approaches often referred to as posthumanism and new materialism have increasingly been calling into question the dominant anthropocentric, dichotomic, and hierarchical thinking. While socio-constructivist research, for example, has provided the possibility of considering all social phenomena as relationally produced among humans, this research stream has simultaneously enacted a separation of culture and nature characteristic of the history of the so-called Western thought. Challenging the former division of labor between natural sciences (focusing on things) and social sciences (focusing on meanings), scholars have called for an interdisciplinary approach emphasizing the interdependency and entanglement of matter and meaning, as well as nonhuman and human (Barad 2007; Haraway 2008; Latour 2009; Pickering 2005). These reworkings resonate with much of indigenous metaphysics in which social and material are perceived as co-constituting and 'cohabiting [with] one another, or feeding one another, or making and re-making one another' (Tallbear 2011; see also Watts 2017) even though these connections have not often been acknowledged (Todd 2016). Thus, there has never been a universal 'we': The ongoing effects of current environmental crises are neither equally produced by all human practices, nor do the crises affect all lives similarly (see Nxumalo 2018)[1]

These more-than-human approaches take relationality as the primary onto-epistemological unit. For instance, Haraway (2008) insists that becoming is always becoming-with, and Ingold (2003, 301) argues for a 'fungal model,' in which ecological relations 'are not really *between* organisms and their environments, as though each were initially 'independently' of one another, but rather constitute the very existential foundation from which organisms grow' (see also Affifi 2017). Critical voices argue that these theoretizations, nevertheless, flatten the power relations and push patterns of exploitation into the background (Pedersen 2013). However, being entangled in relations does not mean that everything is affecting and being affected in the same way. Thus, the potential in the more-than-human turn is in addressing the differential becomings-with or 'versions' (Despret 2016; Prout 2005) that emerge from the natural-cultural intertwinings. As Prout (2005, 81), in the context of childhood studies, has insisted, instead of fixed ideas of what a child or adult is, we should ask, 'how do different versions of child and adult emerge from the complex interplay, networking and orchestration of different natural, discursive, collective, and hybrid materials?' However, the more-than-human turn insists that we do not limit our analyses to human becomings, but 'cultivate attentiveness' (van Dooren, Kirksey, and Münster 2016) toward the ways in which different versions of nonhuman animals, plants, or microbes relationally emerge. In this article, I will attempt to put these insights to work in the context of mold problems in Finnish schools (see also Tammi 2019). While doing this, I will craft some ideas on what a microbial turn might mean in environmental education and ethics.

The humanistic and anthropocentric roots—the emphasis on what it means to be human and to participate in human society—and their consequences in Western education are now being increasingly challenged (Humes 2008; Kahn 2008; Pedersen 2010; Snaza et al. 2014; Taylor 2017). With the emphasis on the more-than-human, schools can be thought of as heterogeneous entanglements extending from their infrastructure to plants and nonhuman animals used in food and clothing, onward to various technologies and their local and global networks (Snaza et al. 2014). Analyses have sought to distract the commonly held binaries by drawing attention to the many ways in which people are constituted by and learn with their more-than-human others (Haraway 2008; Pedersen 2010; Snaza et al. 2014; Spannring 2017; Taylor, Blaise, and Giugni 2013; Rautio et al. 2017). More-than-human studies have also challenged anthropocentrism by showing how learning—understood as interpenetration or 'intra-action' (see Barad 2007) of organism and environment—is ubiquitous across the natural world (Affifi 2017; Dalziell 2017). Thus, environmental education scholars have been crafting 'new ways of theorizing and educating about being with, and in relation to, the planet' aiming at a 'healthy, thriving planet that acknowledges the entangled relationship between human and more-than-human world' (Malone and Truong 2017, 5).

Nevertheless, it has been observed that environment-oriented curricula have had trouble finding their way into schools and when the curricula do, they are at risk of being translated into neoliberal visions of sustainable development (Kahn 2008), practices that objectify animals (Spannring 2017; Tammi et al. 2018) and narrow ideas about the environment (Weston 2004), for example. Investigating how animals are represented in schools, Pedersen (2004) asks who gains and what is hidden in the perceived tendency of schools to prefer small-scale farms over slaughterhouses and pets over laboratory animals. Scholars have contended that critically attuned pedagogies should seek to address and act upon *all* forms of human and nonhuman oppression and injustice near and far, and striving to achieve this aim requires more intimate dialogue between various critical educational discourses (Bell and Russell 2000; Fawcett 2013; Humes 2008; Pedersen 2010). Thus, there has been a call for a holistic approach to respond to the environmental crisis and the many forms of interconnected environmental, human, and animal injustices (Humes 2008; Kalof et al. 2016; Pedersen 2004).

Common worlds at all scales

Recent studies, especially in the context of early childhood education, have taken a 'common worlds' approach[2] in order to move beyond any simplistic developmental and anthropocentric views (Taylor 2017; Taylor and Pacini-Ketchabaw 2015, 2016; Nxumalo and Pacini-Ketchabaw 2017). These studies insist upon 'staying with the trouble' (Haraway 2008, 2016) that our co-implication, encounters, and entanglements with other critters in ordinary and everyday life bring into being. What is common to all of these studies is that they challenge the idea of child–animal relations as cute, innocent, and unproblematic, thus bringing forth how shared vulnerabilities, uncertainties, indeterminacies, and potentialities emerge from the complexities and imperfectness of these encounters (see also Hohti and Tammi 2019).

Still, 'the animal' often implicitly refers to pets or other charismatic animals 'big like us,' while smaller critters, such as insects, have remained 'the unloved others,' unless they are perceived as useful for human endeavors (Lloro-Bidart 2018; Rose and van Dooren 2011). With the more-than-human turn, especially in the context of 'multispecies ethnography,' other than humans previously perceived as 'bare life' or 'that which is killable' are now increasingly considered to appear 'alongside humans in the realm of bios, with legibly biographical and political lives' (Kirksey and Helmreich 2010, 545). These approaches center on 'how a multitude of organisms' livelihoods shape and are shaped by political, economic, and cultural forces' (Kirksey and Helmreich 2010, 545). Thus, the fixed species boundaries are questioned as all living creatures are understood as immersed in and emerging from situated multispecies communities, affecting each others' becoming and knowing (van Dooren, Kirksey, and Münster 2016). In order to explore these comminglings, multispecies scholars have developed a range of sensitizing concepts, such as 'passionate immersion' and 'attentiveness' (van Dooren, Kirksey, and Münster 2016), 'polite visiting' (Despret 2016), 'staying with the trouble' (Haraway 2016), and withling(s) (Tammi et al. 2018). Importantly, these efforts do not merely celebrate the multispecies mingling, but insist asking who benefits when species meet (Haraway 2008; Kirksey and Helmreich 2010).

With an emphasis on the entanglement of humans with their environments *at all scales* (Haraway 2016), I share the efforts put forward by Hird (2009) and O'Malley (2014) among others, of complementing the animal turn with a turn to microbes. Microbes entangle all life in relations and participate in every process of living and dying (Hird 2009, 128) and cause humans to form a rhizome with other animals (Deleuze and Guattari 1987). We are now increasingly aware that we live our posthuman lives in symbiosis with myriad microbes in our guts (our 'second brain'), on our skin, in our mouth, in the food we eat, in the things we touch, in the air we breathe (Affifi 2017; Yong 2016). But often this world within the one we see with naked eyes in our everyday lives remains imperceptible, unless it is interpreted as intervening human activities. Let me quote Bruno Latour's work at length:

> We do not know who are the agents who make up our world. We must begin with this uncertainty if we are to understand how, little by little, the agents defined one another, summoning other agents and attributing to them intentions and strategies. This rule of method is especially important when we are studying a period when the number of agents was suddenly multiplied by millions. [...] When we speak of men, societies, culture, and objects, there are everywhere crowds of other agents that act, pursue aims unknown to us, and use us to prosper [...] Society is not made up just of men, for everywhere microbes intervene and act. (Latour 1993, 35)

I began to include microbes in my thinking about 'the social' and to experiment with the idea of building molds as participants in teaching and learning during my investigations of the phenomenon of school strikes. A few years ago, I read a newspaper article that reported a school strike organized by the parents' association in a small Finnish municipality. The school strike practically meant that parents had collectively decided to homeschool their children for a given period of time (in this case, a week). With the strike, the parents wanted to urge the municipality to take action in renovating their local elementary school that suffered from a mold problem and voiced concern over increasing symptoms, such as runny noses, rashes, and respiratory illnesses. Given that compulsory schooling is highly respected in Finland, parents' associations have seldom engaged in activism related to schooling. Usually, parents' associations engage in activities such as fundraising and organizing events.

Mold, among other types of indoor air pollution, has been referred to as the fifth indoor air epidemic—the others being asbestos, radon, formaldehyde, and cigarette smoke (Salkinoja-Salonen 2016). During the 2010s, at least 20 school strikes, among other political actions, were organized by parents across Finland. The Audit Committee estimated in 2013 that 172,000–259,200 people in Finland work in schools and kindergartens where significant dampness and mold problems have been diagnosed. Daily exposure to moisture and mold damage has been estimated to concern 62,000–94,000 elementary school pupils and 12,000–18,000 high school students (Audit Committee 2013). Overall, the estimated number of people exposed daily is upwards of 600,000, which totals one-tenth of the population. Understandably, the renovation and health-care costs are significant, not to mention the incalculable emotional and social costs for those who have become severely ill (Audit Committee 2013). Moisture and mold in buildings is not only a national problem as the World Health Organization (WHO) has defined it as a considerable health risk in all countries where people spend a considerable amount of their time indoors (WHO 2009, 1). Research on human activity patterns suggests that people have begun to spend their time differently in part due to the relative privilege and luxury indoors. Humans in particular in the global West have become 'indoor species' (Ott 1989).

For me, the school strikes manifested an interesting blurring of a boundary: environmental health issues could now be found inside buildings designed to 'liberate social processes from the temporal and spatial constraints of "nature"' (Ford 2015, 7). In what follows, I first ask, how did moldschools[3] come into being? Here, I employ the concept of 'agencement' (Deleuze and Guattari 1987; Despret 2013) which helps me to think and write about schools as historically specific materializations of more-than-human relations. I consider microbes (here, toxic molds) as participants in heterogeneous agencements, through which the microbes and other participants become rendered capable (Despret 2013). Thus, I am not only 'thinking with microbes' (Hird 2009), but also with buildings and bodies (Murphy 2006), when I explore the complex ways in which they may effectuate each other. Here, I draw inspiration partially from Murphy's (2006) analysis of the emergence of sick building syndrome and chemical exposure in office buildings in the USA during the 1980s and 1990s as it shares similarities with the current mold issue. Her analysis shows that buildings and bodies are co-produced, and there are as many versions of them as there are knowledge practices that seize them. Following Murphy's (2006, 13) definition, assemblages are 'formed of organic and inorganic objects, technologies, bodies, and architecture, and not just of words' none of which can be reduced to others (Pickering 2005). However, agencements are not simply gatherings of objects with predefined powers. These objects are in the state of becoming-with; they effectuate in each other 'the power to be affected'; they 'agence' (Despret 2013, 38). This is why Despret (2013) prefers the original French formulation agencement instead of Massumi's English translation of the term as assemblage. Agencements are open-ended, and when one begins to 'crack open' (Murphy 2006) one agencement, one soon finds it is flowing through many others. One of the benefits in taking a more-than-human turn with the concept of agencement is that it allows me to understand and engage with buildings as ecologies, and to re-situate humans within ecological systems (Gibson, Bird Rose, and Fincher 2015). Humans have not been becoming 'indoor species' (Ott 1989) alone. I will show how the phenomenon of moldschools emerged in the intersections of building and cleaning practices, materials, technologies, microbes, policies, and more.

I then engage with the complexities of this agencement on the mundane level by drawing on my encounters and talks with the teachers, pupils, and their parents in one of the moldschools. Initially interested in how the children navigated during this political situation (Tammi 2017), I contacted the chair of the parents' association and made arrangements for a two-week visit. During this visit I talked with and interviewed 9 parents (60–90 minutes per interview), 12 children (30–60 minutes per interview), and 5 teachers (60–90 minutes per interview) of this school of little more than 100 pupils, gathered documents and reports, and breathed the indoor air in the school and people's homes. I ask, how was the school becoming performed and inhabited differently when entangled with 'material-discursive' (Barad 2007) mold? How were relations with mold becoming perceptible in everyday life, and what were these accounts making imperceptible? I do this by focusing on the fears, hopes, curiosities, inconveniences, confrontations—the flows of forces inside moldschool agencement—that finding one's way within the moldschool brought about. In the final section, I ask, what could moldschools possibly teach us, or ask from us in these troubled times. I will address some questions regarding microbial knowing and becoming thus attempting to craft ideas on how we might consider environmental education and ethics differently.

Emergence of moldschools and indoor climate crisis

As Latour (1993), among others, argues, society cannot be made through the social alone. The emergence of moldschools can be mapped through certain inseparable political, economic, technological, cultural, and natural developments. One crucial entrance point to the phenomenon of moldschools can be located in the Yom Kippur war and the resultant oil crises in the 1970s: The price of oil quadrupled, and Western governments were forced to search for ways to reduce use of fossil fuels. One result was a boom in energy-efficient and super-insulated building and re-building and the corresponding expansion of markets for artificial and chemically processed building materials (Baker 2012; Ionescu et al. 2015; Murphy 2006; Mölsä 2017). However, these changes made buildings more vulnerable to accumulation of moisture and mold growth. For instance, drywall has been found to contain minerals and nutrition (starch, cellulose gum) and the requisite humidity for the toxic species of *Aspergillus* and *Stachybotrys* genera, while the plastic covering that protects the gypsum liner keeps the carbon dioxide and humidity levels favorable for mold (Salkinoja-Salonen 2016, 24). As houses were becoming air-tight, ventilation could no longer be provided with gravitational means. The move from gravitational to mechanical ventilation systems in buildings was expected to enable better regulation of the indoor climate, as well as provide universal standards for optimal performance, but was later found to sometimes boost circulation of mold spores in buildings (Ford 2015; Murphy 2006; Mölsä 2017). Reviewing Salkinoja-Salonen (2016), molds produce vesicles as their metabolic products. The molds exhale humidity. These vesicles contain water-resistant toxins and nutrition for spores to germinate and attach to surfaces. The pressure differences in mechanically ventilated buildings fragment vesicles into airborne nanodrops that can spread across the building and enter the respiratory organs of human inhabitants where the vesicles may cause health effects (Salkinoja-Salonen 2016), depending on the immunological status of the bodies (Mousavi et al. 2016). In addition, mechanical ventilation systems have been found to provide dwellings for various molds. For instance, *Penicillium expansum* has been found in factory-sealed ventilation filters and *Chaetominum* in supply air filters (Salkinoja-Salonen 2016). Further, to save energy, ventilation systems were often shut down for weekends and holidays, or maintenance was neglected, which further allowed moisture to accumulate (Mölsä 2017). Ventilation machines may also cause underpressure that absorbs particles from the walls into the indoor air (Mölsä 2017).

In addition to the attempts to condition indoor air, there are attempts to condition microbial life through seeking super-effective ways of killing (all) bacteria. Salkinoja-Salonen (2016) argues that using boron chemicals in materials, such as soundproofing and thermal insulation, and antibacterial cleaning products in schools and elsewhere killed most microbes, but at the same time gave a competitive advantage to toxic mold genera resistant to these chemicals. An international comparison of 29 moisture-damaged schools and 27 control schools in Finland, Spain, and Holland showed that although the schools in Holland and Spain were significantly more 'dirty' microbiologically, the children in Finnish schools suffered more from symptoms connected to indoor air problems. Thus, the researchers asked whether the schools in Finland have become 'cleaned to death' (Salkinoja-Salonen 2016, 105).

In comparison, when studying the microbiome of hospitals, Kembel et al. (2012) found that hospital rooms were colonized by potentially harmful pathogens originating from the mouths and skins of hospital residents, many of which were rare in outdoor air. The microbial diversity was lower indoors than outdoors, and window-ventilated rooms were richer in microbial diversity than those ventilated mechanically. Opening the window allowed benign or harmless microbes from outdoors to occupy the hospital rooms and exclude potential pathogens as the room's microbiota became more diverse. In many schools, ventilation is left to the machines, and opening a window is prohibited (in some schools, windows are even bolted shut) to assure that these machines work properly. In addition to the ways in which rooms are ventilated, the size and type of the room, the rooms' connectedness to each other, the temperature, and relative humidity, as well as the frequency of use and movements by humans, influence what is found in room dust (Kembel et al. 2014).

Practices of effectivity in building, ventilation, and killing bacteria were coined with the neglect of maintenance and rebuilding (children's places especially) during the deep economic recession of the 1990s (Mölsä 2017). More recently the precarious capitalist economy has further altered the standards of construction and renovation. Responsibilities are often fragmented and blurred between contractors, subcontractors, and sub-subcontractors. Contracts are grounded on low costs, and schedules are made extremely tight due to effectivity pressure (Vesalainen 2018). People may feel that information about the (school) building is being concealed, and their concerns are downplayed, as argued by the parents who organized the school strike. A growing amount of information on human-mold relations is available through the internet, experts, consulting companies, discussion forums, and peer groups, shaping the understanding and engagements with molds. Building molds have also become significant business to some. For instance, apartment sales often involve a building inspection, while the inspectors do not take any responsibility for the results.

Although environmental education often materializes as teaching *about* the environment where interactions with the nonhuman world are mediated through study artifacts within the confines of the built environment, the urgency to enrich traditional disciplines by going out from the school buildings, identifying environmental issues and taking action on them, and enhancing embodied knowledges have been rightly emphasized (Gough 1990; Russell and Bell 1996, see also Fawcett 2013). However, too often the environment is considered to exist outside the school building, and investigating how the indoors might be conceptualized as multispecies sites has been largely absent (however, see Affifi 2017; Weston 2004). While humans have been becoming indoor species, so have a vast number of microbes. After we take microbes seriously, one does not need to leave the school building or bring nature into the classroom to be entangled in multispecies becoming. This multispecies becoming is not separable from historically situated practices of building, cleaning, and other world-making practices through which matter can come to matter differently for humans and more-than-humans alike. Instead of succeeding in purifying culture from nature, the above-mentioned practices generated new and complex natural-cultural agencements and what could in the case at hand be called an 'indoor climate crisis.'

Teaching and learning with material-discursive mold

When I entered the field, I became involved with a messy mixture of affects: hesitations, fears, disappointments, empowerments, curiosities, and more. The situation felt truly troubling and tense. This tension became manifested in the images of a child the adults brought forward during our talks, for example. Although parents and teachers stated that they had talked with the children about mold when questions arose from the children themselves, the adults nevertheless romanticized childhood (Taylor and Pacini-Ketchabaw 2016) by regarding children as 'living in the moment' and 'being relaxed and worry-free by nature.' There was an emphasis in the teachers' and parents' talk that children should play and study instead of needing to worry about their bodies and indoor air pollution. Adults were not only 'protecting children's ears,' but the adults also suggested that children actively 'develop all kinds of ideas in their heads' in relation to the ongoing situation. The teachers let me understand that, given the parents' involvement with the issue, it was the parents' duty to decide whether and how to discuss these issues with the children. Some parents stated that their children had often been present when they had talked about mold in their school. Other parents said that schools were in a better position to deal with the issue with the children as the school could inform all the children at once, and the problem was located inside the building. In all, it seemed that there was no general educational agreement on how to involve children. The children occupied an awkward and leaky middle place.

Thus, when I asked the teachers how mold as an environmental health issue could be taken up with the pupils, the teachers hesitated. Taking up the prevalent microbial life as a subject of study was perceived as pedagogically ambitious but impossible. This was partly due to its politicization in which these teachers were not eager to take part. They repeatedly expressed that their work was, in part, threatened by the parental activism: Some children had started acting disrespectfully and not completing their homework. The parents, according to them, had suggested to the children that 'something else was more important than school work.' To the teachers, the environmental health issue was a disturbance of the work they were supposed to do: Providing all pupils equally the knowledge and skills needed for their lives as future adults.

According to the teachers, their work was tied to the teaching of the learning goals manifested in the national and school curriculum. They repeated the mantra, 'our job is teaching.' The hesitation was partly due to the absence of microbial substance in the curriculum texts and partly due to developmental views that children are not yet capable of dealing with such issues and that these issues were not perceived as relevant to investigate for elementary school pupils. The teachers also positioned themselves as employees of the municipality, a position that required loyalty to the decisions made by the local government. Daily life in schools, no doubt, is routinized most of the time: Learning is understood as interaction between pupils and teachers and is mediated by books and other study materials, a practice that renders the environment something existing for human knowing. Still, the issue of mold materialized occasionally in questions the pupils asked, clashes between pupils and teachers, absences, and study arrangements: Classrooms had been closed, some pupils had a doctor's certificate allowing them to be home-schooled in some subjects, one pupil was permanently studying at home, and another one pupil was studying in a different building than her classmates, for instance.

Although not all pupils (or teachers) experienced symptoms and illnesses commonly connected to mold problems, such as headaches, tiredness, rashes, and respiratory symptoms (Salkinoja-Salonen 2016), the teachers were uncertain about where these effects actually came from. Were the headaches and tiredness merely normal phenomena related to growth or seasonal flu? Did they originate from too little sleep, or were they perhaps psychological caused by the intensification of the mold issue (see also Murphy 2006)? Pupils' differential embodied vulnerabilities had become perceptible in relation to mold and could not be ruled out of teaching and learning. In all, material-discursive mold disturbed the human-centered routines in school, the related dichotomies (e.g. mind and body), and generated new uncertainties.

Still, for the minority, the symptoms were real materializations of the phenomenon of moldschools. When talking with the children, I was told how mold had entangled in their everyday lives, producing new engagements within their environment. For many, the school was not at all the same school it had been earlier. Pupils had hunted and spotted mold from classrooms, hallways, and toilets, and been avoiding certain areas in the school, especially those near the classrooms that were closed due to high volumes of toxin-producing molds. Educational materials also became perceived differently: To avoid getting sick, some pupils were secretly avoiding touching certain materials (e.g. fabrics, books, and xylitol drops[4]) that were brought into their classrooms from moldy parts of the school. Sometimes, the air was difficult to breathe. The noise from the air purifier installed in the classroom distracted some of the pupils.

In children, moldschool sometimes invoked speculation about environmental health outside the school as well. Some of them were finding information on the internet and talking and sharing the 'bits and pieces' they had heard, read, and observed with their friends. Those who had experienced symptoms and sick leave had begun to pay attention to the way in which their body responded or speculated how their body might respond in different places: in different classrooms, outdoors, at their and their friends' homes. For instance, one child told me about her relative who was smoking inside the car that made her nauseated and wondering whether the raccoon dogs who lived nearby might spread diseases. Perhaps those children who were experiencing symptoms were engaged in similar activity as people with multiple chemical sensitivity in office buildings in that the pupils were 'learning new ways to listen to how your body feels, how it connects to the environment' in order to find themselves a safe place (Murphy 2006, 174). One of the children stated, 'If I start to feel bad in school, if I get a headache […], I will see if I get any symptoms at home. If they stop, then it [the reason] must be inside the school.'

As a part of our talks, I asked the children to tell a story titled moldschool. While these stories surely included fiction, they effectively brought up potentialities and vulnerabilities related to mold as viewed from the everyday life in schools. The stories included fears of being exposed and falling sick and descriptions of symptoms and their consequences: 'becoming tired and more tired,' 'having a headache and nausea every day,' 'snapping,' 'having to stay home while friends were at school,' having to visit the doctor, and being prescribed medicine. In addition, the pupils speculated about people's actions: teachers' attempts to solve the issue by re-locating some pupils in different classrooms than their classmates, painting over the mold dots on walls, angry arguments between adults, and fights between children regarding the existence and effects of mold, parents' attempts to make others understand the seriousness of the issue, some teachers' reluctance to admit the problem existed, researchers taking samples, newspapers writing about the issue, talking and wondering with peers 'how can it be so that there is mold in school,' children's investigations of the school building at night, deconstructing the school (e.g. by blowing it up), hopes for building a new one where kids could be 'without all the mold stuff' and symptoms, and some partying and hanging out with friends. In all, these stories mixed hopes, fears, frustrations, and curiosities, and allowed touching some of the ways in which the agencement was working with them. At the same time, they reminded me that the relation to mold was not the only mattering relation to the children of the moldschool.

We can now consider environmental learning as taking place through unfolding of the environment–body relation and the new embodied knowledges and actions that emerge (see also Affifi 2017). The moldschool created a quest for 'finding a safe space' (Murphy 2006) by inviting the children to re-map the school, listen to embodied messages, talk, and demonstrate, among other things. These doings are materialized in a story told by one of the teachers:

> The pupils came to school, and they had art lessons held by another teacher. They started to talk about this mold issue between peers. Apparently, things talked about at home grew so huge in the school that they experienced that 'Why aren't these classrooms closed, and why do we have to be here?' They went to the

> corridor to lie down and complained, 'Now I have a headache, I feel sick.' Then I came and talked with them, and I suggested that the 'school nurse will come in the afternoon and listen to what health issues you have. I don't understand so much about them myself.' And we made a deal. They calmed down and came into the classroom, and we studied the day in the normal manner, and the school nurse came in the afternoon. We have a great school nurse who listens to the children a lot, but in this situation, we didn't clearly plan what she should say, and she said approximately the same calming expressions as I did and had no intention of listening to individual children's symptoms. The children experienced this as downplaying, and four pupils skipped the rest of the classes as a kind of a protest. There were only a few pupils left, and we said that you cannot leave the school in the middle of the day, that it's in no way [appropriate]. And these who stayed at school wondered how they could do it. Then we, of course, contacted the homes that this is not acceptable that pupils leave during the school day as the supervision of the route between school and home is not arranged. The pupil is basically on the school's responsibility. Somehow, a kind of a workplace accident had happened. I had promised them something the school nurse didn't fulfill.

Along with the emergence of the mold issue, the school was becoming practiced differently. Material-discursive mold gave rise to new questions and tensions in the everyday school life. The building, its inhabitants, and air were no longer self-evident. Still, they were attempted to be distanced from the official schooling. Perhaps in the practicing of the official school—the one that agences teachers to attempt to stick with the lesson plans, routinized practices, and familiar interactions—teachers find it next to impossible to allow teaching to 'go wild' (Weston 2004) and to venture into the space of unanticipated intra-actions. The desire to control is present in the design of school buildings, rules, positioning of teachers and pupils, regulations and images of the child, among other things. Mold agencement does not easily fit into this framework as it generates uncertainty. There are no pregiven instructions for dealing with situated phenomena, such as moldschools, as they emerge in the intersections of the particular and the universal, through de-familiarization of the familiar, the in-betweens of humans and more-than-humans. Still, they are right here.

I find it intriguing how the phenomena that are on its skin (and in this case, permeating it and partly constituting it) seem to trouble the official school. Perhaps this is a diffraction of the anthropocentric desire to study the world from a distance: education about environment (Russell and Bell 1996). When there is no distance to begin with and safety is becoming an everyday quest (when our more-than-human becoming is within our bodies, in the air we breathe, in our dwellings), centering on human alone is simply not enough.

While moldschool agencement was effectuating children and school staff and becoming perceptible for them in several ways, there were also important imperceptibilities. First, in everyday life in the moldschools, building molds may appear to have 'nothing good in them' and be 'disgusting,' as expressed by two of the children. Likewise, in the general discourse, molds appear as a unified pathogenic group and an enemy. However, the same molds (e.g. *Aspergillus fumigatus*) that can cause health problems in humans and other animals are also crucial decomposers on which the well-being of the planet partly depends (Mousavi et al. 2016) or are used extensively in the food industry. For instance, *Aspergillus niger* is used to produce citric acid (E330; Salkinoja-Salonen 2016). Capabilities emerge from the relational spacetime. Maybe in buildings, these critters are trying to make an 'earthy living' (Haraway 2008), 'searching' for a livable space, responding within their environments (Affifi 2017). Although health symptoms in humans can distract their living significantly, drifting in the human body, where most building molds cannot live or reproduce (Salkinoja-Salonen 2016), is an undesirable situation for the toxic mold as well. On the other hand, landing on clothing or books may help spores to spread into new environments.

Second, what seems to remain largely imperceptible (especially for pupils) are the manifold processes through which moldschools came into existence and to which they connect and whom they benefit. As argued in the previous section, the attempts to separate and kill may have partly driven buildings into this situation. One child, when talking about the moldy study materials, referred to the cleaning practices: 'They said they should have been ozonized. They were merely disinfected a bit from the top, and it does not help.' Although at first it seems that building molds are not easily harnessed into the service of humans and the capitalist economy, they (among other indoor air problems), nevertheless, have contributed to an increase in building projects and industry (e.g. mold-resistant materials are now available), businesses conducting building inspections, scientific indoor air tests (where rabbit skin, mice brain cells, and boar sperm are used as indicators of toxicity—another example of uneven multispecies communities) and inquiries into experienced symptoms, development of diagnoses (e.g. sensitivity-related illnesses) and related medical treatment, among other things. Although there are some arguments that microbes should be included in compulsory education (Barberán et al. 2016), I am doubtful whether this would help understanding and acting upon the complexities involved in more-than-human agencements. For instance, climate change education research has shown us that increased scientific literacy and numeracy may not be sufficient to increase concern regarding climate change (Kahan et al. 2012). The multifaceted environmental problems we face demand an interdisciplinary approach (Pickering 2005) and consideration of them as agencements (Despret 2013).

I am intrigued to ask, could taking up the phenomenon of moldschools as a study project with pupils work toward a better understanding of the mutual becoming and interdependence of humans and other critters, as well as societal and capitalist processes that generate differential vulnerabilities? As I have pointed out, the teachers may consider taking up these complexities in a classroom too difficult, risky or otherwise not belonging to their work. Yet it appears to me that the children were already engaged in environmental learning through their embodied knowledges, re-mappings, and actions they took in their everyday lives. Living with mold agenced ethical questions regarding how to do well. Material-discursive mold was, in one sense, teaching them. Further, if we move beyond the teacher–pupil relation, it becomes possible to ask how environmental education and learning were taking place in the moldschool all along: through building practices and ventilation, use of chemicals in cleaning, takeover by toxic molds, health effects and absences of pupils and teachers, demonstrations, the re-mappings of the school building, as well as the more playful activities (e.g. mold hunting) the agencement inspired, incited, evoked.

Attuning to microbial life in environmental education and ethics

It is difficult in the current situation not to be involved with building molds. Every day, when I take my son to his daycare, we pass by three empty educational institutions that were closed within the past few years due to mold problems. One day, the city provided my photographer friend the keys to one of these school buildings, and he asked me to accompany him. We walked in from the main doors and found desks and chairs piled on top of each other, waiting for a ride to the junkyard. One floor up, abandoned books and papers were a mess on the hallways and classrooms, cupboards were left open, some full of educational materials. One classroom had a table full of stuffed animals, and another was equipped with microscopes and chemistry equipment and in yet another, there were computers. We went down to the basement and found an archive of educational technologies ranging from cinefilms and boards representing historical events to slide projectors and reel-to-reel machines to abacuses and statues. These were no longer merely instructional devices but devices allied with molds. The air indoors felt heavy, but mold was nowhere to be seen. We hunted for mold with our eyes and noses. Nothing. But when we closed the school's door behind us, we felt the difference between the stuffy air indoors and the cold, fresh air outdoors. Our bodies became differently related, largely beyond our awareness.

Moldschools work as a powerful reminder of situatedness, relationality, and continuous and often conflicting engagement in world-making encounters of diverse associates. As my findings regarding the complexities in the everyday lives in a moldschool suggest, through these encounters the school is no longer what it was thought to be, and its human inhabitants are not quite what was anticipated either. In other words, material-discursive mold troubles anthropocentric epistemologies and ontologies: the commonly held dichotomies (e.g. nature/culture, body/mind), school routines, relations with materials, and more. In many cases, teachers have had to quit their work, sometimes leading to a string of supply teachers, negatively affecting learning processes. Mold-related illnesses are distracting pupils' participation in teaching and learning and sometimes isolating them from their peers. Still, it seems that the official school tends to distance itself from the conflicting multispecies entanglements taking place right here. Unloved microbial companions cannot, however, be wished away. As Lloro-Bidart (2018) concludes in her study on human–invertebrate relations, there is a contradiction in urging pupils to care for the environment while at the same time pushing some critters outside caring practices as 'not only are these nonhuman others also part of our shared ecologies and socioecological histories and should be understood as such, but how we act toward them [...] can have drastic consequences for all living entities.'

Perhaps the pedagogical hesitation is related to the tendency of viewing toxic molds and other indoor air issues as problems themselves and not materializations of the grander problematic of anthropocentric (and economy driven) practices. In talking about global heating, Hird (2009, 129) points out that 'humans might ultimately render the biosphere inhospitable for humans and other animals, but this shifted biosphere will certainly survive our extinction.' Indoor climate crisis is not too different from this. Although certain processes have rendered the buildings uninhabitable for some humans (we now talk about 'mold refugees') and microbial critters, they have not done so to some toxic molds. Apparently, this has happened largely due to the quest for energy- and cost-efficiency coined with the lack of attentiveness in bodies, microbial life, and emerging capabilities.

Moldschools call for attuning to the differential becomings and vulnerabilities not only in humans but also in buildings and microbiota, among others. For instance, pupils are becoming differently vulnerable within their schools, and schools may differ in their microbiota, producing differential vulnerabilities in societal and global scales. Further, the microbial communities are becoming differently affected in schools, in relation to the practices of building, killing, ventilation, and other actions affecting microbial diversity. These nested practices of ignorance have fed back to humans as illnesses, but at the same time, could these illnesses be regarded as new sensibilities, new response-abilities? If we accept this provocative suggestion, we may ask, what are these symptoms making available for us, what are they asking from us, what are they making urgent?

Re-thinking the environment as not the surroundings of given critters but instead as the emergent phenomenon of their commingling (Haraway 2008; Ingold 2003; Tallbear 2011) allows us to entertain an idea of environmental education as being not only about the environment (Edwards and Cutter-Mackenzie 2011; Russell and Bell 1996) but also thinking of it as being of the environment. The environment may be characterized as being always already educational in the way it affects and is affected by given human and nonhuman becomings and processes constituting it (Affifi 2017). Thus, multispecies scholars have called for cultivating 'arts of attentiveness' by asking what matters to those we live with, including the awkward, unloved and loathed animals, plants, fungi, and microbes (Despret 2016; van Dooren, Kirksey, and Münster 2016).

While studies on toxic molds' sociability are very limited, I would like to speculate and get wild with the recent studies on slime mold (*Physarum polycephalum*). This critter has been found to be able to interpret environmental cues, memorize, practice forethought, and respond to the unpredictabilities of its circumstances (Dalziell 2017). As *Physarum* does not have any center (e.g. brain) in which to locate its 'cognitive' capabilities, Dalziell (2017) suggests that it challenges the idea that consciousness is produced by the body. Instead, consciousness could be regarded as 'written into the repertoire of materiality, its biological possibilities, such that matter *is* the capacity to think, to cognise' (Dalziell 2017, 170). Interestingly, depending on the environmental conditions, slime mold sometimes appears as singular (single celled spores) and sometimes as plural (plasmodium), provoking hesitation on how to address it/them and reminding Deleuze and Guattari's (1987) notion that monism = pluralism. While there seems to be a tendency to think of *Physarum's* capabilities as inherent and not arising from the particular material practices (see also Barad 2007), these studies allow important speculations for environmental education. What if we entered places as if they were always already knowledgeable? What if we considered buildings and classrooms as always already alive, already mattering, beyond humans? What if we acted 'as if everything matters' (Carruthers 2006)? Could such questions sensitize us to better recognize practices of both suppression and flourishing in human and more-than-human worldings?

Although moldschools are in many ways sad occasions of multispecies becoming, these schools may also work as reminders that even when there is no detected indoor air problem, microbes are there. In reference to the recent more-than-human turn in environmental education, I thus suggest that the (school's) interior is always already a lively multispecies community in-the-making. Attuning to microbial life in buildings and elsewhere could bring to our attention the interrelated materialities, such as walls, machines, cleaning liquids, weather, air, dirt and dust, and their potential significance in environmental education and ethics. By asking, how do microbes come to know, be and do in the world, the attention is drawn to the myriad ways in which world can come to matter, to the ways in which the world can come to know itself or make itself intelligible (Barad 2007). Cultivating attentiveness to all modes of knowing, becoming, and doing, both human and more-than-human, is to give worth to different co-existing versions of mattering. This is one of the most crucial, yet troubling exercises for cultivating critical and caring thought (Haraway 2016, 116). The ethical questions of how to care (e.g. how to meet, eat, breathe, and kill well) are continuously present in the ordinary and everyday life in schools even when there is no detected indoor air problem. The quest for caring needs to be extended from recycling and other practices of consumption to the attentiveness toward the onto-epistemologies in our surroundings, and within our bodies, for humans are environments too. Of course, we have no certainty of how bodies may affect and become affected. But maybe this uncertainty is what allows us to question 'our responses and accountabilities' and remain hesitant and 'curious about the ethical implications of certain acts of caring rather than taking them for granted' (Nxumalo and Pacini-Ketchabaw 2017, 3). I hope that my efforts in conceptualizing educational practices and places as always already more-than-human agencements as well as the sensitizing questions I have posed regarding the difficulties in attuning to microbial life further inspires such hesitation and curiosity.

Disclosure statement

No potential conflict of interest was reported by the author.

Funding

This research was supported by grants from Ella and Georg Ehrnrooth foundation, Emil Aaltonen foundation.

Notes

1. I cannot help thinking about the recent news about the toxic dumping by the oil company Chevron in the Amazon in Ecuador that has had drastic effects on the indigenous people and more-than-humans, for example (see Donziger 2016).
2. The common worlds approach 'merges Latour's (2009) notion of common worlds as a collective of human and non-human constituents with Haraway's (2008) grounded and relational "worldly" ethics' (Taylor and Pachini-Ketchabaw 2016).
3. 'Moldschool' is a vernacular term used to refer to schools (and other educational institutions) where an indoor air problem often caused by toxic molds has been detected or assumed. Whereas 'sick building' in the 1980s signaled confusion between buildings and the bodies in them (Murphy 2006), moldschools signal confusion between nature (mold) and culture (school), already teasing out images of the school as a multispecies site.
4. I count xylitol-drops as materials for (dental) health education as pupils are obliged to take them after each meal in order to prevent acid attack.

Acknowledgments

I thank my colleagues at the *Animate* project, for the ongoing feedback, discussions, and experimentations in the more-than-human world. I also thank the article's anonymous reviewers and the editors of this Special Issue for their comments and suggestions that were invaluable for the development of this article.

Funding

This research was supported by grants from Ella and Georg Ehrnrooth foundation, Emil Aaltonen foundation.

ORCID

Tuure Tammi http://orcid.org/0000-0002-7582-8887

References

Affifi, R. 2017. "The Metabolic Core of Environmental Education." *Studies in Philosophy and Education* 36 (3): 315–332. " doi:10.1007/s11217-016-9555-y.

Audit Committee. 2013. *Rakennusten kosteus- ja homeongelmat* [*The Moisture and Mold Problems in Buildings*]. Finnish parliament: Audit Committee report 1/2013.

Baker, L. 2012. *A History of School Design and Its Indoor Environmental Standards, 1900 to Today*. Washington, DC: National Clearinghouse for Educational Facilities.

Barad, K. 2007. *Meeting the Universe Halfway: Quantum Physics and the Entanglement of Matter and Meaning*. Durham & London: Duke University Press.

Barberán, A., T. Hammer, A. Madden, and N. Fierer. 2016. "Microbes Should Be Central to Ecological Education and Outreach." *Journal of Microbiology & Biology Education* 17 (1):23–28. doi: doi:10.1128/jmbe.v17i1.984.

Bell, A., and C. Russell. 2000. "Beyond Human, beyond Words: Anthropocentrism, critical Pedagogy, and the Poststructuralist Turn." *Canadian Journal of Education* 25 (3):188–203. doi:10.2307/1585953.

Carruthers, B. 2006. "Praxis: Acting as If Everything Matters." MA thesis, Lancaster University.

Dalziell, J. 2017. "Microbiology as Sociology: The Strange Sociality of Slime." In *What If Culture Was Nature All along?* edited by Vicky Kirby, 153–178. Edinburgh: Edinburgh University Press.

Deleuze, Gilles, and Felix Guattari. 1987. *A Thousand Plateaus*. Minneapolis, MN: University of Minnesota.

Despret, V. 2013. "From Secret Agents to Interagency." *History and Theory* 52 (4):29–44. doi:10.1111/hith.10686.

Despret, V. 2016. *What Would Animals Say If we Asked the Right Questions?* Minneapolis, MN: University of Minnesota Press.

van Dooren, T., E. Kirksey, and U. Münster. 2016. "Multispecies Studies." *Cultivating Arts of Attentiveness." Environmental Humanities* 8 (1):1–23. doi:10.1215/22011919-3527695.

Donziger, S. 2016. "Chevron's "Amazon Chernobyl" in Ecuador: The Real Irrefutable Truths About the Company's Toxic Dumping and Fraud." Huffington post, May 27, 2016. https://www.huffingtonpost.com/steven-donziger/chevrons-amazon-chernobyl_b_7435926.html?guccounter=1.

Edwards, Susan, and Amy, Cutter-Mackenzie. 2011. "Environmentalising Early Childhood Education Curriculum through Pedagogies of Play." *Australasian Journal of Early Childhood* 36 (1):51–59.
Fawcett, L. 2013. "Three Degrees of Separation: Accounting for Naturecultures in Environmental Education Research." In *The International Handbook for Research on Environmental Education*, edited by Robert Stevenson, Michael Brody, Justin Dillon, and Arjen Wals, 409–417. New York, NY: Routledge.
Ford, D. 2015. "The Pneumatic Common: Learning in, with and from the Air." *Educational Philosophy and Theory* 47 (13–14):1405–1418. doi:10.1080/00131857.2014.930683.
Gibson, K., D. Bird Rose, and R. Fincher. 2015. *Manifesto for Living in the Anthropocene*. Brooklyn: Punctum.
Gough, N. 1990. "Renewing Our Mythic Links with Nature: Some Arts of Becoming Ecopolitical in Curriculum Work." *Curriculum Perspectives* 10 (2):66–69.
Haraway, D. 2008. *When Species Meet*. Minneapolis, MN: University of Minnesota Press.
Haraway, D. 2016. *Staying with the Trouble*. Durham & London: Duke University Press.
Hird, M. 2009. *The Origins of Sociable Life: Evolution after Science Studies*. London: Palgrave Macmillan.
Hohti, R., and T. Tammi. 2019. "The Greenhouse Effect: Multispecies Childhood and Non-innocent Relations of Care." *Childhood* https://doi.org/10.1177/0907568219826263
Humes, B. 2008. " Moving toward a Liberatory Pedagogy for All Species: Mapping the Need for Dialogue between Humane and anti-oppressive Education." *Green Theory & Praxis: The Journal of Ecopedagogy* 4 (1):65–85.doi: 10.3903/gtp.2008.1.4
Ingold, T. 2003. " Two Reflections on Ecological Knowledge." In *Nature Knowledge: Ethnoscience, Cognition, and Utility*, edited by Glauco Sanga and Gherardo Ortalli, 301–311. Oxford: Berghahn Books.
Ionescu, C., T. Baracu, G.-E. Vlad, H. Necula, and A. Badea. 2015. "The Historical Evolution of the Energy Efficient Buildings." *Renewable and Sustainable Energy Reviews* 49:243–253. doi:10.1016/j.rser.2015.04.062.
Kahan, D., E. Peters, M. Wittlin, P. Slovic, L. Ouellette, D. Braman, and G. Mandel. 2012. "The Polarizing Impact of Science Literacy and Numeracy on Perceived Climate Change Risks." *Nature Climate Change* 2 (10):732–735. doi: 10.1038/nclimate1547.
Kahn, R. 2008. " Towards Ecopedagogy: Weaving a Broad-Based Pedagogy of Liberation for Animals, Nature, and the Oppressed People of the Earth." In *The Critical Pedagogy Reader*, edited by Antonia Darder, Martha Baltodano, and Rodolfo Torres, 522–540. New York, NY: Routledge.
Kalof, L., J. Zammit-Lucia, J. Bell, and G. Granter. 2016. "Fostering Kinship with Animals: animal Portraiture in Humane Education." *Environmental Education Research* 22 (2):203–228. doi:10.1080/13504622.2014.999226.
Kembel, S. W., E. Jones, J. Kline, D. Northcutt, J. Stenson, A. M. Womack, B. J. M. Bohannan, G. Z. Brown, and J. L. Green. 2012. "Architectural Design Influences the Diversity and Structure of the Built Environment Microbiome." *The ISME Journal* 6 (8):1469–1479. doi:10.1038/ismej.2011.211.
Kembel, S., Meadow, J. T. O'Connor, G. Mhuireach, D. Northcutt, J. Kline, Maxwell Moriyama, G. Brown, B. Bohannan, and J. Green. 2014. " Architectural Design Drives the Biogeography of Indoor Bacterial Communities." *PloS One* 9 (1):e87093. https://doi.org/10.1371/journal.pone.0087093.
Kirksey, E., and S. Helmreich. 2010. "The Emergence of Multispecies Ethnography." *Cultural Anthropology* 25 (4):5: 545–576. doi:10.1111/j.1548-1360.2010.01069.x.
Latour, B. 1993. *The Pasteurization of France*. Cambridge, MS, London: Harvard University Press. doi:10.1086/ahr/95.4.1215-a.
Latour, B. 2009. "A Collective of Humans and Non Humans: Following Daedalus's Labyrinth." In *Readings in the Philosophy of Technology*, edited by David Kaplan, 156–172. Plymouth: Rowman & Littlefield.
Lloro-Bidart, T. 2018. " "Cultivating Affects: A Feminist Posthumanist Analysis of Invertebrate and Human Performativity in an Urban Community Garden." *Emotion, Space, and Society* 27:23–30. doi:10. 1016/j.emospa.2018.02.006.
Malone, K., and S. Truong. 2017. "Sustainability, Education, and Anthropocentric Precarity." In *Reimagining Sustainability in Precarious Times*, edited by Karen Malone, Son Truong, and Tonia Gray, 3–16. Singapore: Springer.
Mousavi, B., M. Hedayati, N. Hedayati, M. Ilkit, and S. Syedmousavi. 2016. "Aspergillus Species in Indoor Environments and Their Possible Occupational and Public Health Hazards." *Current Medical Mycology* 2 (1):36–42. doi:10.18869/acadpub.cmm.2.1.36.
Murphy, M. 2006. *Sick Building Syndrome and the Problem of Uncertainty: Environmental Politics, technoscience, and Women Workers*. Duke University Press: Durham and London.
Mölsä, S. 2017. *Näin Suomi homehtui – hyvä rakentamistapa sai aikaan pahaa jälkeä.* [*This is How Finland Molded – The Good Practice of Building Gone Awry.*] https://www.rakennuslehti.fi/2016/06/nain-suomi-homehtui-hyva-rakentamistapa-sai-aikaan-pahaa-jalkea/.
Nxumalo, F. 2018. " Situating Indigenous and Black Childhoods in the Anthropocene." In *Research Handbook on Childhoodnature*, edited by Amy Cutter-Mackenzie, Karen Malone, and Elisabeth Barratt Hacking. Cham: Springer International Handbooks of Education.
Nxumalo, F., and V. Pacini-Ketchabaw. 2017. "Staying with the Trouble' in Child-insect-educator Common Worlds." *Environmental Education Research* 23 (10):1414–1426. doi:10.1080/13504622.2017.1325447.

O'Malley, M. 2014. *Philosophy of Microbiology*. Cambridge: Cambridge University Press.

Ott, W. 1989. *Human Activity Patterns: a Review of the Literature for Estimating Time Spent Indoors, outdoors, and in Transit*. US Environmental Protection Agency. In: Proceedings of the Research Planning Conference on Human Activity Patterns, EPA/600/4-89/004, 3-1 to 3-38. Las Vegas, NV: EPA National Exposure Research Laboratory.

Pedersen, H. 2004. " Schools, speciesism, and Hidden Curricula: The Role of Critical Pedagogy for Human Education Futures." *Journal of Futures Studies* 8 (4):1–14.

Pedersen, H. 2010. "Is 'the Posthuman' Educable? On the Convergence of Educational Philosophy, animal Studies, and Posthumanist Theory." *Discourse: Studies in the Cultural Politics of Education* 31 (2):237–250. doi:10.1080/01596301003679750.

Pedersen, H. 2013. "Follow the Judas Sheep: materializing Post-Qualitative Methodology in Zooethnographic Space." *International Journal of Qualitative Studies in Education* 26 (6):717–731. doi:10.1080/09518398.2013.788760.

Pickering, A. 2005. "Asian Eels and Global Warming: a Posthumanist Perspective on Society and the Environment." *Ethics & the Environment* 10 (2):29–43. doi:10.1353/een.2005.0023.

Prout, A. 2005. *The Future of Childhood: Towards the Interdisciplinary Study of Children*. London: Routledge Falmer.

Rautio, P., R. Hohti, R.-M. Leinonen, and T. Tammi. 2017. "Reconfiguring Urban Environmental Education with 'Shitgull' and a 'Shop." *Environmental Education Research* 23 (10):1379–1390. doi:10.1080/13504622.2017.1325446.

Rose, D. B., and T. van Dooren. 2011. "Unloved Others: Death and Disregard in the Time of Extinctions." *Australian Humanities Review* 50:1–178. doi: 10.22459/AHR.50.2011

Russell, C., and A. Bell. 1996. "A politicized Ethic of Care: Environmental Education from an Ecofeminist Perspective." In *Women's Voices in Experiential Education*, edited by Karen Warren, 172–181. Dubuque, IA: Kendall Hunt.

Salkinoja-Salonen, M. 2016. *Diagnostisia Työkaluja Rakennusten Patologiaan* [*Diagnostic Tools for the Pathology of Buildings*]. Helsingin yliopisto: Mikrobiologian julkaisuja 50.

Snaza, N., P. Appelbaum, S. Bayne, D. Carlson, M. Morris, N. Rotas, J. Sandlin, J. Wallin, and J. Weaver. 2014. "Toward a Posthumanist Education." *Journal of Curriculum Theorizing* 30 (2):39–55.

Spannring, R. 2017. "Animals in Environmental Education Research." *Environmental Education Research* 23 (1):63–74. doi:10.1080/13504622.2016.1188058.

Tallbear, K. 2011. "Why Interspecies Thinking Needs Indigenous Standpoints." *Theorizing the Contemporary*, Cultural Anthropology *website*, April 24. https://culanth.org/fieldsights/260-why-interspecies-thinking-needs-indigenous-standpoints.Accessed 8.1.2018

Tammi, T. 2017. "Poliittinen prosessi ja demokratiakokeilujen tartunnat koulun arjessa." (The political process and the contagions of democratic experiments in everyday life in school). PhD diss., University of Helsinki: Studies in Education.

Tammi, T. 2019. "Breathing Well at the Wastelands? Indoor Climate Change in Schools and the Daily Lives of Arctic Children." In *Social, Material and Political Constructs of Arctic Childhoods – An Everyday Life Perspective*, edited by Pauliina Rautio and Elina Stenvall. Singapore: Springer Nature, pp.17-34.

Tammi, T., Rautio, P., Leinonen, R.-M., and R. Hohti. 2018. Unearthing withling(s): Children, tweezers, and worms and the emergence of joy and suffering in a kindergarten yard. In: Research Handbook on Childhoodnature (Springer International Handbooks of Education), edited by Amy Cutter-Mackenzie, Karen Malone and Elisabeth Barratt Hacking, 1-13. Cham: Springer. https://doi.org/10.1007/978-3-319-51949-4_68-1

Taylor, A. 2017. "Romancing or Re-configuring Nature in the Anthropocene? Towards Common Worlding Pedagogies." In *Reimagining Sustainability in Precarious Times*, edited by Karen Malone, Son Truong and Tonia Gray, 61–76. Singapore: Springer.

Taylor, A., M. Blaise, and M. Giugni. 2013. "Haraway's 'bag Lady Story-telling': Relocating Childhood and Learning within a 'post-human Landscape." *Discourse: Studies in the Cultural Politics of Education* 34 (1):48–62. doi:10.1080/01596306.2012.698863.

Taylor, A., and V. Pacini-Ketchabaw. 2015. "Learning with Children, ants, and Worms in the Anthropocene: Towards a Common World Pedagogy of Multispecies Vulnerability." *Pedagogy, Culture and Society* 23 (4):507–529. doi:10.1080/14681366.2015.1039050.

Taylor, A., and V. Pacini-Ketchabaw. 2017. "Kids, raccoons, and Roos: Awkward Encounters and Mixed Affects." *Children's Geographies* 15 (2):131–145. doi:10.1080/14733285.2016.1199849.

Todd, Z. 2016. "An Indigenous Feminist's Take on the Ontological Turn:'Ontology'is Just Another Word for Colonialism." *Journal of Historical Sociology* 29 (1):4–22. doi:10.1111/johs.12124.

Tammi, T., P. Rautio, R.-M. Leinonen, and R. Hohti. 2018. "Unearthing Withling(S): Children, Tweezers and Worms, and the Emergence of Joy and Suffering in a Kindergarten Yard." In *ChildhoodNature Child–Animal Relations*, edited by Tracy Young and Pauliina Rautio. Springer Nature.

Vesalainen, S. 2018. "Tuore raportti sisäilmaongelmista paljastaa: Pääsyyt ongelmiin ovat riskinotto, välinpitämättömyys ja raha" [Recent Report on Indoor Air Problems: Main Reasons are Risk Taking, Neglectance and Money]. Yle, March 19. https://yle.fi/uutiset/3-10116480. Accessed 19.3.2018.

Watts, V. 2017. "Indigenous place-thought & agency amongst humans and non-humans (First Woman and Sky Woman go on a European world tour!)." *Re-visiones*, 7. doi:http://re-visiones.net/index.php/RE-VISIONES/article/view/228/409.

Weston, A. 2004. " What If Teaching Went Wild?" *Canadian Journal of Environmental Education* 9:31–46.

WHO. 2009. Who Guidelines for Indoor Air Quality: Dampness and Mould. World Health Organization. http://www.euro.who.int/data/assets/pdf_file/0017/43325/E92645.pdf.WHO Regional office for Europe: Copenhagen.

Yong, Ed. 2016. *I Contain Multitudes: The Microbes within us and a Grander View of Life*. New York, NY: Harper Collins.

'An atmosphere, an air, a life:' Deleuze, elemental media, and more-than-human environmental subjectification and education

Marcelina Piotrowski

ABSTRACT

Subjectification in environmental movement education comprises an influx of more-than-human 'others,' including the classical elements: air, water, earth, fire. In this conceptual article, I consider what environmental movement education research, which includes inquiring into processes of political subjectification, might entail, when thinking with the elements. As part of this focus, I identify and propose thinking alongside an 'elemental Deleuze,' by attuning to how the elements thread through Deleuze's many works. Thinking with the elements alongside the field of elemental media studies, I turn to a series of examples from research conducted on subjectification and education in anti-oil pipeline movements in British Columbia, Canada, to suggest that the elements are productive media of atmospheres and affects that generate political 'life.' My objectives in this article are to articulate implications and ways of engaging with more-than-human forms of subjectification in environmental movements, and the implications for environmental education research.

Introduction

> It is the wolf itself, and the horse, and the child, that cease to be subjects to become events, in assemblages that are inseparable from an hour, a season, an atmosphere, an air, a life (Deleuze and Guattari 1987, 261).

Subjectification in environmental movement education comprises an influx of more-than-human 'others,' including the classical elements: air, water, earth, fire. In this conceptual article, I consider what environmental movement education research, which includes inquiring into processes of political subjectification, might entail, when thinking with the elements. As part of this focus, I identify and propose thinking alongside an 'elemental Deleuze,' by attuning to how the elements thread through Deleuze's many works. Thinking with the elements, I turn to a series of examples from research I conducted on subjectification and education in anti-oil pipeline movements in British Columbia, Canada, to suggest that the elements are productive media of atmospheres and affects that generate political 'life.' My objectives in this article are to articulate ways of engaging with more-than-human forms of subjectification in environmental movements, and their implications for environmental education research. This approach to subjectification moves beyond individuation, personal agency, or identity, towards examining 'subjectification' as a process emerging from a confluence of material forces (Deleuze 1988, 1990).

Material (Sonu and Snaza 2015; Payne 2016) and post-critical (McKenzie 2005; Nakagawa and Payne 2014; Payne 2016) approaches emerging in the field of environmental education orient towards post-anthropocentric ways of thinking about environmental education research and pay particular attention to the relation between processes of inquiry and assertions about ontology. This article engages with the new materialisms and the post-critical turn in environmental education to recognize the mutual constitution and entanglement between human and non-human elements in the field of environmental movement education in particular. In doing so, my aim is to contribute to post-anthropocentric undoing of the humanist subject (St. Pierre 2004) that has operated so prevalently as a focus of research in environmental movement education. In this article, thinking with an elemental Deleuze about environmental movement education complements and forges important affinities with the 'elemental turn' in media studies (Peters 2015; Parikka 2015).

More-than-human environmental movement education

Climate change, oil spills, water toxicity, and smog, are but a few examples of some of the world's urgent environmental problems stretching ecological resilience and challenging human and other animal life, and environmental movements are considered to be important forces of public pedagogy and political education aiming at political subjectification, presumably aimed to engage with, 'resolve' or delay ecological devastation (Kovan and Dirkx 2003; Hall 2004; Walter 2007, 2011, 2013). Environmental movement education research on informal, non-formal, and self-directed learning contexts (Walter 2013) has typically been informed by critical and transformative learning theories focusing on adults' environmental education and subjectification through the process of individuation, emancipation, actualization, and humanization (Clover 2002, Kovan and Dirkx 2003; Hall 2004; Walter 2007, 2011, 2013). This approach emphasizes that ecological ethics and humanization are both entwined in a critical project that grapples with the oppressive mechanism of consumer Capitalism and its impact on critical consciousness and the environment.

Departing from such humanist orientations and contending with the implications of posthumanism and matter in environmental education research is not easy. Environmental education scholars often consider what role learners, students, 'a public,' and teachers might occupy with new refigurations of 'being' in the post-Anthropocene. Gough (2016) notes this difficulty, suggesting that environmental education frequently inadvertently resuscitates the humanist political subjects each time it proposes an anthropocentric gaze to environmental concerns, despite articulating anthropocentrism as damaging.

More-than-human subjectification in education research, and in my case, in environmental movement education research, responds to what Wallin (2016) has called a necrophilosophic approach that these fields have been invested in. Necrophilosophic approaches to environmental education and research, resuscitate 'dead figures' with the aims to 'freeze life into epistemic states hence evading the encroaching death of the Humanist subject as a figure of rationality, progress, and perfectibility above and beyond the contingency of material reality' (Wallin 2016, 4). The question in environmental education research is whether 'the material' can be seen as forceful, affective, and productive of political becomings, in ways that does not privilege human anthropocentrism, positions the elements as a 'milieu' or 'background material' that help to cultivate ecological sensibility, or treats the elements as objects primarily in need of conservation.

Elements appeared to exert a force that continuously generated wonder (MacLure 2013) in my recent study about political subjectification and environmental education in two anti-oil pipeline movements in Canada, where I observed elemental forces as profoundly affective in the emergence of political ontologies and the role of learning in their ongoing becoming. Protests, workshops, community gardens, sit-ins, documentary films, speaker events, online, and social media communication organized by environmental activist groups involved in the oil pipeline debates—all conventional sites and modes of environmental movement education—often took secondary place to the swarming and often surprising force of wind, ocean water, oil, and other elements that emerged as vital to political life.

Weaver and Snaza (2016) characterize the move to more-than-human research in education as a risky, but necessary one, in order to show how 'we are always exposed, materially and vulnerably, to the more-than-human world in which we dwell' (p. 90). This approach moves away from examining how to educate for an ecologically concerned human subjectivity through an 'educational project grounded in rational empiricism' (Sonu and Snaza 2015, 261) and instead it shifts attention to mapping how 'human agency is an effect of previous inhuman actions and material entanglements' (p. 266). In his research on outdoor education, for example, Clarke (2017) drew attention to the idea that a 'flat' environmental education ought to consider 'material intra-relations that constitute their [learners'] current thisness—their haecceitical self' (p. 313). Such a 'flatness' speaks to the dissolution of hierarchies between nature and culture and refocuses on the monist continuity of the material world and its various expressions (Clarke 2017). In this article, I engage with this turn to materialism in environmental education, which draws attention to learners and learning as relationally entangled with matter, and which questions to what extent humans are the initiators of thought, thinking, and knowledge (Gough 2016), but first, I turn to 'the elements' which have increasingly become a focus of attention in fields as different as geography (Adey 2015, McCormack 2008, 2010, 2015, 2016) and medieval studies (Cohen 2011, 2014).

Elemental folds

The material fold, or 'the material part of ourselves which is to be surrounded and enfolded'(Deleuze 1988, 104), is a key force affecting subjectification in environmental movements. Deleuze (1993) noted that elements enfold to produce a human 'figure' that comes to express the world. A political subject is but an instance of subjectification that becomes an expression of a chaotic world that far exceeds the confines of any humanist ontology, and emerges as a fold of the elements in their various forms: water, air, dust, waves, particles, fog, glaciers, sunshine, etc. As a fold, the political subject emerges as a momentary pleat in a material 'chaosmos,' which Deleuze (1993) defines as a 'dark backdrop' (p. 87) consisting of a multiplicitous motley crew of possible bifurcations that elements take. Here, the classical elements, air, water, earth, fire are not distinct from each other, but are pleated in new ways through recursive and erratic geological and biological cycles and processes.[1]

The elements are an integral force in the production of more-than-human, non-individuated subjectivities. Elements and humans are connected as a pliable matter that bends and folds as though in a sheet that can be stretched and re-moulded—i.e. a topology (de Freitas 2016). Deleuze (1993) strikingly illuminates a 'world [that] exists only in its representatives as long as they are included in each monad. It is a lapping of waves, a rumor, a fog, or a mass of dancing particles of dust' (p. 98) that enact the vitality of the world. In other words, for Deleuze, the chaos of the world expresses itself through elements, which in turn affect, enfold and refold into other expressions. While Deleuze (1995) claims that there are individuations without 'subjects,' he refers to the need to imagine forms of subjectification that are event-like, and which overspill the thresholds of human ontologies. In these 'event-type individuations,' where 'there's no subject: a wind, an atmosphere' (Deleuze 1995, 115) we are invited to view the elements folding a material world (an outside) into a 'subject' (an inside) through sense encounters. This is explicitly a move *away* from drawing attention to how the elements, and 'weather' for example, affect identity or culture (see for example Sturken 2001, or Vannini et al. 2012), and a move towards viewing the elemental morphology and subjectification as part of the same continuum.

More-than-human forms of subjectification, which Deleuze and Guattari (1987) also refer to as a haecceity, is an individuation that 'no longer constitutes persons or "egos"' (p. 111). Instead, it requires a refolding of political 'life' along lines that do not correspond a body, but that are constituted by non-human forces, and particularly elemental ones

> There is a mode of individuation very different from that of a person, subject, thing, or substance. We reserve the name haecceity for it. A season, a winter, a summer, an hour, a date have a perfect individuality lacking nothing, even though this individuality is different from that of a thing or a subject. They are haecceities in the sense that they consist entirely of relations of movement and rest between molecules or particles, capacities to affect and be affected. When demonology expounds upon the diabolical art of local movements and transports of affect, it also notes the importance of rain, hail, wind, pestilential air, or air polluted by noxious particles, favourable conditions for these transports (p. 261).

It is not trivial that Deleuze and Guattari draw attention to the movement of rain, hail, wind, and air, when they focus on the transport of affects in the passage above. Deleuze and Guattari's ecological politics aims at the dissolution of Platonic representational boundaries between 'essences' and 'appearances' (Hayden 2008). Air infects and is infected, transports and is transported, is atmospheric (sky), and produces atmospheres (moods). In turning to the elements, I attend to, experiment with, and blur the relations between such different approaches to atmospheric bodies (McCormack 2010). Deleuze's politics here consist of attributing elements their affective capacities, that is, acknowledging their vitality as well as their capacity to produce life by generating difference.

Atmospheric affects

> Gale force winds propelled a Russian tanker to lose control off the coast of Western British Columbia near Haida Gwaii in the fall of 2014. The winds drifted the tanker around 40 km off shore, unsecured, defying rescue attempts of the Canadian Coast Guard to attach a cable to the ship and secure it in place. The wind intensified the dense atmosphere of waiting for the implementation of an oil tanker traffic ban and the cancellation of the Enbridge Northern Gateway oil pipeline project. The gale's playfulness with the ship, the ineffectiveness of efforts to contain it in the wind, and the slow news reporting about the event generated an atmosphere of frustration and mobilization. Communication between non-indigenous activists on the central coast of British Columbia with Indigenous people living on the nearby island of Haida Gwaii who were first to report on the drifting ship on social media, intensified. People became apprehensive of the Coast Guard's capacity to use tugboats to assist in manoeuvring ships in the Douglas Channel in the future, as the wind evoked memories of the Exxon Valdez oil spill, a boat that defied human intentionality in containing the elements.

Air, wind, and waves have ambitious agencies (Gannon 2016), in that they make people sense, anticipate, and want to move and do things. Atmospheres are in this sense meteorological *and* affective (McCormack 2008). Atmospheres, which have lost their grounding in the meteorological and material realm in research need to become re-enchanted to their material connections (Ingold 2011). To affirm the connection between material and affective atmospheres it is important to draw attention to the more-than-human forces which comprise a political 'life.' Atmospheric fieldwork (McCormack 2010) and atmospheric methods (Anderson and Ash 2015) call for attunement to the relations between atmospheric bodies, and the way in which atmospheres in the material sense, can produce atmospheres in the affective sense which emerge before comprehension or thought. Atmospheric fieldwork is a 'non-representational approach to [diagramming] atmospheres' coexistence [that] involves a flattening and breaking down of distinctions between living and dead matter to suggest that all objects have the potential to equally impact or weigh upon an atmosphere' (Anderson and Ash 2015, 42).

Air, earth, water, and fire are elemental media (Peters 2015) that enact infrastructural agency (McCormack 2016) which enables movements and affects to form. Air for example operates in its affective (having the capacity to affect, to move, and to create mood) and meteorological sense (and not in the metaphorical sense, i.e. cloud computing) (Adey 2015; Anderson 2009; Ingold 2011; McCormack 2008, 2016). Dust (Parikka 2015), earth (Maddalena and Russill 2013), icebergs (Ruiz 2014), and clouds (Peters 2015) are also media that move—not only particles, plastics, and hurricanes, but affects that mobilize affect and subsequently thought. We might think of the elements as media. Peters (2015) argues that 'media are vessels and environments, containers of possiblility that anchor our existence and make what we are doing possible' (p. 2). For humans, the elements are important media because they have long been used to move affects and signals (e.g. astrology, aviation, sea navigation, and now satellites all depend on the air and atmospheric conditions) and they have been equally important for non-human animals in enabling their communication (e.g. oceans filter out light to enhance sound conductivity, enabling whales to communicate) (Peters 2015). Elements in the form of air and water emerge as weather that mediates many elements of living on the West Coast of Canada (Vannini et al 2012). Thinking about the elements' infrastructural agency is different from relegating them to a 'backdrop' or 'background' material of the world in which they are ontologically distinct, and which a fully-formed human encounters or tries to conserve.

Smog for example, is a medium of atmospheric information that not only moves particles of dust from fires or manufacturing sites into people's bodies, but that also enfolds and enters into more-than-human political subjectivities involved in anti-smog activism. The materiality of smog and how it is inhaled through breath by human bodies, is a specific material that enfolds political subjectification in environmental movements in places where smog poses a significant health hazard (Parikka 2015). In this approach, smog is not an object/ive of people's subjectification i.e. their activism. Smog is literally consumed by people, and is expressed in the figure of a 'political subject,' producing a 'them' because of the affective and biological sensations and perturbations it invokes.

Elemental media might therefore be seen at 'work in and through bodies, or, more widely, through materials and things' (Parikka 2015, 93). For example, deteriorated quality of tear-gas-infused air at Occupy protests produced distinct activism atmospheres. Air became more than a forceful element that produces havoc in the form of hurricanes, and became a texture of political 'life' that circulated affect in and through specific human bodies (including their tears, anger) (Parikka 2015). The point is that the elements operate as media that produce and move affects that are productive of political subjectivities. As I show next, drawing from an example based on research I conducted on environmental movement campaigns and public education in oil pipeline conflicts in Canada, elements such as air, produce significant affects.

Elemental life

> The hierarchy of substances is abolished: a single one replaces them all: the whole world can be plasticized, and even life itself since, we are told, they are beginning to make plastic aortas (Barthes 1957, 111).

Imagine a beating, working, plastic heart. Such devices (silicone hearts can already be printed on 3D printers) would be complex and elegant expressions of the collapse of human/elemental life thresholds. In thinking about 'a life,' including 'a political life' as never solely of a human subject, but operating outside of, in excess of anthropocentric boundaries, life becomes conceptualized as exceeding that of the individual (Deleuze 2001). 'Life' flows through human subjects, albeit temporarily, as it actualizes in specific bodies. Materials have the propensity to ask questions and animate new 'life.' Even if they are not of the beating heart variety that Barthes imagined, they swarm together to create affections that might be materialized in a 'specific life'—such as a person active towards an environmental cause. They enfold human bodies with forces that move them and unravel their solidity, challenging them to examine their 'bounded' porosity.

Material folds do not produce immediate or direct effects, and especially not ones that directly materialize in an individual's 'feeling.' In this sense, the elements produce 'atmospheres' that are pre-individual and which circulate outside of familiar categories of emotions. Stewart (2011) highlights the material force that acts on the production of atmospheres when she notes that 'the atmospheric' is expressive of newness: an atmosphere 'buzzes with the resonance of nascent forms quickening or sloughing off, materialities pressing into the expressivity of something coming into existence' (p. 446). Atmospheres are not constant, but are expressions of vitality, or something stirring, or coming into being.

In my study on environmental movements related to oil-pipeline conflicts in British Columbia, Canada, I observed several ways in which the elements enfold political becomings. People concerned about oil-pipelines emerge as expressions of 'subjectivity' that is more-than-human. Not being able to swim in the water on certain days inflects the way in which bodies ingest and enfold toxins and bacteria, as they themselves are enfolded in the waves. Water pollution challenges culture/nature, human/non-human binaries that position water sports as 'outdoor' or 'wilderness' adventures and an 'escape' to Nature. One example from this study shows how political subjects emerge as more-than-human:

> I remember coming around the bend in Bondi [Beach] on a bus and it was the first time I had been in [Sydney] Australia. I come around this crest, and I remember vividly. 'Holy shit, this is just fantastic.' Just blew my mind from suburbia Ontario. I'm getting really tingly thinking about that moment. I jumped in the water, and it progressed to, 'Why are people polluting this water? Why is the city dumping?' You know, Sundays we weren't supposed to go into the water because of the 'whatever counts' that were bacterial. Long story short is that my passion for surfing started and just continued from there in terms of enjoying hiking and going in the mountain for snowboarding. You know, snowboarding is an extension of surfing. I think surfing's kind of led the path to how I think about treating the world, really.

I read the excerpt above: 'the path to how I think about treating the world,' not an expression of a humanist participant or transformative learning moment leading to actualization, but an assembled voice that contained plastic surfboards/waves/trash/bacteria undulating in the water—the singularization of 'a life' in a human expression of joy and abjection. Put differently, a human becomes a fold, or an expression of a more-than-human subjectivity that contains many virtualities in which it can be expressed. While environmental education literature has argued for the importance for people to be 'immersed' in 'the elements' by going 'outdoors' (practices with which I frequently engage in myself) as vital to developing an ethical orientation to the environment, I agree with Clarke and Mcphie (2014) who argue that such approaches are inadequate in highlighting the vitalism of elements in daily life, and produce humans as being 'in relation to' the environment, rather as being 'of' the environment, and emerging 'of' elements that comprise them. That very subtle linguistic distinction constitutes an ontological struggle to define the role elements have in producing subjectivity, and is indicative of a broader struggle to redistribute new models of vitality and life in the efforts launched through new materialisms, and most notably the work of Karen Barad (2007, 2014) and Jane Bennett (2010), as well as previous process philosophers and vitality theorists such as Henri Bergson (1907) and Alfred Whitehead (1927). This is particularly important, as 'life' features ever more prominently in environmental conflicts.

For example, energy and oil pipeline companies vie with environmental movement groups to define the relation between 'life' and the elements. In environmental conflicts related to oil pipelines, the focus on the materiality of life, and the life of matter is amplified and highlighted by oil pipeline companies that attempt to articulate 'energy literacy' in the form of defining what constitutes a good 'human' life, while maintaining the oil industry 'at life.' Enbridge, a pipeline company that planned to develop the Northern Gateway Pipeline in Canada, launched a campaign focused on the material foundations of 'life.' In their 2014 advertising campaign, *Life Takes Energy*, the company turned to the politics of 'life itself' after unsuccessfully gaining social license and buy-in for the Northern Gateway Pipeline by focusing on the economic benefits of the oil pipelines and their safety in earlier campaigns.

The *Life Takes Energy* campaign circulated through print, television, social media, and digital display advertising, focusing on the theme of 'E=___,' in which energy was equated with the means for a child to go to school on a school bus (E = Awestruck), the ability to heat up food for the family (E = Dinner with dad), or a means to go for a road trip (E = Amazing journey). In *Life Takes Energy*, the focus of the campaign shifts from 'death' (an oil spill) to 'life' and particularly celebrating the role of the elements in sustaining human life. The elements are not a 'dark matter' but media of humanness, i.e. celebration of birthdays, going to school, reading, and so on. As conversations about risk are rendered invisible, the company draws attention to the benefits of oil in sustaining 'life.'

The campaign appeared to draw attention to the material, and particularly fossil fuel basis of life, producing 'life itself', and what constitutes the essence of human life (presented through depictions of family, curiosity/learning, and adventure), as a modality through which to reaffirm the value of continuing life in arrangements of petro-subjectivity. In refusing to let petro-subjectivity 'die,' oil pipeline corporations engage in discursive biopolitics that attempt to bind 'life' to oil. Through this, they position political subjectification as having to continuously resuscitate itself to 'life' through enactments of environmental 'authenticity' as people concerned about oil pipelines are positioned within hypocrisy micropolitics that try to render them in/authentic, and in/credible, and as not understanding the energy needs for sustaining 'a life' or 'their life' (Piotrowski 2017).

The politics of 'life itself' might then be understood in relation to Braidotti's (2010) discussion of the shifting boundaries of life and death in light of a need to think about Foucault's concept of biopower differently for the post/Anthropocene, a shift that she defines as an understanding of biopower 'not only in the sense of the government of the living but also with relation to practices of dying' (p. 201). Braidotti (2010) argues that the politics of 'life itself' demands 'reconsideration of the concept of subjectivity in terms of 'life forces' (204). This reconstitutes the relation between human and non-human life as symbiotic, less dualistic, and less oppositional. As Braidotti shows, this refigures biopolitics a concept with an 'ecophilosophical dimension of reflection and inaugurates alternative ecologies of belonging both in kinship systems and in forms of social and political participation'(p. 204). As activist groups and environmental non-governmental organizations in environmental movements conceptualize themselves in educational terms and grapple to educate for new forms of subjectivity that might counteract petro-subjectivities, environmental education has been engaging with its own re-definitions of what constitutes 'being' and vitality.

New materialism emerged because there was a need to re-orient 'being' and what counts as vitality (St. Pierre, Jackson and Mazzei 2016). Whenever there is a rupture in the 'apparent order of infinity' (Deleuze 1988, 128) entire programs for accounting 'life' differently emerge: 'there are spatio-temporal programmes of organization which are irreducible in themselves, and on the basis of which living beings are disseminated' (p. 128). In other words, the turn to posthumanist and new materialist thought and methodologies that engages with the failures of the Anthropocene operates as a modality for refolding 'life' into more-than-human subjectivities. This is not to suggest that new materialist environmental education scholarship aims to 'save' the human subject 'by some other means' in light of the Anthropocene. 'Death' distributes force, or 'life itself,' along other, not-just-human (and sometimes animal) lines, in ways that genuinely contend the dissolution of Anthropocentric ways of life. Deleuze (1995) notes how subjectification is dependent on relations to death, in addition to relations to life. Subjectification is a process of

> establishing different ways of existing, depending on how you fold the line of forces, or inventing possibilities of life that depend on death too, on our relations to death (p. 92)

'Death' does not refer to the end of human beings or ignoring humans altogether, as might sometimes be suggested with the 'post' of 'posthumanism,' but to de-centering the attending to their becoming recombinant in new folds of subjectification, by way of reconfiguring new conceptions of 'life' with other 'finite' materials, and not in order to 'save' the human subject through these new configurations, but to attune to the vitality which has always figured and produced the human subject in the first place. By finite materials, I refer both to ecologically destroyed habitat, but also to the inability of any 'organism' to be independent, stand outside of, or transcend the planet. Hence, Deleuze and Guattari's (1987) call for 'becomings-elementary, -cellular, [and] -molecular' (p. 248) point to an understanding and engagement in an ecological topology based on the possibility of new folds and new life.

A humanist subject cannot possibly hold the 'life' of political subjectivity in what overflows what it claims to be its anthropocentric thresholds. This more-than-human subjectification is concerned with sustaining possibilities of new life forms, so that even if a human/person/subject seizes to become political, political 'life' emerges elsewhere. 'Life' beyond the anthropocentric subject, is concerned with a plurality of becoming—of being born in another place, at another time, in another, new, configuration. A life, is always more-than-one (Manning 2010), even if it is expressed in the form of 'a subject'—such as a learner, an activist, and 'its' capacity to think. Individuation becomes one momentary actualization of 'a life' that exists in virtual possibilities through which more-than-human and human affects come together.

> Every body, every thing, thinks and is a thought to the extent that, reduced to its intensive reasons, it expresses an Idea the actualisation of which it determines. However, the thinker himself makes his individual differences from all manner of things: it is in this sense that he is laden with stones and diamonds, plants 'and even animals' (Deleuze 1994, 255).

Residents living in oil pipeline conflict areas use the elements to describe their political subjectification as though being a surfer, a naturalist, an environmentalist was a result of 'their' intentional engagements with 'nature,' when conversely it is 'a life' that becomes expressed and actualized through 'them' in those very practices. In other words, elements operate without 'us' (Cohen 2014)—a scary prospect for environmental educators in environmental movements whose public education campaigns attempt to steer away from imaging what Thacker (2011) calls 'a-world-without-us.' By saying that elements act without us, I do not attribute any kind of teleological directionality to 'nature' that may be related to the term 'Gaia' as a self-regulating entity (Clarke and McPhie 2014). Larger constellation of elemental forces act despite, or whether or not a human subject comes to be expressed, but this is not necessarily harmonious or symbiotic. If one can speak of an ecological ethic traversing 'elemental Deleuze,' is rather in the realization that 'the subject' emerges as part of or from 'nature' that is grotesque rather than divine, and immanent rather than transcendent (Hayden 2008).

Rautio (2017) has shown that it is through 'diluting of a clear, identifiable one-to-one, that often causal relations, through which humanist approaches to viewing the world, to viewing education, can be dismantled' (p. 7). More-than-human subjectification thereby best be disassociated with the intention of a clear 'outcome' or macropolitical 'effect' as well as I would argue, with any intention or telos. The concept of a 'subject' in environmental movement education research might be examined in relation to different configurations of 'a political life' by attuning to how elements move and fold to produce political affects, and at times actualize in distinct 'bodies.' Elemental education research might hence focus on 'the material, sensory labor of attending to an emergent and enduring hum' (Stewart 2011, 447) and chaos of the elements that enlivens the world, but which has been relegated to 'non-verbal' communication (Peters 2015) for lack of words to describe the vitality of the elements as media.

Dromology of sea/sky/earth media

If 'something in the world forces us to think' (Deleuze 1994, 139) the effect on thought is not direct. Thought emerges only after affective encounters with the elements in the form of sense: 'wonder, love, hatred, suffering' (p. 139) rather than direct comprehension or recognition. The fold is the process through which materials affect sensations, styles, manners, and spaces in which subjectivity is expressed, first as impersonal affective force, then as emotion or a project, and subsequently as thought. Deleuze (1993) argued that 'folded forms are styles or manners. We go from matter to manner; from earth and ground to habitats and salons' (p. 40), from a material texturology, to a manner of thinking, i.e. a logology. It appears that Deleuze (1990) suggests that the elements, and 'the earth' in particular appear to shape subjectivities. These 'manners' appear as different textures, depending on the particular problem/solutions, bottlenecks, knots, boiling: echoes, reflections, traces, prismatic deformations, perspectives, and thresholds. In other words, the elements do not make direct impact, but become evident in a sensed encounter out of which the emotional, evaluative, project, conscience emerges in react to changes in atmosphere, rearrangements of matter, the presence of fire, chemicals.

For example, elements might enfold political subjectification as they relate to conundrums, and frustrations related to temporality, and more specifically in relation to dromology, or the science of movement and speed (Virilio 2006). What I mean by that is that an elemental dromology might attune to the way in which specific elements are associated with speeds in environmental movement that have distinct implications for knowledge practices and epistemology. In the oil pipeline movement, I examined people living in small towns on the central coast of British Columbia lamented the presence of journalists from larger urban centres, who would fly in to 'write up' Northern protests and share stories about 'Northern resistance' (or perceived lack of) to the proposed oil pipelines. Thinking expansively with the idea that the air mediates affects, I was intrigued by the way in which movements by air produced an atmosphere of outrage about knowledge geopolitics and the hierarchies of voice, documentation, and storytelling that had impact on people's engagements in political projects. With each new arrival of journalists, environmental NGO's, and even researchers (me included) by air, people could sense a growing atmosphere of frustration with becoming targets of 'Southern' education, and 'objects' to be 'written up,' whether in the form of research or news stories.

Meanwhile water travel was embraced as a slow medium that was perceived to be a practice that circulated trust and respect for local waters. Films like *Stand* (Bonello & Teichrob, 2012) about a stand-up paddle boarder that went to Kitimat, and *Tipping Barrels* (Gauthier & Reid, 2011) about surfers who wanted to explore the waves in the Great Bear Rainforest which the oil pipelines would endanger, and *Pipedreams Project* (Vandecasteyen & Des Roches, 2011) about three friends on a kayaking trip along British Columbia's west coast, all highlight slow aquatic engagements with elements that produce distinct affects, both in their films, as well as to the communities to which they travelled. Environmental education research hence might engage with the specific elemental forces that make up knowledge practices, including how the elements facilitate or prevent knowledge/mobilities.

In closing, I want to recall a conversation with one of the people I spoke with in the North of British Columbia. I learned that he had developed cancer, allegedly from working at the local aluminum smelter for twenty years. He had been politically active his entire life largely because of his work at the smelter, and I asked him if he was as active in the oil-pipeline movement as he had been before he became ill. He was, but not as much. Cancer re-prioritized different aspects of life for him. This had nothing to do with his 'energy literacy' or his resistance to being mobilized by public pedagogy campaigns organized by local and regional environmental groups. Aluminum (earth and fire) intervened to singularize in 'the life' of a subject, but it also unfolded and re-folded that subject in other ways. The elements are part of 'a political life' that produces distinct affective atmospheres and which act as media which help move affects that enable 'a life' to materialize in multiple expressions.

Political subjectivities cease to function as beings, and need to be examined as composites of folds, as process-ontologies that are vital because of elements' process-power (Parikka 2012). A challenge and opportunity for environmental movement education research will be to avoid the creep of the humanist voice, and to hear the materiality in voice if they engage in empirical research with elements/people. In this article I illustrated and made a few suggestions for how to do that by turning to the elements and their enfolding of more-than-human political subjects in environmental education. Elemental atmospheres, new biopolitics of life itself, and elemental dromology articulate three entry points into the focus on education in/as/through elemental/more-than-human subjectification. In particular material approaches to elements as media of affect and 'life' might focus on 'dirty matter' at the level of grime, earth minerals, dust, and smog (Parikka 2012). This darker turn to matter in environmental movement education really matters because within environmental education literature and amongst people involved in environmental movements, the tendency has been to celebrate being and 'going outside' with positioning the political subject as though they were *against* such dark matters when they are actually expressed, enfolded, and refolded by them.

Notes

1. It is outside the scope of this article to consider where one element stops and another resumes or to engage in discussion of precise definition of 'the elements.' I am grateful to one of the reviewers for drawing my attention to this point, and for suggesting an excellent overview of changing conceptions of 'the elements' in ecological philosophy written by David Macauley (2005) 'The flowering of environmental roots and the four elements in Presocratic philosophy From Empedocles to Deleuze and Guattari' in *Worldviews: Environment, Culture, Religion*.

ORCID

Marcelina Piotrowski https://orcid.org/0000-0003-4863-6975

References

Adey, P. 2015. "Air's Affinities: Geopolitics, Chemical Affect and the Force of the Elemental." *Dialogues in Human Geography* 5 (1): 54–75.

Anderson, B. 2009. "Affective atmospheres." *Emotion, Space and Society*, 2 (2): 77–81.

Anderson, B. and J. Ash. 2015. "Atmospheric Methods." In *Non-Representational Methodologies: Re-Envisioning Research*, edited by P. Vannini, 34–51. New York: Routledge.

Barad, K. 2007. *Meeting the Universe Halfway*. Durham, NC: Duke University Press.

Barad, K. 2014. "Diffracting Diffraction: Cutting Together-Apart." *Parallax* 20 (3). doi:10.1080/13534645.2014.927623

Barthes, R. 1957. *Mythologies. Collection Pierres Vives*. Paris: Seuil.

Bennett, J. 2010. *Vibrant matter: A Political Ecology of Things*. Durham, NC: Duke University Press.

Bergson, H. 1907. *Creative Evolution*. Mineola, New York: Dover Publications, Inc.

Bonello, A. & Teichrob, N. 2012. Stand [Motion Picture]. Canada: b4apres Media & Nicolas Teichrob Photography. Retrieved from http://standfilm.com/

Braidotti, R. 2010. "The Politics of "Life Itself" and New Ways of Dying." In *New Materialisms: Ontology, Agency, and Politics*, edited by D. Coole and S. Frost, 201–218. London, UK: Duke University Press.

Clarke, D. A. G. 2017 "Educating Beyond the Cultural and the Natural: (Re)Framing the Limits of the Possible in Environmental Education." In *Reimagining Sustainability in Precarious Times*, edited by K. Malone, S. Truong, and T. Gray, 305–319. London: Springer.

Clarke, D. A. G., and J. Mcphie. 2014. "Becoming Animate in Education: Immanent Materiality and Outdoor Learning for Sustainability." *Journal of Adventure Education & Outdoor Learning* 9679: 1–19.

Clover, D. E. 2002. "Traversing the Gap: Concientización, Educative-Activism in Environmental Adult Education." *Environmental Education Research* 8 (3): 315–323.

Cohen, J. J. 2011. "An Abecedarium for the Elements." *Post-medieval: A Journal of Medieval Cultural Studies* 2 (3): 291–303.

Cohen, J. J. 2014. "Elemental Relations." *O-Zone: A Journal of Object-Oriented Studies*, 1 (1): 51–53.

de Freitas, E. 2016. "The new empiricism of the fractal fold." *Cultural Studies: Critical Methodologies*, 16 (2): 224–234.

Deleuze, G. 1988. *Foucault*. Translated and edited by S. Hand. Minneapolis: The University of Minnesota Press.

Deleuze, G. 1990. *The logic of sense* (M. Lester & C. Stivale, Trans.). New York, NY: Columbia University Press. (Original work published 1969)

Deleuze, G. 1993. *The fold: Leibniz and the Baroque*. New York, NY: Continuum.

Deleuze, G. 1994. *Difference and Repetition*. New York, NY: Columbia University Press.
Deleuze, G. 1995. *Negotiations: 1972–1990*. Translated and edited by M. Joughin. New York, NY: Columbia University Press. (Original work published 1990)
Deleuze, G. 2001. *Pure Immanence: Essays on a life*. New York, NY: Zone Books.
Deleuze, G., and F. Guattari. 1987. *A Thousand Plateaus: Capitalism and Schizophrenia*. Translated and edited by B. Massumi. Minneapolis: University of Minnesota Press. (Original work published 1980)
Gannon, S. 2016. "Ordinary Atmospheres and Minor Weather Events." *Departures in Critical Qualitative Research* 5 (4): 79–90.
Gauthier, R. & Reid, M. (Producers) & Gulliver, B. (Director). (2011). Tipping barrels [Motion Picture]. Canada: Sherpas Cinema. Retrieved from https://vimeo.com/33234007
Gough, N. 2016. "Postparadigmatic Materialisms: A "New Movement of Thought" for Outdoor Environmental Education Research?" *Journal of Outdoor and Environmental Education* 19 (2): 51–56.
Hall, B. L. 2004. "Towards Transformative Environmental Adult Education: Lessons from Global Social Movement Contexts." *Counterpoints* 230: 169–191.
Hayden, P. 2008. "Gilles Deleuze and Naturalism: A Convergence with Ecological Theory and Politics." In *An [Un]Likely Alliance: Thinking Environment[s] with Deleuze|Guattari*, edited by B. Herzogenrath, 23–45). Newcastle upon Tyne, UK: Cambridge Scholars Publishing.
Ingold, T. 2011. *Being Alive*. London: Routledge
Kovan, J. T., and J. M. Dirkx. 2003. ""Being Called Awake": The Role of Transformative Learning in the Lives of Environmental Activists." *Adult Education Quarterly* 53 (2): 99–118.
Nakagawa, Y., and P. G. Payne. 2014. "Critical Place as a Fluid Margin in Post-Critical Environmental Education." *Environmental Education Research* 21 (2): 149–172
Macauley, D. 2005. "The Flowering of Environmental Roots and the Four Elements in Presocratic Philosophy: From Empedocles to Deleuze and Guattari." *Worldviews: Environment, Culture, Religion* 9 (3): 281–314.
MacLure, M. 2013. "The Wonder of Data." *Cultural Studies: Critical Methodologies* 13 (4): 228–232.
Maddalena, K., and C. Russill. 2013. "Is the Earth an Optical Medium? An Interview with Chris Russill." *International Journal of Communication* 10: 3186–3202.
Manning, E. 2010. "Always More Than One: The Collectivity of *a Life*." *Body & Society* 16 (1): 117–127.
McCormack, D. P. 2008. "Engineering Affective Atmospheres on the Moving Geographies of the 1897 Andree Expedition." *Cultural Geographies* 15: 413–430.
McCormack, D. P. 2010. "Fieldworking with Atmospheric Bodies." *Performance Research* 15 (4): 40–48.
McCormack, D.P. 2015. "Devices for doing atmospheric things." In *Non-Representational Methodologies: Re-Envisioning Research*, edited by P. Vannini, 89–111). New York: Routledge.
McCormack, D. P. 2016. "Elemental Infrastructures for Atmospheric Media: On Stratospheric Variations, Value and the Commons." *Environment and Planning D: Society and Space* 35: 1–20.
McKenzie, M. 2005. "The "Post-Post Period" and Environmental Education Research." *Environmental Education Research* 11 (4): 401–412.
Parikka, J. 2012. "New Materialism as Media Theory: Medianatures and Dirty Matter." *Communication and Critical/Cultural Studies* 9 (1): 95–100.
Parikka, J. 2015. *A Geology of media*. Minneapolis: University of Minnesota Press.
Payne, P. G. 2016. "What Next? Post-Critical Materialisms in Environmental Education." *The Journal of Environmental Education* 47 (2): 169–178.
Rautio, P. 2017. ""A Super Wild Story" Shared Human–Pigeon Lives and the Questions they Beg." *Qualitative Inquiry*. 23 (9): 722–731. doi:10.1177/1077800417725353
Peters, J. D. 2015. *The Marvelous Clouds: Toward a Philosophy of Elemental Media*. Chicago, IL: University of Chicago Press.
Piotrowski, M. 2017. ""Authentic" Folds: Environmental Audiences, Activists and Subjectification in Hypocrisy Micropolitics." *Continuum: Journal of Media & Culture* 6: 1–13.
Ruiz, R. 2014. "Iceberg Media." *International Journal of Communication* 8 (1): 2525–2530.
Snaza, N., & J. Weaver. 2016. "Introduction: Education and the Posthuman Turn." In *Posthumanism and Educational Research*, edited by N. Snaza and J. Weaver, 1–14. New York, NY: Routledge.
Sonu, D., and N. Snaza. 2015. "The Fragility of Ecological Pedagogy: Elementary Social Studies Standards and Possibilities of New Materialism." *Journal of Curriculum and Pedagogy* 12 (3): 258–277.
St. Pierre, E. 2004. "Deleuzian Concepts for Education: The Subject Undone." *Educational Philosophy and Theory* 36(3): 283–296. doi:10.1111/j.1469-5812.2004.00068.x
St. Pierre, E. A., A. Y. Jackson, and L. A. Mazzei. 2016. "New Empiricisms and New Materialisms: Conditions for New Inquiry." *Cultural Studies: Critical Methodologies* 16 (2): 99–110.
Stewart, K. 2011. "Atmospheric Attunements." *Environment and Planning D: Society and Space* 29 (3): 445–453.
Sturken, M. 2001. "Desiring the Weather: El Niño, the Media, and California Identity." *Public Culture* 13 (2): 161–189.
Thacker, E. 2011. *In the dust of this planet: Horror of philosophy Vol. 1*. Washington, DC: Zero Books.

Vandecasteyen, R. & Des Roches, F. (Producers and Directors). (2011). Pipedreams [Motion Picture]. Canada. Retrieved from https://www.cinemapolitica.org/film/pipedreamsproject

Vannini, P., D. Waskul, S. Gottschalk, and T. Ellis-Newstead. 2012. "Making Sense of the Weather: Dwelling and Weathering on Canada's Rain Coast." *Space and Culture* 15 (4): 361–380.

Virilio, P. 2006. *Speed and Politics*. Los Angeles, CA: Semiotext(e).

Wallin, J. 2016. "Pedagogy at the Brink of the Post-Anthropocene." *Educational Philosophy and Theory*, 1857 (July): 1–13. doi:10.1080/00131857.2016.1163246

Walter, P. 2007. "Adult Learning in New Social Movements: Environmental Protest and the Struggle for the Clayoquot Sound Rainforest." *Adult Education Quarterly* 57 (3): 248–263. doi:10.1177/0741713606297444

Walter, P. 2011. "Dead Wolves, Dead Birds, and Dead Trees: Catalysts for Transformative Learning in the Making of Scientists-Environmentalists." *Adult Education Quarterly* 63 (1): 24–42.

Walter, P. 2013. "Theorising Community Gardens as Pedagogical Sites in the Food Movement." *Environmental Education Research* 19 (4): 521–539.

Whitehead, A. N. 1927. *Process and Reality an Essay in Cosmology*. New York, NY: The Free Press.

Re-assembling environmental and sustainability education: orientations from New Materialism

Greg Mannion

ABSTRACT

A growing number of scholars call for the use of New Materialist frameworks for research across social and natural sciences. In general, however, there is little rigorous, in-depth or detailed advice on how postqualitative research is to be empirically conducted. Also, what the implications might be for environmental and sustainability education remain unclear. In response, drawing on data from a place-responsive heritage education project, employing theory from Deleuze and Guattari, I provide orientations for Assemblage Pedagogy and Assemblage Research. Assemblage Pedagogy involves educating for more sustainable ways of life through: (1) Interrupting existing education assemblages and experimenting with new approaches, (2) Practicing, relating, and entangling 'from the middle', involving the human and more-than-human to actualise the capacities and relations needed, and, (3) Evoking and performing new practices and expressions designed to create more sustainable ways of life. Wider implications for researching environmental and sustainability education are considered.

Introduction

Within ontological, relational, animal, and material turns (Castree 2004; Coole and Frost 2010), 'New Materialist' research is emerging across the social sciences (Barad 2003, 2007; Braidotti 2013; Deleuze and Guattari 1987; Deleuze 1994; Haraway 2003, 2008; Latour 2005; Pickering 2005). The application of New Materialism to pedagogy and educational research has also gained traction (Alaimo and Heckman 2008; Lather and St. Pierre 2013; Masny 2013; St. Pierre 2013, 2016; Snaza and Weaver 2015). These perspectives are diverse but often draw on Deleuze (1994) and Deleuze and Guattari (1987) and theories of assemblage. This article picks up these debates and extends them through applying them to environmental and sustainability education ('ESE' hereafter) and considering the associated challenges and benefits.

An assemblage is not just a collection or compilation of things. A DeleuzoGuattarian assemblage must be a collection of material, human and nonhuman parts with expressive capacities to affect and be affected by each other, and by other assemblages (Deleuze and Guattari 1977). The parts of an assemblage and the assemblages themselves are produced through their relations or their connectivity. These arrays of entities are decomposable and are not seamless – this means they can be put together and taken apart. Assemblages can comprise 'corporeal, technological, mechanical, virtual, discursive, and imaginary' elements that are in interaction (Renold and Ivinson 2014). Deleuze and Parnet's (1977) definition draws attention to how various things come together – how they are composed (rather than merely randomly compiled). Indeed, assemblages work the way they do despite and because of their *differences*. Like musical notes on a composer's score, rather than being haphazardly scattered, it is the particular arrangement of the notes, their durations, and other mark-ups that bring about expressive affect in a listener when played in a performance by a sympathetic musician.

> [An assemblage] is a multiplicity which is made up of many heterogeneous terms and which establishes liaisons, relations between them across ages, sexes and reigns – different natures. Thus, the assemblage's only unity is that of co-functioning: it is a symbiosis, a 'sympathy'. (Deleuze and Parnet 1977, 52)

Across the social sciences, Buchanan (2015a) suggests that there has been something of a failure to be clear about the implications of assemblage theory. In part, Buchanan shows why it has been difficult to be true to the ideas of Deleuze and Guattari; the contemporary view also suggests not overly specifying research practices (St. Pierre 2013). Another problem comes from how meanings are perhaps getting lost in translation as terms, such as 'agéncement' are translated from French. Unlike the noun 'assemblage', the verb, agéncement, perhaps better captures the process of coming together with openness and sensitivity to what is different and what might become. Hence, there is a need to find a viable way to apply Deleuze and Guattari's ideas adequately to ESE practice and research and explore their benefits.

Using an assemblage approach can help us make post-humanism viable in pedagogy and research but why should we do this? This is an important contribution since humanistic approaches to education in general appear to have failed us to date in many ways to adequately address issues such as climate change and wider global precarity. More widely, arguments against humanism are becoming more ubiquitous and compelling. Arguments include (Malone, Truong, and Gray 2017) the need for a de-centring of the human in our ontologies and epistemologies (for example, in solely looking to the needs of humans in definitions of sustainability or in child-centred approaches), a countering of human exceptionalism in general towards a relational orientation (in terms of, for example, human-plant, human-animal, human-place), a revision of 'stewardship' views of the environment (with its paternalistic associations of sympathy, mastery and control), a greater acceptance of environmental crisis of the Anthropocene (particularly climate change, but in connected ways food insecurity, migration, and so on), and a foregrounding of the importance of alternative ways of knowing (via, for example, indigenous knowledge, and the embodied, affective and ethical dimensions of experience). On these grounds, it is timely in ESE to begin now to explore and apply various forms of post-humanist and new materialist theory in rigorous but creative ways.

New Materialist ESE

Assemblage theory, like other New Materialist perspectives, share a desire to work in post-paradigmatic ways to challenge linguistic and social constructivist approaches and require a rethinking of how materials and discourses relate (Gough 2016). New materialists argue that since people, places and materials are entangled and emergent, a relational approach is required. Another feature is how approaches decentre the human and to take a more distributed view of agency. A third is a desire to unpack the binaries seen as undergirding Western thought around themes such as nature and culture, human and non-human, idealism and materialism. With these themes in mind, New Materialism should be attractive to and have consequences for environmental and sustainability programmes and their underpinning ontologies. However, with some exceptions, these efforts remain under-represented, less securely founded, and only latterly critiqued in ESE (Buchanan 2015, 2016; Gough 2016).

Some progress in New Materialist ESE is apparent when it comes to *describing* learner experience. Fawcett (2002) is credited with an early call for collective learning 'with' other species. Taylor (2017), drawing on Deleuze and Haraway, explicitly seek to account for more-than-human agency. Others emphasise the significance of the turn to flatter ontologies for learning within the age of the Anthropocene (Taylor and Pacini-Ketchabaw 2015). Early childhood researchers have challenged the binary of culture/nature and the idea of human stewardship of nature (Pacini-Ketchabaw and Taylor 2015; Malone, , Truong, and Gray 2017) where the significance of learner embodiment is common (Hackett and Somerville 2017). Lloro-Bidart (2016) asks us to consider non-human actions and human lived experiences in a relational way suggesting that other species and the human can be a community of knowers. Assemblage researchers seek to align epistemology, ontology and methodology within a linked approach to reality, representation and subjectivity (St. Pierre 2016). However, there is a noticeable reticence in this literature to do more than describe life in new ways. Less common, is the provision of advice for educators or exemplification for researchers.

In other assemblage-oriented ESE research, there are some directions for travel (*inter alia,* Clarke and McPhie 2016; Mannion and Gilbert 2015; Ross and Mannion 2012; Van Poeck and Lysgaard 2016). Mannion, Fenwick and Lynch (2013) analysed teachers' experiences to argue for place-responsive pedagogy requiring planning for reciprocity between humans and the learning setting. Clarke and McPhie (2016) suggest all education regardless of whether indoors or out should move to become more place-responsive. 'Assemblage Pedagogy', in this article, can be understood as a development of an expression of place-responsive pedagogy and research (Mannion, Fenwick and Lynch 2013). Often outside of ESE, a small but increasing number of authors in different subfields employ Deleuzo-Guattarian ideas around assemblage (Fox and Alldred 2015, 2018; Green and Duhn 2015; Hultman and Lenz Taguchi 2010; Leander and Rowe 2006; Lenz Taguchi 2013; Rautio 2013a, 2013b; Taylor and Pacini-Ketchabaw 2015). Smith and Dunkley (2018) employ assemblage theory to usefully *describe* child-nature relations in a study of children's roaming in outdoor settings using GPS systems.

Notably, this journal's special issue and a previous one (see the editorial from Duhn, Malone, and Tesar 2017) seek to continue to 'trouble' (*inter alia*) the categories of child/nature/urban, the nostalgic view of children's upbringing via nature experience, and the anthropocentrism of the idea of stewardship. Across these sources, it is commonly argued that a post-human reading of ESE requires us to stay with this trouble. To boot, it appears to require serious shifts in thinking and action. The result, inadvertently perhaps, is a positioning of post-human transformative pedagogies a bit beyond the grasp of the ordinary teacher or researcher:

> Changing the entrenched habits of modern western humanist thought, which are so adept at dividing humans off from nature, requires persistence, vigilance and a preparedness to take risks. It is hard work. It requires us to continually interrogate what it means to be human, to resituate humans firmly within the environment, and to resituate the environment within the ethical domain. (Taylor 2017, 3)

Assemblage theory as toolbox

We can expect that any good research will not usually exactly follow a pre-given formula for data collection, analysis and write-up and New Materialist approaches are not different in that respect. In order to deploy Deleuze and Guattari's ideas in research, we are required to rework contemporary ontological models of what counts as methods, data, and their analysis. Yet, if valid, an ontology that asks us to link discursive and material experience will surely marry with our sensate experiences of living in the world and make some aspects of researching in this way more viable. Indeed, over a number of projects, once the shift is made to accept and work within a new onto-epistemological frame, it does seem possible to deploy Deleuze and Guattari's ideas in different ways. There are also now many good sources for support albeit often located outside of ESE literatures. There is also the attractive advantage with this way of working of having scope to create and use new concepts which could be of value to a reinvigorated ESE. Assemblage theory in ESE can be seen as a useful way forward:

> A theory is exactly like a box of tools. It has nothing to do with the signifier. It must be useful. It must function. And not for itself. If no one uses it [...] the theory is worthless. [...] It is an instrument for multiplication and it also multiplies itself. (Deleuze speaking in conversation with Foucault. [Foucault 1977, 208]).

Perhaps core to the use Deleuze and Guatarri's philosophy, is a need to focus on pragmatic use of theory alongside questions of immanence (what can ESE do and what can learners-in-place become), rather than questions of identity (for example, what is sustainability?). In that spirit, I offer some (but not comprehensive) tool-like orientations and implications of 'Assemblage Pedagogy' and 'Assemblage Research' within ESE.

The goal of this article is both to describe and show *how* programming in ESE can be understood via assemblage theory. A key move is to understand how an ontology of becoming shapes the business of data collection, their analysis, and onward presentation in a research paper such as this. In showing how this works in research processes, the reader is invited to generate their own significance of these learners' lived experience in an outdoor environmental education programme. This empirical evidence is assembled here for its effects – the extent to which both photographs and transcribed evidence help us see or palpate how data collection and analysis occurred while still setting them to work now for readers. Assemblage researchers set out to produce effects, something new. Here, I set out to provide an expressive and productive account – to evoke young people's lived experiences as enfolded with, in and of a place. After Springgay and Truman (2017, 4), assemblage-oriented research, and in an important related way, assemblage pedagogies, will have an immanent quality commencing in the speculative middle 'where things grow, expand and pick up speed' [...] 'You have to be "in it", situated and responsive. You are not there to report on what you find or what you seek, but to activate thought'. This call for attention to unpredictable immanence applies to assemblage pedagogues, assemblage researchers and readers of their outputs alike (Jackson 2010).

Firstly, I describe the origin and purposes of the project, Stories in the Land and our approach to assemblage curriculum making. Thereafter, we describe the approach taken to data collection before offering three general orientations for Assemblage Pedagogy. Lastly, I draw out some exemplifications and consequences for ESE.

The project – Stories in the Land

The data used in the article come from a project entitled *Stories in the Land* (http://storiesintheland.blogspot.co.uk) which was led by Dr Joyce Gilbert who then worked for the Royal Scottish Geographical Society. I was the funded adviser and researcher over a three-year period. In this work, we took 'droving' as a key eco-social practice with which schools and youth groups could engage (Gilbert and Mannion 2014).

The project can be conceived as a place-responsive, arts-based, outdoor environmental education programme that connected participants from different generations from schools, youth groups and their communities with local places through journeys on the widespread network of 'drove roads' across Scotland. A 'drove road' is a route that was used – mostly in the last century – to bring livestock from one place to a market town or to bring cattle up into the higher ground to avail of summer pastures. A.R.B. Haldane (1900–1982), a social historian and writer, became interested in drove roads and the history of droving. His book, *The Drove Roads of Scotland* (1952) included a linen map showing an extensive arterial network of routes linking communities from the highlands to the lowlands (Figure 1).

At the outset, Haldane's map offered a perspective of some of the entanglements we considered potentially significant. Through taking an assemblage approach to pedagogy, we looked for actual and virtual entry points into this rhizome of roads, related practices and folklore, stories, and songs (Deleuze and Guattari 1987). The pedagogical approach of journeying on these routes

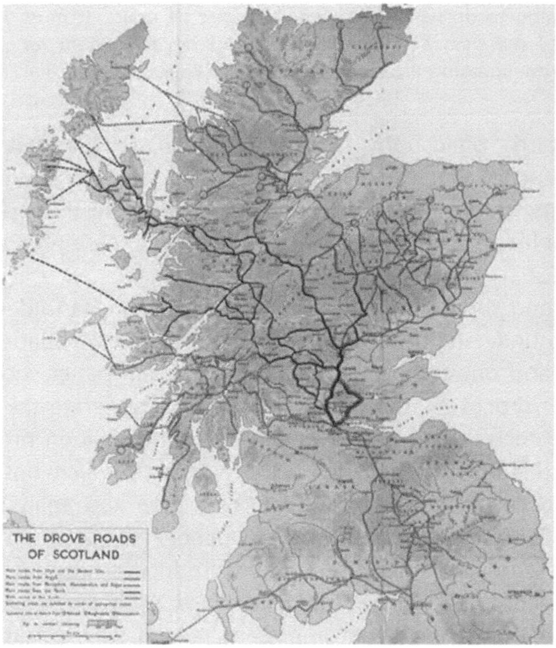

Figures 1 and 2. Haldane's map of the main drove routes whereupon the seasonal flow of thousands of cattle made their way from highland and coastal regions to market towns and on to England (reproduced by kind permission of Birlinn Publishers, Edinburgh). A school community on a community drove walk. Source: © (2014), the author. Reprinted with permission.

and investigating practice was, for us, reminiscent of Honeyford and Watt's (2018) notion of pedagogical burrowing or the need to 'plug into' (Jackson and Mazzei 2012) rhizomes in affective ways.

Nearly every community in Scotland has a local drove route but, despite their once vital importance to Scotland's culture and economy, droving remains a largely forgotten way of life in Scotland. The *Stories in the Land* approach was not only a form of *local* history pedagogy. Our approach asked participants to respond to past generations and the places inhabited by them and to consider the use of contemporary places inhabited by us now. Starting with the local lived experience (Duhn 2012a, 2012b), and drawing on traditional oral knowledge and practice, we invited learners to affect and become affected by these practices. We saw the potential for a situated, interdisciplinary, and intergenerational form of place-responsive pedagogy (Mannion, Fenwick and Lynch 2013). The project pedagogy was initially based on two principles arising from our reading of Deleuze and Guattari (Mannion and Gilbert 2015): (1) people and places are reciprocally enmeshed and co-emergent, and (2) people learn through making embodied responses to differences. We asked: How might our ways of life be prompted to be more sustainable in the light of the drover experience?

Assemblage curriculum

The key idea of 'assemblage', (Deleuze and Guattari 1987; Deleuze 1994) captures the ethos of our curriculum design work. We took from Lorimer (2005) a sense of the importance of the 'more-than-human' – for example, the place elements that mattered included cattle but also local plants and their uses by drovers. We also sought to provide lived experiences of drover life that would be sensory, embodied and affective for learners (Figure 2). In effect, we sought to devise 'more-than-human' pedagogies which would evoke and provide opportunities for catalytic witnessing of the significance of landscape, story, material practices and the movements of

Figures 3 and 4. As part of the project, normal school practices of teacher delivered lessons in classrooms are de- and re-territorialised as pupils were encouraged to walk on old drove routes and were assisted in setting up impromptu drovers' camps often in sites of droving historical significance. They also met local crofter, Ronnie Campbell, who took part in the last drove in 1949 as a boy who told them stories of his past experience (Figure 4). Source: © (2014), the author. Reprinted with permission.

animals and people in past times with a view to informing new ways of life today. We sought to maximise affective intensity for learning with, through, and in relation. Mulcahy (2012) captures the core business of the assemblage pedagogue as working to create significant 'conditions' for learners to engage and respond through making affective connections with others – human and more-than-human others. 'Good' teaching is in contact with the moving dimensions of experience that allow for affective connection. 'Basically the "good" is affectively defined as what brings maximum potential and connection to the situation' (Mulcahy 2012, p 23).

Our assemblage approach included an oral storytelling element. Learners were to 'become collectors, creators and tellers of old and new stories inspired by these epic journeys' (project documents). Following Somerville, Power, and de Carteret (2009), we addressed our relationship with place via re-storying across certain contact zones which were for this project (inter alia) Now-The Past, School-Community, Adult-Child, Farming-Wild Space, Nature-Culture. While traditionally, these relationships are policed or in Deleuzian terms, 'territorialised' along certain relations, our desire was to de-territorialise some of these relations through reconnecting communities with their schools, learners with their pasts, and indoor learning with outdoor settings in new ways (Figures 3 and 4).

As well as hearing about past experiences, learners valued the opportunity to engage in practices that were affectively impactful.

> Primary Pupil: You actually get to experience things. [...] And you feel it. [...] Rather than just watching it happen, you can actually do it.

The kind of curriculum making took 'droving' as a significant practice-based process where the person, animals and environment are in movement-based relation and are co-emergent (Ross and Mannion 2012). In this project, learners were invited to become similarly *attuned* to the environment *through* the practices of droving (Figures 5 & 6). Some activities allowed for intense memorable and sensory experiences.

> Teenage girl: And we learnt like all about like the things that we ate. And even, like we made butter and oatcakes and all that kind of thing. [...]

Our curriculum as assemblage approach countered transmission and constructivist views of learning. We followed Ingold's view that learners 'do not superimpose meaning on a world ("nature" or "physical reality") that pre-exists apart from ourselves, for to live we must dwell *in* the world, and to dwell we must already relate to its constituents. Meaning inheres in these relationships' (Ingold 1993, 222). Ingold's eco-social take on learning is profoundly Deleuzian: the

Figures 5 and 6. Attunement to a place comes through action in that place. On route, learners were engaged in learning through practices that were made possible by the way place-based elements came together or were brought into relation. For example, they learned about plant identification, collection and their uses in cooking using traditional recipes. Source: © (2014), the author. Reprinted with permission.

social and the natural are together coming into being in an on-going way. Within an anthropological frame, the less cultured novice as an emergent 'point of growth' (Ingold 2003, 304) apprehends by accompanying more experienced people into practice assemblages. Learners 'attend to the movements of others' to become more skilled (Ingold 2000). By his view, organisms that grow and develop can exercise agency but only in an entangled way within environments embodying skills in ways that are connected to perception and experience in that environment. Action, in contrast to mentally considered agency, includes the flow-like interplay of many diverse kinds of entities and forces in relation (weather, gravity, human and non-human agency). Acting in the world is experienced through *undergoing* the effects of the world's processes, one's own and others' actions (Ingold 2015). Dealing with unforeseen problems was part of undergoing droving as an educational experience:

> Teenage girl: Well I think I would say [...] the best way [to learn] is to go out there and do it and not talk about it and explain like in theory what could go wrong. But go out and experience what could go wrong and learn how to overcome it.

Ingold (2012) reminds us that whenever we encounter material things in a place (whether it be cattle, fire, water, or food), it is in flux with the consequence that 'this matter-flow can only be followed' (Deleuze and Guattari 1987, 451) as they move, grow, decay, and get transported. Taking account of how this was for the drovers makes more apparent how it is different for us now:

> Teenage girl: We had a lot more luxuries than the drovers would have had. So, it kind of gives you [a] perspective on how difficult it would have been for them.
>
> We found out how people used plants in their day to day lives
>
> We found out about life long ago
>
> We found out how people managed with what they had round about them (they couldn't go to shops to buy things)
>
> Respondent aged 12 (Questionnaire. Open Question)

Assemblage research

My own role in this project was as a contracted researcher. Research data collection included extensive participant observation fieldwork (across days spent planning, community walking, reflection and dissemination), sound recordings of stories told, researcher and respondent-taken

Table 1. Focus groups and respondents.

Focus group interview data drawn upon	No. of focus groups
Primary school pupils: 7–12 years of age	7
Secondary School age pupils: 13–17 years of age	7
Teachers: 7+ schools, primary and secondary schools	5
Facilitators and Assisting Adults	12
Total	**31**

Figures 7 and 8. The facilitation team and their practices can be seen as an assemblage of humans with capacities to engage with the non-human elements of place in different ways which afforded interdisciplinary learning within the team. Members had backgrounds in storytelling, singing, basket weaving, geography, history, anthropology and drama and spanned a number of generations. The oldest, storyteller in the traveller tradition, was Essie Stewart. Several of the team were fluent Gaelic speakers. Source: © (2014), the author. Reprinted with permission.

photographs, and over 30 transcribed focus group and other interviews with participants over a two-year period. Focus group interview data drawn upon here are summarised in Table 1. Each interview mostly had 3–7 respondents and lasted 15–60 minutes each. Participating interviewees (and their facilitators, teachers and parents [if under 16 years of age]) provided on-going written and oral informed consent for sound recording and photography (for onward use as included in this publication).

After Deleuze, a research-assemblage (Fox and Alldred 2015) seeks to bring into affective relation composable and decomposable human and non-human entities. For Fox and Alldred (2014), a research-assemblage should be defined in terms of the affective relations among the 'events' to be researched, the research tools and sources, the 'data' generated, and the researcher. (They also remind us that the physical spaces where research takes place, research culture, cultures, ethics committees; research outputs, editors, reviewers and readers are also part of this relational set). Within this immanent ontological frame, data are to be collected and analysed whilst being in the midst of these entities. Research assemblages are interventionist experiments orchestrated for producing data through the fieldwork and methods employed. The approach to researching the project also involved various different kinds of components – humans, events, technologies, frameworks, and theories. All of these together can be considered as a research assemblage (Fox and Alldred 2015).

Fox and Alldred (2018) suggest data collection involves identifying assemblages in the eco-social world, exploring how elements affect and are affected, and identifying particular and more aggregate flows that are generated by these assemblages. In that vein, working as a participant observer, I took part in witnessing camping practices, journeyed as a member of the story tribe (Figures 7 and 8), and captured data on the hoof as events unfolded as well as in project dissemination events (Figures 9 and 10). While participating in project events and during interviews and field note writing, assemblage researchers attempt to apprehend points of growth as significant for the respondents involved. Hence, lived experience contains actual and virtual or immanent events with affective, sensory and embodied dimensions. Data analysis takes the onward

Figures 9 and 10. In our assemblage-oriented research, interviewing happened mostly while events themselves were unfolding, fieldwork involved travelling with the team as they moved between schools and at community celebration days (Figure 9). Figure 10 shows the Duke of Edinburgh Award group of young people on stage being interviewed in front of a live audience in the Storytelling Centre in Edinburgh, Scotland. Source: © (2014), the author. Reprinted with permission.

immanent effects of an assemblage as primary; assemblages produce rather than, in hindsight, attempt to mirror or objectively represent reality.

Reporting the research – as in this article and other outputs – involved creating assemblage communications that are themselves 'eventful'. This meant attending in accounting to ways of crating affective hybrids from the research assemblage to impact on research consumers. The use of video (on-line) and images here helps to exemplify this aspect as does the dissemination celebrations and exhibitions (Figures 9 and 10). Axiologically, these research practices are key to enacting new materialist eco-sociologies that seek 'not only to analyse but to improve the world' (Fox and Alldred 2016, 52).

The stories young people created were often told to local audiences. Within an assemblage ontology, the agéncement of a research output too aims to produce an evocative account and calling for a response from readers.

In summary, we offer three orientations to Assemblage Pedagogy which were produced from the analysis of data in this project.

Assemblage pedagogy

Orientation 1: Practice, facilitate, and entangle from the middle with the more-than-human

Honeyford and Watt (2018, drawing on Preston, 2015) see 'assemblage pedagogies' as a form of 'burrowing' that is deeply connected to local settings, fundamentally productive, but also as dynamic and improvisational since they plug into and pull from other contexts in uncertain ways. Assemblage pedagogies demand reciprocal place-responsiveness from participating entities. As environmental and sustainability educators, assemblage pedagogies must take a purposeful way of planning and enacting with and through encounters with the more-than-human elements of place. The educational goal is to generate ESE learning as a result of significant coming together of people, places and practices. Assemblage pedagogies are designed and experienced through practicing, relating and becoming entangled with more-than-human elements and processes.

One project action involved four groups of teenagers using horses as a mode of travel which formed part of their Duke of Edinburgh Award. Each group travelled on horseback for several days on drove routes. Ahead of these trips, training was required, including navigation, campcraft, natural horsemanship, and first aid. The role of the facilitator was key to understanding

Figures 11 and 12. One DofE group on a practice ride arriving on the shores of Lochaline (Morvern, Scotland) and coming to the end of one of their journeys. Source: © (2014), the author. Reprinted with permission.

how educational experience was enabled through connecting people, place, other species, and practices in open-ended ways, as Bonnie (natural horsemanship teacher) explains:

> Bonnie: Being in an outside space and travelling from one place to another brings up thoughts and feeling that help younger people and children just explore where they are in themselves. [...] When we looked at the routes with the children we really did explore the past. [.] It started them asking lots of questions about it. [...] We talked about the post routes and the drovers and about hard life would have been for them. (Figures 11 and 12)

One of the young people noted the significance of horse-rider relations (Renold and Ivinson 2014) in actualising learning to respond to more-than-human aspects such as terrain and their companion animal:

> DofE Member: And we learnt what each different horse is like. [...] Gwen was probably the bravest one. So if there was something that the horses weren't sure about, like... like a bridge that didn't look very safe or something then we could always rely on Gwen to be the first one to cross so that the other horses would follow. So, it was good to know which horse we could rely on to do things like that.

Sheila, facilitator in the programme, also noticed the reciprocal nature of response making with the more-than-human from which learning arose (Figures 13 and 14).

> Sheila: They have worked so hard to get to this point. [...] I am not sure they actually really knew what they were taking on. So, it could have gone either way. [...] I think if they had given up, the ponies would have given up, 'cause the animals take their cues from them. If they had got stressed, the ponies would have got stressed. But they didn't.

ESE assemblage pedagogues need to actively seek out which key features in the biophysical and social environment are to be co-mingled and harnessed as features in a pedagogical way. They need to second guess which practices and interactions might afford or lead to learning about environmental issues, processes and problems, how to solve them, or at least lead to motivations to work towards solving them (after Stapp 1969). When well chosen, the encounters will be memorable, affective, significant and involve affective intensity.

In assemblage ESE, practitioners will also likely need to consider the ethico-political nature of this task. On the one hand, we seek to catalyse, provoke becoming but these becoming cannot all be fully determined in advance in educational programme aims and objectives. Assemblage ESE desire for the emergence of new person–place relations that are more sustainable but heretofore are by definition and within an assemblage ontology as yet unexpressed. Within an assemblage approach to pedagogy, facilitators sustain relative yet important autonomy for the pedagogue as a socio-historically contextualised and connected agent. It is the ESE facilitator – alongside other participants (human and more-than-human) – who plays a crucial role in orchestrating, sustaining and, later, dissolving pedagogical assemblages over time. In this project,

Figures 13 and 14. Across project activities, horses, as more-than-human agencies, played different roles in making journeying and the transportation of materials possible. The project participants also valued the opportunity to get to know particular horses by name and by character and which could be relied upon to lead over bridges or on uneven terrain. Source: © (2014), the author. Reprinted with permission.

educational facilitation came in many guises. Youth workers, horse trainers, artists and others helped catalyse opportunities for people of all ages to take part in local journeys of varying lengths from short walks of an hour or so, to expeditions lasting several days on horseback.

Orientation 2: Interrupt existing assemblages and experiment within new ones

Assemblage learning cannot be fully predetermined because it requires the on-going connections of agencies across the assemblage which are themselves in flux in unpredictable ways in pedagogical assemblages. Therefore, the assemblage pedagogue must interrupt, intervene and experiment. As Anderson et al. (2012) explain, new assemblages are needed to experiment with contemporary orderings. Understanding existing assemblages is a good place to start but this is not sufficient for an assemblage approach to pedagogy. Being open to change is clearly a related requirement (Anderson et al. 2012). Another is having a desire to target what we see as ESE problems – these may be human–environment relational problems such as the biodiversity loss, climate change or where our food has come from in the past. Bazzul and Kayumova's (2016) analysis suggests that using environmental issues in assemblage pedagogies might create the kinds of ontological and epistemological openings for the wished-for changes we have yet to imagine.

Assemblage pedagogies commence by being purposefully unsure about what might become but uses problems to create openings that allow for new relations among people and place so that more sustainable ways of life might emerge. These new ways of life are, in Deleuze and Guattari's terms, the 'lines of flight' (Bazzul and Kayumova 2016) that are needed to escape current norms and structures. The application of assemblage thinking to our contemporary context of unsustainability also usefully provides a way to link actual if emergent realities (material and discursive) with virtual possibilities into one frame. This is a challenge faced but not always addressed or theorised in his way in contemporary transformation and sustainability educational models.

A key feature of assemblage pedagogies is captured by the Deleuzian notions of de-territorialisation and re-territorialisation. Assemblage pedagogues actively seek to interrupt the contemporary flow of school life through the various project initiatives through re-purposing what is available, to evoke and promote responses in learners. We were interested in assemblages that had the capacity affect and be affected by their constituent elements. To be educational, I suggest assemblage educators work will involve facilitating some coalescing of elements that result in desired emerging differences or 'lines of flight' (Deleuze and Guattari 1987, 338) for participants. Lines of flight can take elements of an existing assemblage and recompose them in new

Figures 15 and 16. The role of the ponies in altering the relations between children, the outdoors, and the community was palpable throughout the project. In the 'Travelling Tales' project action, a pony and trap was used by the facilitation team to walk between primary schools over a three-day period. Stopping off at schools allowed for human-non-human encounters, time for parents to walk with pupils, and share knowledge about tack and saddles. Source: © (2014), the author. Reprinted with permission.

relationship to change an assemblage's expressions, norms and/or political realities. These lines of flight may be obstructed or reterritorialised (back into the pre-established assemblage) or become successful in more completely 'de-territorialising' existing assemblages and result in the creation of newly established assemblages (Nail 2017). These ideas allow us to see how assemblage pedagogies have a socio-historical, political and transformative dimension but this is needs to be understood as a relational, human and non-human, expressive and embodied, discursive and material activity.

In our project, we actively sought to interrupt the contemporary flow of school life through the various project initiatives. As Figures 15 and 16 (above) show, this involved appropriating spaces in new ways, working with other species (in this case horses, often Rosie, the dog, and at times, cattle, plants and other wildlife). The project at least temporarily de-territorialised school timetables, norms and routines to open up possibilities for learning across new boundaries: home-school, indoor-out, human and non-human, now and the past. Through camp craft and storytelling practices we sought to perform curriculum making differently evoking and promoting 'becoming-drover' sensibilities leading potentially to a reconsideration of life now, and, ultimately, in the future.

Orientation 3: Evoke, story and perform new expressions while becoming with the world

After Deleuze, we can say we come to know at the same time as *enacting* a world through inhabiting it and becoming attuned in our senses, skills and dispositions to its 'differences, positions and juxtapositions' (Anderson and Harrison 2010, 9). By this view, all 'representations' (such as written texts, maps, or artworks) are really better understood as presentations that coalesce to enact a world. Teaching and learning cannot be about content delivery alone or be solely based on mental constructivism. Assemblage educators will support and address learners' everyday performative embodiments in place collaborating in 'experimenting with the corporeal sensibilities that are enfolded into how we learn to affect and be affected by the world, including relations with the non-human' (Anderson and Harrison 2010, 25). For Stories in the Land, we took a view that stories could be retold and recreated by children and adults in Gaelic and English for different community audiences as part of an outdoor landscape experience (Figures 17–19).

Researcher: In what way do you think stories were in the land for you?

Primary Pupil: [They are] kinda tied into the land and everything.

Other Pupil: Yeah, yeah.

Researcher: Stories are 'tied into the land'?

Figure 17, 18 and 19. Storytelling in action. Pupils also constructed and told their own stories. Source: © (2014), the author. Reprinted with permission.

> Primary Pupil: They're about places [...] and what they actually believed.
>
> Primary Pupil: Stories help you learn about the past of your own place.
>
> Primary Pupil: It's important to get more stories told in Scotland so they don't get forgotten.

Many eventful pedagogical assemblages arose and fell away across the life of the project. These educational events are challenging to evidence in non-representational research. They happened through the ways in which humans and non-humans were involved in harnessing together and dispersing various entities such as animals (cattle, horses, wildlife), people (children, parents, community members), place elements (drove roads, weather), materials (saddles, food, tents and other heterogeneous elements of journeying). In addition to these, there was a harnessing of discursive elements (e.g. the social history of droving, drovers' songs and stories, children's created stories, curricular framings, and policies around, for example, outdoor learning). Our intentional actions as educators can be, 'compatible with the idea that the human being is both immersed in a world of nonhuman forces and inseparable from affective relations with nonhuman animals and other things' (Bowden 2015, 78). We become with the world (Figures 20–22).

A core idea of Deleuze and Guattari (1994) is that it is always difference that gets repeated. This project action (see figures above) perhaps exemplifies this ontological idea in action. In this task, the participants were invited to place unspun sheep's wool inside their footwear. As the days journeying progressed, the effect of the children's walking was to create felt in the shape of their own feet inside their boots. In the process, wool becomes felt with the action of the person and the activity in a place. This experience shows how the heterogeneous material entanglements (of feet, different steps taken, local sheep, and footwear) produces each different boot with each different step imprinting a new difference. The tasks exemplify how difference gets repeated both in how each learner walked that day, and in how each felt boot emerges as a result. The boots display (Figure 22) is an evocative visualisation of the expression of the linked goals of learning *about* droving, learning via *making*, *journeying* and learning through becoming-with-materials. In assemblage pedagogies, we are called to teach, learn and research through acting in an immersed way within a cultural and material world, at one and the same time acting and undergoing changes in response to differences encountered. Teaching, learning, and researching are forms of becoming with the world.

Summary

ESE assemblage pedagogues can and must work in local and beyond-local ways to interrupt traditional education to experiment with bringing human and more-than-human entities into

Figures 20, 21 and 22. Making and displaying the 'felt feet'. The 'felt feet' are created by agitating unspun woollen fibres with soapy water by placing the material on the feet within a bag inside one's footwear. Through walking, the fibres interlock and tangle into a new, firm, shaped piece of felt. Source: © (2014), the author. Reprinted with permission.

relation to create conditions of affective encounter for learners. We have shown that assemblage pedagogies can be about experimenting with/in place, producing affective expressions, reordering, de-territorialising and re-territorialising, tracing and interrupting the relations of heterogeneous elements into new purposeful assemblage for pedagogical ends.

Because not every collective, compilation or gathering is an assemblage, pedagogues and participants need to be 'assemblage convertors' (Bennett 2010; Buchanan 2015) who activate and co-ordinate specific types of gatherings and beings into consequential arrangements with purposes.

> The elements in an assemblage are never contingent or purposeless. This is why I have suggested that 'arrangement' is better translation of agéncement than assemblage. It foregrounds more obviously the strong element of agency that underpins Deleuze and Guattari's original conception of the term. We cause our assemblages to be the way they are inasmuch that if they were arranged differently we would not be able to endure them. (Buchanan 2016, online)

We have used our case analysis to derive the following more general orientations for *Assemblage Pedagogy* and *Assemblage Research*. We summarise them here as a framework for practitioners in ESE seeking to follow this approach in other settings.

To actualise *Assemblage Pedagogy* in ESE we might:

1. **Practice, facilitate, and entangle from the middle** with the human and **more-than-human** to actualise the capacities for more sustainable ways of life,
2. **Interrupt existing assemblages and experiment with new ones** to create more purposeful ESE pedagogies and research practices. Assemblages are pedagogical in that they can involve an intervention designed to describe, interrupt, and recast existing relations including existing educational relations. that involves a 'resistance to closure and encourages a stress on open-endedness around the emergent and the already formed' (Anderson et al. 2012, 176), and,
3. **Evoke, story, and perform new expressions while becoming with the world** creating more sustainable ways of life. Educational assemblages have a purpose: to affect and be affected by the participants and to create new capacities.

For *Assemblage Research*, four practices reflect our approach here:

1. **Analyse and collect from the middle.** Researching what is 'becoming vital' in the midst, analysing from the middle of flows and movements is a norm. Methods of collecting and analysing are themselves a form of reciprocal response making in the midst of materials, practices, people and other entities within environments.
2. **Evoke significance.** Act to disclose local and wider significance via the data collection and analysis. This means attending to practices and performances of how desires, affect and influences impact in a given moment in time recognising that this moment is part of a wider flow of desires and influence. Yet, we must not lose sight of how local and extra-local, contemporary and past practices relate and influence how what is assembled in the here and now.
3. **Engage with the more-than-human.** These may be entities, practices and processes: the capacity to influence or be influenced is not ascribed solely to people. A third orientation asks us to attend to the more-than-human entities, practices and processes, at times, perhaps, noticing what the role of other species in learning.
4. **Attend reflexively to our own capacities.** We reflexively draw upon our experience as researchers within research assemblages. We too are often implicated as participants within the very assemblages we research and the place-responsive research methodologies (Lynch and Mannion 2014) we might employ.

The (above) lists for pedagogy and for research overlap to some degree and they are not comprehensive and need to be further responded to and situated or place-plugged via one's own ethico-political judgements. Despite their partial nature and these limitations, they may be starting points to provoke further assemblage approaches in ESE.

Discussion and implications

Staying open

In this project, the purposeful pedagogical tasks required assemblages to bring key relevant entities into new affective relationships, often creating intense learning for participants in affecting and being affected by more-than-human others. Through coming to know practices experientially, learners can be prompted to create new forms of dwelling, journeying and storying. Assemblage pedagogies are based on a key idea that assemblages are undergoing on-going composition across human/more-than-human and other boundaries. In education, because of this radical relationality, our facilitation processes required an openness to outcomes that cannot all be predetermined. From ESE, the approach allows for a tentative open-ended exploration rather than a prescriptive approach to telling learners what living more sustainably might mean on a vulnerable and changing Earth. We need to 'examine reality from within the intractabilities and ambiguities of the human perspective [and] push towards an ecological politics: new ways of imagining living with the earth' (Saldahna and Stark 2016, 434–437).

Encountering differences

In assemblage ESE, places are not backdrops to the social action, and generational relations are linked to place-person relations. This way of working means educators, learners, and their collaborators will actively seek out *place-based intergenerational differences* and work with these in *more-than-representational expressions.* Findings suggest that schools and other educational settings can benefit from enrolling local community members from diverse generations *and* harnessing places into emergent processes of curriculum making. By this view, learning is an all-age, emplaced process derived from encountering differences within social and material worlds. Because education is traditionally generationally niched and indoor, intergenerational and outdoor educational practice are seen as particularly relevant (Mannion 2017).

Entangled educator

Not just any entanglement will form an assemblage with pedagogical reach. It is important here to notice that curriculum designers and pedagogues are not insignificant in the process of shaping educational throwntogetherness (Massey 2005). Creating an educational agéncement means *purposefully* selecting out some elements enrolling them into new relationships. Within ESE, the role of educators in activating and participating in assemblage pedagogies remains important. Bowden (2015) argues that entities, including pedagogues, when harnessed into any assemblage do maintain their autonomy. However, this autonomy must be understood as a form of on-going entangled association where we undergo experiences in the setting whilst acting too: 'agency is inseparable from the different 'assemblages' of heterogeneous affective bodies that a given body enters into' (Bowden 2015, 61). Certainly, assemblage thinking means we must accept our contextualised embeddedness in settings, accepting that our bodies and practices are always entangled with the more-than-human, conditioning what happens, but not fully determining the on-going emergence. In outdoor ESE, this might include the roles of educators amid other agencies such as that of animal, amid other non-agential forces or effects coming from weather and terrain. An outdoor ESE facilitator will for sure be swayed in their decision making and participants will be affected by changing conditions in the environment. Our intentional will as educators – what we do and want to do – is always dependent on the forces and agencies of other entities. Different elements in an assemblage have different capacities to act, affect and be affected whilst participating. In fact, it is these differences that give rise to the educational and other effects of an assemblage. Assemblage theory allows us to consider the role of ESE pedagogues in intervening intentionally within educational processes of all kinds and not falling into a trap of human exceptionalism. Whether wearing the wool of a sheep on our feet or eating foraged food, participants were being and learning together (Snaza et al. 2014, 50) with non-humans.

Becoming sustainable

To change educational practice, and to address the wicked problems of sustainability we desperately need new assemblages to interrupt the contemporary status quo. Agyeman, Bullard, and Evans (2003, 2) defines sustainability as 'the need to ensure a better quality of life for all, now and into the future, in a just and equitable manner, while living within the limits of supporting ecosystems'. In this project, quality of life for all included humans and more-than-humans, whose relationality in local and wider trans-local environmental connections was raised as significant in place-responsive assemblage pedagogies. Walking and responding with/in place across the generational boundary had the potential to catalyse change in our ways of life and work as starting points for considering what it means to live a more sustainable life. Sustainability pedagogies need to be rooted in an awareness of the pasts of our current unsustainability. Understanding our own local histories in this interdisciplinary, intergenerational way seemed to be a viable if tentative exploration of what living more sustainably in the land might mean (Haluza Delay, DeMoor, and Peet 2013).

In sum, at a time when ESE should be striving for the ways to actualise how to live a sustainable life, assemblage thinking and pedagogy has a lot to offer. There is the orientation towards new embodied, emergent interconnections, understanding the formation of new planes of immanence between groups (learners and their communities, people and animals, communities and outdoor places), and the voicing of heretofore unvoiced (animals and other entities). New human and more-than-human educative assemblages can create new becoming (Goodley 2007). Ontologically, assemblage ESE reminds us that what things, people and places are and what they are becoming are connected and open-ended. In Assemblage ESE, we can be less concerned with essences and more fully embrace a relational and processual approach. ESE then becomes an experiment based on evocative productive pedagogies of desire for change. Within assemblage pedagogies, the shifts required include from transmission to active inhabitation, experiencing with our bodies, witnessing and evoking through practice, from representation to creating more-than-representative communicative expressions with new audiences, and from conforming to staying open to transforming, performing and responding to differences found in all kinds of places.

Acknowledgements

Thanks especially to Dr. Joyce Gilbert who led the *Stories in the Land* project as education officer at Royal Scottish Geographical Society at that time. (See http://storiesintheland.blogspot.co.uk). Thanks also to participating schools and their communities, Coal Youth Centre, and Storyline Scotland (Sally Harkness). Facilitators included: Alasdair Taylor, Essie Stewart, and Claire Hewitt (storytellers), Barney Strachan (sound artist), Alastair Davidson (bushcraft) Sarah Hughes and Richard Bracken (Room 13), Jane and Derry Wilkinson (willow), Alistair Strachan (historical re-enactment), Kate Langhorne (song and anthropology), Sheila Ryan (Duke of Edinburgh link), Bonnie Mealand (Ardgour Riding School), Ruaraidh Ormiston (Newtonmore Riding School). Thanks to the funders: Robertson Trust, Abernethy Trust, Ganochy Trust, Heritage Lottery, Ernest Cook Trust. For further information on the felt feet task, see https://bellacouche.com/feet-felt-walking-wool/

Disclosure statement

No potential conflict of interest was reported by the author.

ORCID

Greg Mannion http://orcid.org/0000-0003-2233-9278

References

Agyeman, J., R.D. Bullard, and B. Evans, eds. 2003. *Just sustainabilities: Development in an unequal world*. London: Earthscan.

Alaimo, S. and S. J. Hekman, eds. 2008. *Material feminisms*. Bloomington, IN: Indiana University Press.

Anderson, B., and P. Harrison. 2010. "The promise of non-representational theory". In *Taking-Place: Non-Representational theory and human geography.*, edited by A. Anderson and P. Harrison, 1–34. London: Ashgate.

Anderson, B., M. Kearnes, C. McFarlane, and D. Swanton. 2012. "On Assemblages and Geography." *Dialogues in Human Geography* 2(2): 171–89. doi: 10.1177/2043820612449261

Barad, K. 2003. "Posthumanist Performativity: Toward an Understanding of How Matter Comes to Matter." *Signs* 28 (3): 801–831. doi: 10.1086/345321

Barad, K. 2007. *Meeting the universe halfway*. Durham: Duke University Press.

Bazzul, J., and S. Kayumova. 2016. "Toward a Social Ontology for Science Education: Introducing Deleuze and Guattari's Assemblages." *Educational Philosophy and Theory* 48 (3): 284–299. doi: 10.1080/00131857.2015.1013016

Bennett, J. 2010. *Vibrant Matter: A Political Ecology of Things*. Durham, NC: Duke University Press.

Bowden, S. 2015. "Human and non-human agency in Deleuze." In *Deleuze and the non/human.*, edited by Rolfe, J. and H. Stark, 60–80. London: Palgrave Macmillan.

Braidotti, R. 2013. "Posthuman Humanities." *European Educational Research Journal* 12 (1): 1–19. doi: 10.1177/0263276407085156

Buchanan, I. 2015. "Assemblage Theory and Its Discontents." *Deleuze Studies* 9 (3): 382–92. doi: 10.3366/dls.2015.0193

Buchanan, I. 2016. "What Must We Do about Rubbish?" *Drain Magazine* 13 (1). Available on-line: http://drainmag.com/what-must-we-do-about-rubbish/ accessed 17 April 2017.

Castree, N. 2004. *Nature*. London: Sage.

Clarke, D. A. G., and J. McPhie. 2016. "From Places to Paths: learning for Sustainability, Teacher Education and a Philosophy of Becoming." *Environmental Education Research* 22 (7): 1002–1024. doi: 10.1080/13504622.2015.1057554

Coole, D., and S. Frost. 2010. *New materialisms: Ontology, agency and politics*. Durham, NC: Duke University Press.

Deleuze, G. 1994. *Difference and Repetition*. P. Patton. NY: Columbia.

Deleuze, G., and C. Parnet. 1987. *Dialogues*. Columbia: Columbia University Press.

Deleuze, G., and F. Guattari. 1987. *A thousand plateaus: Capitalism and schizophrenia*. Minneapolis: University of Minnesota Press.

Duhn, I. 2012. "Places for Pedagogies, Pedagogies for Places." *Contemporary Issues in Early Childhood* 13 (2): 99–107. doi: 10.1177/1463949115585658

Duhn, I. 2012. "Making 'Place' for Ecological Sustainability in Early Childhood Education." *Environmental Education Research* 18 (1): 19–29. doi: 10.1080/13504622.2011.572162

Duhn, I., K. Malone, and M. Tesar. 2017. "Troubling the Intersections of Urban/Nature/Childhood in Environmental Education." *Environmental Education Research* 23 (10): 1357–68. doi: 10.1080/13504622.2017.1390884

Fawcett, L. 2002. "Children's Wild Animal Stories: Questioning Interspecies Bonds." *Canadian Journal of Environmental Education* 7 (2): 125–39. doi: 10.1080/13504622.2017.1325452

Foucault, M. 1977. "Intellectuals and Power - A Conversation Between Michel Foucault and Giles Deleuze". In *Language, Counter-Memory, Practice: Selected Essays and Interviews by Michel Foucault*, edited by D. Bouchard, 205–217. Oxford: Basil Blackwell.

Fox, N. J., and P. Alldred. 2015. "New Materialist Social Inquiry: designs, Methods and the Research-Assemblage." *International Journal of Social Research Methodology* 18 (4): 399–414. doi: 10.1080/13645579.2014.921458

Fox, N., and P. Alldred. 2016. *Sociology and the New Materialism*. London: Sage.

Fox, N. J., and P. Alldred. 2018. "Mixed Methods, Materialism and the Micropolitics of the Research-Assemblage." *International Journal of Social Research Methodology* 21 (2): 191–204. doi: 10.1080/13645579.2017.1350015

Gilbert, J., and G. Mannion. 2014. "Stories in the Land: In the Tracks of the Highland Drovers." *Scottish Local History. Spring* 88:32–41. ISSN: 0266-2027.

Goodley, D. 2007. "Towards Socially Just Pedagogies: Deleuzoguattarian Critical Disability Studies." *International Journal of Inclusive Education* 11 (3): 317–34. doi: 10.1080/13603110701238769

Gough, N. 2016. "Postparadigmatic Materialisms: A 'New Movement of Thought' for Outdoor Environmental Education Research?" *Journal of Outdoor and Environmental Education* 19 (2): 51–65. ISSN: 2206–3301.

Green, M., and I. Duhn. 2015. "The Force of Gardening: investigating Children's Learning in a Food Garden." *Australian Journal of Environmental Education* 31 (01): 60–73. doi: 10.1017/aee.2014.45

Hackett, A., and M. Somerville. 2017. "Posthuman Literacies: Young Children Moving in Time, Place and More-than-Human Worlds." *Journal of Early Childhood Literacy* 17 (3): 374–391. doi: 10.1177/1468798417704031

Haldane, A. R. B. 2008. *The drove roads of Scotland*. Edinburgh: Birlinn Ltd.

Haluza-Delay, R., M. J. DeMoor, C. Peet. 2013. "That We May Live Well Together in the Land ... : Place Pluralism and Just Sustainability in Canadian and Environmental Studies." *Journal of Canadian Studies* 47 (3): 226–258.

Haraway, D. J. 2003. *The Companion Species Manifesto: Dogs, People, and Significant Otherness*. Chicago: Prickly Paradigm Press.

Haraway, D. J. 2008. *When species meet*. Minneapolis: University of Minnesota Press.

Honeyford, M. A., and J. Watt. 2018. "Burrowing and Becoming: teaching Writing in Uncertain Times." *Pedagogies: An International Journal* 13 (3): 260–279. doi: 10.1080/1554480X.2017.1399797

Hultman, K., and H. Lenz Taguchi. 2010. "Challenging Anthropocentric Analysis of Visual Data: A Relational Materialist Methodological Approach to Educational Research." *International Journal of Qualitative Studies in Education* 23 (5): 525–542. doi: 10.1080/09518398.2010.500628

Ingold, T. 1993. "The Art of Translation in a Continuous World." In *Beyond Boundaries: Understanding, Translation and Anthropological Discourse*, edited by G. Palsson, 210–230. Oxford: Berg.

Ingold, T. 2000. *The Perception of the Environment: Essays on Livelihood, Dwelling and Skill*. London: Routledge.

Ingold, T. 2003. "Two Reflections on Ecological Knowledge." In *Nature Knowledge: Ethnoscience, Cognition, Identity.* edited by G. Sanga and G. Ortalli, 301–310. New York: Berghahn.

Ingold, T. 2012. "Toward an Ecology of Materials." *Annual Review of Anthropology* 41 (1): 427–442. doi: 10.1146/annurev-anthro-081309-14592

Ingold, T. 2015. *The Life of Lines.* London: Routledge.

Jackson, A. Y. 2010. "Deleuze and the Girl." *International Journal of Qualitative Studies in Education* 23 (5): 579–587. doi: 10.1080/09518398.2010.500630

Jackson, A., and L. Mazzei. 2012. *Thinking with Theory in Qualitative Research: Viewing Data across Multiple Perspectives.* New York: Routledge.

Lather, P., and E. A. St. Pierre. 2013. "Post-Qualitative Research." *International Journal of Qualitative Studies in Education* 26 (6): 629–633. doi: 10.1080/09518398.2013.788752

Latour, B. 2005. *Reassembling the social: An introduction to Actor-Network-Theory.* Oxford: Clarendon.

Leander, M., and D. Wells Rowe. 2006. "Mapping Literacy Spaces in Motion: A Rhizomatic Analysis of a Classroom Literacy Performance." *Reading Research Quarterly* 41 (4): 428–460. doi: 10.1598/RRQ.41.4.2

Lenz Taguchi, H. 2010. *Going Beyond the Theory/practice Divide in Early Childhood Education: Introducing an Intra-active Pedagogy.* London: Routledge.

Lenz Taguchi, H. 2013. "Images of Thinking in Feminist Materialisms: Ontological Divergences and the Production of Researcher Subjectivities." *International Journal of Qualitative Studies in Education* 26 (6): 706–716. doi: 10.1080/09518398.2013.788759

Lloro-Bidart, T. 2017. "A Feminist Posthumanist Political Ecology of Education for Theorizing Human-Animal Relations/Relationships." *Environmental Education Research* 23 (1): 111–130. doi: 10.1080/13504622.2015.1135419

Lorimer, H. 2005. "Cultural Geography: The Busyness of Being 'More-than-Representational." *Progress in Human Geography* 29 (1): 83–94. doi: 10.1191/0309132505ph531

Malone, K., S. Truong, and T. Gray. 2017. *Reimagining Sustainability in Precarious Times.* Singapore: Springer.

Mannion, G. 1999. *Children's Participation in Changing School Grounds and Public Play Parks in Scotland*, Stirling: University of Stirling. Available online: http://dspace.stir.ac.uk/bitstream/1893/88/1/Children%27sParticipation%20inChangingSchoolGroundsAnd PlayParks.PDF Accessed 1st Feb 2018.

Mannion, G. 2017. "Intergenerational Education: We are in a new place." In *Family, Intergenerationality and Peer-Group Relations.*, edited by S. Punch and R. Vanderbeck. London: Springer.

Mannion, G., A. Fenwick, and J. Lynch. 2013. "Place-Responsive Pedagogy: Learning from Teachers' Experiences of Excursions in Nature." *Environmental Education Research* 19 (6): 792–809. doi: 10.1080/13504622.2012.749980

Mannion, G., and J. Gilbert. 2015. "Place-responsive Intergenerational Education." In *Intergenerational Space.*, edited by R. Vanderbeck and N. Worth, 228–241. London: Routledge. ISBN: 978-0-415-85531

Masny, D. 2013. "Rhizoanalytic Pathways in Qualitative Research." *Qualitative Inquiry* 19 (5): 339–348. doi: 10.1177/1077800413479559

Massey, D. 2005. *For space.* Thousand Oaks, CA: Sage.

Mulcahy, D. 2012. "Affective Assemblages: body Matters in the Pedagogic Practices of Contemporary School Classrooms." *Pedagogy, Culture and Society* 20 (1): 9–27. doi: 10.1080/14681366.2012.649413

Nail, T. 2017. "What Is an Assemblage?" *SubStance* 46 (1): 21–37.

2015. Pacini-Ketchabaw, V. and A. Taylor, eds. *Unsettling the colonialist places and spaces of early childhood education.* New York: Routledge.

Pickering, A. 2005. "Asian Eels and Global Warming: A Posthumanist Perspective on Society and the Environment." *Ethics and the Environment* 10 (2): 29–43. doi: 10.1353/een.2005.0023

Rautio, P. 2013. "Children Who Carry Stones in Their Pockets: On Autotelic Material Practices in Everyday Life." *Children's Geographies* 11 (4): 394–408. doi: 10.1080/14733285.2013.812278

Rautio, P. 2013. "Being Nature: Interspecies Articulation as a Species-Specific Practice of Relating to Environment." *Environmental Education Research* 19 (4): 445–57. doi: 10.1080/14733285.2013.812278

Rautio, P., and J. Winston. 2015. "Things and Children in Play – Improvisation with Language and Matter." *Discourse: Studies in the Cultural Politics of Education* 36 (1): 15–26. doi: 10.1080/01596306.2013.830806

Renold, E., and G. Ivinson. 2014. "Horse-Girl Assemblages: Towards a Post-Human Cartography of Girls' Desire in an Ex-Mining Valleys Community." *Discourse: Studies in the Cultural Politics of Education* 35 (3): 361–376. doi: 10.1080/01596306.2014.888841

Ross, H., and G. Mannion. 2012. "Curriculum Making as the Enactment of Dwelling in Places." *Studies in Philosophy and Education* 31 (3): 303–313. doi: 10.1007/s112

Saldanha, A., and H. Stark. 2016. "A New Earth: Deleuze and Guattari in the Anthropocene." *Deleuze Studies* 10 (4): 427–439. doi: 10.3366/dls.2016.0237

Smith, T. A., and R. Dunkley. 2018. "Technology-Non-Human-Child Assemblages: reconceptualising Rural Childhood Roaming." *Children's Geographies* 16 (3): 304–318. doi: 10.1080/14733285.2017.1407406

Snaza, N., P. Appelbaum, S. Bayne, D. Carlson, M. Morris, N. Rotas, J. Sandlin, J. Wallin, and J. A. Weaver. 2014. "Toward a Posthumanist Education." *Journal of Curriculum Theorizing* 30 (2): 39–55.

Snaza, N., and J. Weaver. 2015. *Posthumanism and Educational Research.* London: Routledge.

Somerville, M., K. Power, and P. de Carteret. 2009. *Landscapes and Learning: Place Studies for a Global World*. Rotterdam: Sense Publishing.

Springgay, S., and S. E. Truman. 2018. "On the Need for Methods beyond Proceduralism: speculative Middles, (in)Tensions, and Response-Ability in Research." *Qualitative Inquiry* 24 (3): 203–214. doi: 10.1177/1077800417704464

St. Pierre, E. A. 2013. "The Posts Continue: becoming." *International Journal of Qualitative Studies in Education* 26 (6): 646–657. doi: 10.1080/09518398.2013.788754

St. Pierre, E. A. 2016. "The Empirical and the New Empiricisms." *Cultural Studies ⟨—⟩ Critical Methodologies* 16 (2): 111–124. doi: 10.1177/1532708616636147

Stapp, W. B. 1969. "The Concept of Environmental Education." *Environmental Education* 1 (1): 30–31. doi: 10.1080/00958960903210031

Taylor, A. 2017. "Beyond Stewardship: Common World Pedagogies for the Anthropocene." *Environmental Education Research* 23 (10): 1448–1461. doi: 10.1080/13504622.2017.1325452

Taylor, A., and M. Giugni. 2012. "Common Worlds: reconceptualising Inclusion in Early Childhood Communities." *Contemporary Issues in Early Childhood* 13 (2): 108–119. doi: 10.2304/ciec.2012.13.2.108

Taylor, A., and V. Pacini-Ketchabaw. 2015. "Learning with Children, Ants, and Worms in the Anthropocene: Towards a World Pedagogy of Multispecies Vulnerability." *Pedagogy, Culture and Society* 23 (4): 507–529. doi: 10.1080/14681366.2015.1039050

Thomas, G. 2011. "The Case: generalisation, Theory and Phronesis in Case Study." *Oxford Review of Education* 37 (1): 21–35. doi: 10.1080/03054985.2010.521622

Fieldnotes and situational analysis in environmental education research: experiments in new materialism

Andy Ruck and Greg Mannion

ABSTRACT

This article details the approach taken to a recent study in an environmental education context, with a focus on the writing and analysis of ethnographic fieldnotes. This approach drew upon aspects of new materialist theory and multi-species ethnography for the writing of fieldnotes, and Situational Analysis for their analysis. These approaches represent a 'site of experimentation' that arose through attempts to carry out research in a manner sensitive to new materialist theories, whilst operating within a time-bound, collaborative study. As well as highlighting potential synergies, this article also explores the constant tensions that arose when attempting to use existing qualitative research methods in combination with new materialist theories. It is not intended as a guide to conducting research in a 'new materialist' or 'post-qualitative' manner, but rather as an insight into the tensions, synergies and on-the-ground methodological struggles within this site of experimentation, and what was produced through the 'research assemblage' thereby created. The article aims to demonstrate the ways in which existing qualitative research methods were re-oriented through this approach, as well as the onward effects of these re-orientations on what then amounted to 'the data'.

Introduction

The growing use of new materialist theories in environmental education research (Clarke and Mcphie 2016, Green and Duhn 2015) is indicative of a wider relational and material turn across the social sciences (Coole and Frost 2010, Fox and Alldred 2015). These theories, drawn from the work of Deleuze and Guattari (1987), Haraway (2008), Barad (2007) and others, have given rise to a number of proposed methodologies that are sensitive to new materialist theories. Prominent among these are 'post-qualitative' methodologies (St. Pierre 2011), a term that encompasses a wide range of critiques of 'conventional humanist qualitative methodologies' drawn from postmodernism, poststructuralism and posthumanism – collectively referred to as the 'posts' (St. Pierre 2014, 2). Importantly for research in both environmental education contexts and elsewhere, post-qualitative methodologies rest on a number of key ontological shifts which, for their key proponents, render them incompatible with many of the 'normalising humanist concepts' (St. Pierre 2014, 10) central to qualitative research. These shifts have given rise to Taylor's (2016, 18) subsequent question, 'if the "usual" methodological procedures are no longer possible in the posthuman… then how to proceed?'.

In partial response, this article details the approach taken to a recent doctoral study in an environmental education context, with a focus on the writing and analysis of ethnographic fieldnotes. The approach taken was a 'site of experimentation' (Bridges-Rhoads 2015, 704) that arose through attempts to learn and enact new materialist theories whilst in the midst of a funded doctoral study that was embedded in a wider funded project. I[1] detail how I drew upon aspects of new materialist theory and multi-species ethnography (Pacini-Ketchabaw, Taylor, and Blaise 2016) for the writing of fieldnotes, and Situational Analysis (SA) for their continuous analysis.

There is, as Bridges-Rhoads (2015, 705) notes, 'no guarantee that such experimentation will produce a qualitative research ... that can claim to "do something different"', and every risk of it collapsing back into the 'conventional humanist' approach critiqued by St. Pierre and others. Indeed, the approach taken here created constant tensions when used in combination with new materialist theories, not least with the drawing up in advance of a 'research design' (Taylor 2016), the use of an apparently essentialising practice such as SA (Jackson 2013), and the use of fieldnotes in the first place (MacLure 2013a). I acknowledge and explore these tensions throughout this article. This article, then, is not intended as a guide to conducting research that is sensitive to new materialist theories, but rather as an insight into the tensions, synergies and on-the-ground methodological struggles within this site of experimentation, and what was produced through the 'research assemblage' thereby created (Fox and Alldred 2018). I hope to highlight the ways in which existing qualitative research methods came to be re-oriented through attending to aspects of new materialist theories, as well as the onward effects of these re-orientations on what then amounted to 'the data'[2], and on myself as researcher.

Polli: Nation and post-qualitative research

This article draws upon a recent doctoral study linked to Polli:Nation – a UK-wide environmental education initiative that engaged young people from 260 primary and secondary schools in the transformation of their school grounds into pollinator-friendly habitats. Funded by the UK's Heritage Lottery Fund and run by the school grounds charity Learning through Landscapes (LtL), Polli:Nation had the stated aims of restoring pollinator populations deemed vital for the maintenance of natural heritage and food production (Goulson 2014), and 'creat(ing) a network of knowledgeable and enthused young conservationists' with an improved understanding of their local environment and of wider environmental issues (LtL 2014, 13). The project involved young people mainly in practical conservation tasks such as planting flowers and building 'bug hotels', and in conducting surveys of the biodiversity in their school grounds.

For the doctoral element of this research, I conceived of a study that took an ethnographic approach (Brewer 2000) to addressing the question: 'What processes and features within the Polli:Nation project appear significant in terms of young people's response-making to environmental issues?. This stemmed from growing calls for environmental education researchers to focus on what participation in environmental education programmes 'looks and feels like for the learners concerned', as well as/instead of the aims and outcomes of such programmes (Rickinson, Lundholm, and Hopwood 2009, 97).

I was drawn to new materialist theories in the early stages of this PhD through literature highlighting the complex sets of factors interacting to determine the outcomes of environmental education programmes (Higgins, Nicol, and Ross 2006; Mannion, Fenwick, and Lynch 2013), as well as by their strong focus on more-than-human elements (Bennett 2010; Tsing 2013) – of particular interest given the study's focus on a project built around a particular inter-species relationship. Perhaps most importantly, however, I saw the potential utility of the new materialist project of re-thinking humanity as inextricably bound up with matter, through which new materialist theorists are seeking to re-frame the relationship between humans and our environments. I agreed with Coole and Frost (2010, 4) that such a re-framing is especially important in light of current environmental challenges, and with Clarke and Mcphie (2014) that this may have many implications for outdoor and environmental education practice. This, in turn, led me to an initial exploration of the literature on post-qualitative research.

St. Pierre (2014, 3) explains that the term 'post-qualitative' is a 'large and ambiguous' one, and that her use of it began largely in response to 'positivist' qualitative research methodologies that had persisted despite the influence of postmodernism, poststructuralism and posthumanism during the 1990s and early 2000s. These approaches have been collectively referred to as the 'posts' (St. Pierre 2014), and have concerned themselves principally with the deconstruction of concepts central to humanist qualitative methodology, including 'validity', 'voice' and 'authenticity' (Lather 1993, 2000). For St. Pierre (2014, 2), the resultant ontological shifts render thinking in 'the posts' incompatible with what she labels 'conventional humanist qualitative methodology'. This incompatibility has, in turn, led to her adoption of the term 'post-qualitative'.

Lather (2016, 125) has recently written of a 're-turn or second coming of postmodernism', and draws upon a hugely heterogenous range of theories and methodological shifts to summarise the learnings from this ontological turn. To me, two such fundamental shifts from conventional humanist methodologies stood out as being key to adopting a methodology sensitive to new materialist theories for the study in question. The first is to resist the tendency to focus on identifying order, regularity, and a linear relationship between cause and effect (Fox and Alldred 2015), instead 'rethinking … causality as entanglements with surprising effects' from which 'unpredictable novel possibilities are always emerging' (Fenwick et al. 2015, 123–124). The second is a de-centring of the researcher to focus instead on material and more-than-human entanglements, of which humans are a part. For Jackson (2013, 741), the privileging of human experience in conventional qualitative methodologies is rooted in Cartesian dualist thought, which 'affirms the existence of reality in the mind and therefore the thinker's existence *outside the realm of the material*'. In fact, she continues, materials and humans are inseparable: 'The material is not purely produced by human intention, nor does human agency pre-exist or transcend the material: *they mutually constitute one another*' (744).

For Taylor (2016, 18), accepting the co-constitution of humans and materials means that research can no longer be thought of as an 'individualized cognitive act of knowledge production', and is instead 'an enactment of knowing-in-being that *emerges in the event of doing research itself*' (Emphasis added). Crucially, then, planning such research in advance is highly problematic. Although a number of wider orientations are offered for conducting research in a manner sensitive to new materialist theories (Lenz Taguchi 2012; Vannini 2015; Mannion 2019), this key ontological shift accounts for the lack of prescriptive guidance on how such research ought to be carried out. Indeed, for St. Pierre (2014, 10), 'there's no recipe, no textbook that explains, step-by-step, how to 'do' a Foucaultian power-knowledge reading or genealogy; there's no 'research design' or 'research process' for how to 'do' Derridean deconstruction'. The notion that humans and materials 'mutually constitute one another' also calls into question other accepted norms within qualitative research, such as the idea of data as existing separately from the researcher (Brinkmann 2014), and as I will explore later in this article, essentialising practices such as coding (St. Pierre and Jackson 2014).

'The rush to application'

Despite its sound ontological basis, carrying out research that was truly post-qualitative proved problematic in this particular research context. This was, in the first instance, due to my relative lack of familiarity with the new materialist theories underlying it when commencing my doctoral studies. By the time I had gained a fuller understanding of the ontological shifts described above, I had already been caught up in what St. Pierre (2017, 2) calls 'the *rush to application*, to methodology'. This includes a perceived need among doctoral students to quickly decide upon a research design before they have fully considered how a given theory might be applied to their specific research context.

In this study, the temptation of this 'rush to application' was compounded by both the structured, time-bound nature of doctoral research in general, and the instrumental requirements of this funded project in particular. In most UK universities, doctoral researchers are required to undergo various systematic and formal assessment processes during their studies. This was a typical PhD in this sense, with the first year being largely structured around the ten-month Progress Review – a written and oral assessment that required a clear outline of my research design and proposed methods. Furthermore, of significance to this study in particular was the parallel requirement to provide an evaluation of Polli:Nation as a whole for the funders and partner organisations. This created a restriction in terms of time and scope: the research had to take place in Polli:Nation-participating schools, and had to be completed during the 2016–17 academic year. Moreover, the stated purpose of my University's involvement in the project was to 'ensure evaluation of behavioural change outcomes as well as natural heritage outcomes' (LtL 2014, 3), implying a need to somehow 'evidence' this within the available timeframe. These factors combined to create a strong felt need for early clarity around data collection and analysis, and a subsequent unwillingness to simply wait for 'situations of breakdown, surprise, bewilderment, or wonder' (Brinkmann 2014, 722) to occur.

Ahead of my ten-month Progress Review, I had designed a study that would eventually comprise thirty sessions of participant-observation in twelve schools in Scotland, followed by focus groups with pupils in twenty participating schools across the UK. Nonetheless, I was well aware of a tension that had not been fully resolved between new materialist/post-qualitative thinking and the apparently 'humanist qualitative study' (St. Pierre 2018, 2) I was embarking upon. Before beginning participant-observation, then, I drew upon two approaches to the writing and analysis of fieldnotes that I felt would increase the study's sensitivity to new materialist theories. Firstly, I would draw upon multi-species ethnography (Pacini-Ketchabaw, Taylor, and Blaise 2016, Taylor and Pacini-Ketchabaw 2015) for orientations on the writing of fieldnotes. Secondly, I would create situational and relational maps, both drawn from SA (Clarke, Friese, and Washburn 2017), throughout the research process. In the following sections, I outline these methods, their links to new materialist theories, and the tensions that arose when employing them within the research context described here.

Could I call this a post-qualitative approach? St. Pierre (2018, 8), while asserting the impossibility of pinning down post-qualitative research to any defining features, is insistent that '*a study cannot be made post qualitative after the fact.* To repeat, one begins post qualitative inquiry without a methodology' (Emphasis in original). Perhaps, then, it was already too late. Perhaps choosing even *more* methodological approaches to draw upon only exacerbated the extent to which this study was not post-qualitative. Instead, then, I characterise what follows as a site of experimentation (Bridges-Rhoads 2015) – an attempt to learn new materialist theories whilst in the midst of doctoral research. As well as exemplifying the tensions that come with pursuing research that is sensitive to new materialist theories, I hope to go beyond discussions as to what this approach 'is', to instead consider what was nonetheless *produced* through this unique research assemblage (Fox and Alldred 2018).

Theoretically sensitive fieldnotes

During early participant-observation sessions, my approach had been to follow a well-known formula for taking fieldnotes – taking short-hand 'jottings' (Emerson, Fretz, and Shaw 1995) while participating in Polli:Nation activities, then expanding these into a detailed descriptive narrative as soon as possible after these notes were made (Mack et al. 2005). Later, however, I made two key changes to the way I collected fieldnotes. These changes, I felt, served to both focus my attention more fully on the more-than-human elements present, and to encourage me to 'think with and write with philosophical texts' (Bridges-Rhoads 2017, 2).

The first of these changes was to leave my pen and notebook behind. It felt to me that the perceived obligation to constantly take notes had taken me further away from the role of 'participant' than I had intended, and more towards being an 'observer'. No longer taking notes, I felt, enabled me to be better attuned to more-than-human elements and inter-species encounters, and to better develop what Pink (2009) calls 'embodied knowing'. I shifted instead to writing notes as soon as possible after each session. This shift was inspired by an article by Pacini-Ketchabaw, Taylor, and Blaise (2016, 155) in the emergent field of multispecies ethnography, which also provided orientations for the second key shift I made.

The second change I made with regard to fieldnote-writing was to develop a series of questions specific to this study, to prompt my note-taking after each session. These, I felt, served to connect the fieldnotes more clearly with new materialist theories and focus my attention on more-than-human elements. I drew upon a range of questions, drawn from two key sources. These were, firstly, the emergent literature on multi-species ethnography – a form of relational ethnography (Desmond 2014) that has been applied principally in early-years contexts (Ogden, Hall, and Tanita 2013; Taylor and Pacini-Ketchabaw 2015). The central feature of multi-species ethnography, for Pacini-Ketchabaw, Taylor, and Blaise (2016, 151), is to avoid being human-centric in research encounters, instead 'tracing how our lives, children's lives and the lives of other animals in our common worlds are entangled, interconnected (and) mutually dependent'. Secondly, I developed questions based on key ideas from new materialist theory. These drew most closely upon the idea of 'concepts', 'percepts' and 'affects'. The idea of directing one's thinking through the selected use of particular theories has clear resonances with the 'diffractive' approach to data analysis drawn from Barad (2007). Recent articles advocating this approach (Mazzei 2014; Lenz Taguchi 2012), however, are concerned with using theory to analyse *existing* data such as audio recordings of interviews. Here, I conceived instead of an approach that enabled me to explore those theories whilst in the midst of carrying out the research itself.

Devising questions based on new materialist concepts was accompanied by a slight sense of unease at using terms that I had only recently begun to engage with. Was my understanding of them 'correct'? I proceeded, however, with something like the approach advocated by Strom (2018), who stresses that 'it did not matter if I did not understand every single word of what I was reading. What mattered was if I found anything that *worked for me*' (106). This claim is based on Deleuze's own recommendation to avoid trying to read too deeply into his work, because '(t)here is nothing to explain, nothing to understand, nothing to interpret. It is like plugging in to an electric circuit' (1990, 8). I also kept in mind Springgay and Truman's (2018, 4) assertion that '(y)ou are not there to report on what you find or what you seek, but to *activate thought*' (emphasis added).

Concepts, percepts and affects were, in this case, ideas that *worked for me* as ways to think about how young people experienced and came to understand the activities making up the Polli:Nation project, and how these were bound up with the more-than-human elements present. In very brief terms, following Semetsky (2015), I understood affect to be a *force* that is present within an assemblage, potentially giving rise to feelings. This differs from a percept, which is the perception of the feeling itself, after it has been 'felt' (Colebrook 2002). A concept, broadly speaking, is an idea. For Semetsky (2015), it is an idea that is 'invented in practice' – that is, arising through a given situation and serving to facilitate understanding of the nature or purpose of that situation. Importantly, these concepts are *becomings*, rather than fixed entities. In Polli:Nation activities, I came to label the concepts in circulation as, for example, 'young people as (becoming) conservationists', 'young people as (becoming) community activists/active citizens' and 'Polli:Nation as (citizen) science project'.

The box below (Figure 1) contains extracts from a set of fieldnotes taken on a day of participant-observation in a Polli:Nation-participating secondary school, where a group of seven pupils

What human/more-than-human encounters were there?

- While gathering dry leaves from the ground, Liam finds a large woodlouse, and lets it crawl over his hand. He comes and shows it to me. He calls it a "slater" - I'd forgotten that's what people call them in this part of the country. Unlike some of the boys in the group, who display high levels of energy, Liam is quiet, thoughtful, and does his own thing within the group. He tells me how slaters are actually crustaceans, because they have a shell. I complement him on his knowledge and ask him where he's learned it. "I just like animals and stuff", he shrugs. At this point, one of the other boys, Brian, quickly takes the slater off him and throws it towards one of the girls, who is scared of insects. She screams and runs away. Liam waits until the commotion dies down, then goes and finds the slater, and places it carefully at the side of the path, among the leaves.

How do human and more-than-human elements respond to these encounters? How might they have experienced them?

- What to say about the slater? It was clearly in its usual environment – damp, dark places. Being in amongst those dead leaves was exactly that sort of place. And having looked up a bit about slaters… They are indeed "isopod crustaceans" that feed on dead plant material (so probably searching for food when Liam found it?), and are usually active at night (so are they sort of drowsy this time of day?) They use their antenna to feel their way around. They still breathe through gills as their ancestors lived underwater. As for how it might have experienced today, though… Crawling over new surfaces (people's hands), then being chucked through the air, then picked up again. It's hard to know if any of that registered beyond a brief sensation…

- Brambles: Are they growing at this time of year? We noticed how they'd been "re-rooting", ie, growing from one spot, then sticking into the ground again in another. The parent helper pointed that out – don't think I knew they did that. It's amazing how long they are too. You pull one up at one end, and are left with a long spiky string of it, tangled with other bits and sometimes re-rooting. Now they're being uprooted and chucked into a waste pile over by the polytunnel. Same with the weeds in the sensory garden.

Figure 1. Extract from fieldnotes from a participant-observation session in a Polli: Nation-participating secondary school, using a series of theoretically sensitive questions.

were engaged in the commonplace activity of building a 'bug hotel'. The pupils were part of the school's department of Additional Support Needs (ASN), and participated in the project for the same two periods every week. It is not the full set of fieldnotes taken on that day. The extracts have instead been selected in order to exemplify some of the questions I drew upon, to give an impression of how the questions operated, and the shifts made from conventional fieldnote writing to an approach re-oriented within a new materialist framing.

What was affectively palpable today? (Affect)

- Again, the human/ non-human encounters – or, for the young people, the chance to see, touch, handle insects (or crustaceans in this case) up close. Usually they are spontaneous/ unexpected encounters.

- Mrs C as an "activist" teacher/ unique character: This is only the second time I've met Mrs C, and her attitude and demeanour really come across as she tells me about how she's been running the Polli:Nation project so far. She puts a great emphasis on doing things her way, being herself, and not caring what impression she gives. I have noted already how a school's level of activity within Polli:Nation seems to depend on a passionate and outgoing teacher. She is a strong character to carry a project like this forward. It seems to depend on that a lot.

- That woodland really is unique. It offers the chance for free exploration, out of sight from the teachers. From the teachers' point of view, it is small enough that the pupils can be let out of sight – it's a fenced-off area, and no real harm can come to them. Stretching several hundred metres from one end to the other, the woodland is bounded at one end by a small stream (which the group call "the Burn") and several gardens belonging to neighbouring houses, and at the other, a fence and locked gate which separate it from the rest of the school grounds. Mrs C says she really appreciates the woodland as a valuable resource for outdoor learning, but it's only her that uses it for most of the year. It being behind a fence and locked gate means it can only be used as part of a timetabled class like this. On the way into the woodland today, I also notice some rhododendrons, a few silver birches, a carpet of leaf litter, and the pungent smell coming from the Burn, which had some of that froth in it that forms in eddies on burns and rivers.

What (new?) ideas/ "concepts" are in circulation? (Concepts)

- Giving a "home" to slaters, other bugs, bees, etc. Anthropomorphism when thinking about the bug hotel. As I've said before, this seems to be an activity that young people really get into, perhaps because of the relatability of giving something a "house" or "home". "So this is the Presidential Suite?", Liam grinned, pointing at an area at the top of the bug hotel, after Mrs C had joked about the hotel getting its "first customer" when they hadn't finished it yet.

Figure 1. Continued.

'Language practices' and 'performative privilege'

The changes outlined above represent small steps aimed at better attuning myself to more-than-human elements and inter-species encounters, and developing a higher degree of 'embodied knowing' (Pink 2009). Clearly, however, there are ongoing tensions when applying these orientations alongside new materialist theories, mainly owing to ontological concerns with an

- Different ideas/ reactions to other species: What was interesting today was the contrasting responses to the slater from Liam and Brian. Picking it up and throwing it, vs carefully placing it among the leaves. That caring attitude (a "commonworld" perspective?) is also quite different from the "stewardship"/ "other species as 'useful'" point of view the project is founded on, too. So could I say that other species such as the slater (and more obviously for this project, the bees) carry the capacity to provoke different affective responses in young people? Something to be scared of, something to take care of, something that's "useful" to humans? And Polli:Nation seems to create the spaces for these different kinds of "othering" to happen

- Idea of "weeds" vs "plants", or more widely, "nature vs non-nature". Or even "types of nature". As the groups were pulling out weeds from the planters in the sensory garden, Brian noticed Danny pull out something that wasn't a weed, and exclaimed "Haha, that's a plant!"

- Young people/ citizens as conservationists. The idea that we can "help" other species ("environmental stewardship") through building this protective space for them. And that the other species need "protection".

Figure 1. Continued.

over-reliance on 'language practices' (MacLure 2013a). The first tension was simply the continued use of fieldnotes. As MacLure (2013a, 664) reminds us, any data gathered from fieldnotes, interviews, focus groups and scholarly papers are inherently humanist, and have serious limitations when dealing with elements outside of 'the ideational and cultural aspects of utterances (spoken or written)'. Whilst I had stopped writing fieldnotes whilst actually *in* the research situation, I nonetheless continued to rely on notes written afterwards as key data from these participant-observation sessions (although these sat alongside photos and situational maps). Furthermore, directing my note-taking with questions arguably exacerbated the separation between my own participation in Polli:Nation activities, and the 'inner mental activities inside a separated human being' (Hultman and Lenz Taguchi 2010, 536) when interpreting them afterwards. One could, then, experiment with not writing fieldnotes at all, or at least without the use of pre-conceived research questions to guide our analysis. Yet we would, of course, still encounter a reliance on textual representation at some point – whether through writing for publication, or a PhD thesis. As Bridges-Rhoads (2015, 709) reflects, although she may write as a 'situated speaker' (after Richardson and St. Pierre 2005), the expectation is still that 'I am the researcher. I have to produce a text about that research. That is what qualitative research is'.

A second, related tension is an unease with what Petersen (2018, 11–12) describes as the danger of 'speak(ing) on behalf' of someone or something – in the case of the example above, of the woodlouse. 'The power to narrate somebody's (or in this case something's) story', she says, and re-producing it through language practices such as writing, is in danger of engendering 'another form of oppression' on the part of humans over other species. Pacini-Ketchabaw, Taylor, and Blaise (2016, 157) stress that it is more a case of 'sensing and following … rather than rushing in to interpret and represent them'. In their paper, one of the authors, Mindy Blaise, recounts an experience of conducting multi-species ethnography with dogs. Here (similarly to my experience with the woodlouse), she admits that 'I have no idea what dog knowledgeabilities might be or how they will emerge because I am not a dog', but stresses that the key is in 'suspending my pull towards meaning making for long enough to sense dog worlds and dog agencies on dog terms' (Pacini-Ketchabaw, Taylor, and Blaise 2016, 156).

As shown in the fieldnote extract, however, such suspension of meaning-making proved difficult in this particular study. To me, anything I wrote about the woodlouse felt at least like some kind of attempt to interpret, or to write *about* the woodlouse from a human perspective. This rush to interpretation was, as previously discussed, undoubtedly encouraged by the questions I had set myself when writing fieldnotes. It was also, however, most likely exacerbated by the difficulty in sensing anything of the knowledgeabilities of woodlice (compared to, say, those of dogs), and the comparative ease in considering the significance of different pupils' responses to the woodlouse: one throwing it, one voicing a care for its life. The fieldnotes therefore primarily focused on how relations between pupils and more-than-human elements were altered via these encounters, but crucially, placed their emphasis predominantly on the pupils.

It is possible that in these early attempts to think with more-than-human elements, I am only at the beginning of what Taylor (2017, 1455) describes as a 'practice that requires a dedicated apprenticeship'. Even so, I am persuaded by Petersen's (2018) argument that that '(a)s long as researchers operate in the realm of the discursive and affective… we uphold a significant performative privilege, which cannot and should not be denied through rhetorical denouncements of privilege' (11). With 'performative privilege', Petersen is referring to the aforementioned ability to 'speak on behalf' of someone or something, and the 'inevitable' tendency for practices such as writing to lend a particular authority to the researcher's interpretation (12). She illustrates this with reference to another article of which Blaise is an author (Banerjee and Blaise 2013). In this article, the authors describe a study carried out in Hong Kong focused on air, in which 'we did not set out to research air, rather Hong Kong air found us' (241). Petersen (2018), however, refutes this attempt to denounce the 'privilege' of the researcher, suggesting that in fact, 'participants were on the lookout for some non-human actor to walk and think with, and therefore were under particular conditions of openness to being found' (10). In other words, despite these attempts at open-ness, the identification of and decision to include air as the subject of the study was still, ultimately, a human one. Likewise, through thinking and writing about it, the researchers 'inadvertently (came) to produce a-i-r, 'air', on their own terms' (12). This was my distinct sense with regard to any attempts to think with more-than-human elements during this study.

'Produc(ing) something other'

Whilst the fieldnote extract above makes clear some of the tensions discussed in the previous section, this new approach to taking fieldnotes was one that had clearly activated thoughts relating to new materialist theories, thereby producing new ways of thinking in the midst of my research encounters. The following two points illustrate this with reference to the fieldnote extract above.

Firstly, before the Polli:Nation project began, I had identified some of the key discourses underlying it. These included what I had labelled a 'utilitarian view of other species' and a somewhat anthropocentric 'stewardship' perspective towards the conservation of pollinators (Taylor 2017). Drawing upon the idea of concepts, however, and keeping in mind that these are 'invented in practice' (Semetsky 2015), encouraged me to pay attention to *new* ideas that were in circulation, rather than simply looking for manifestations of these underlying discourses. In the example above, whilst there *were* clearly moments that pertained to these pre-identified discourses, I also noted the woodlouse's 'capacity to provoke different affective responses in young people'. Such responses include the beginnings of what, following Taylor (2013, 2017), I labelled a 'commonworld perspective'. For Taylor (2017, 1456–1457), this refers to a perspective that goes beyond 'the rational quest to know about the world from a distance that characterises western

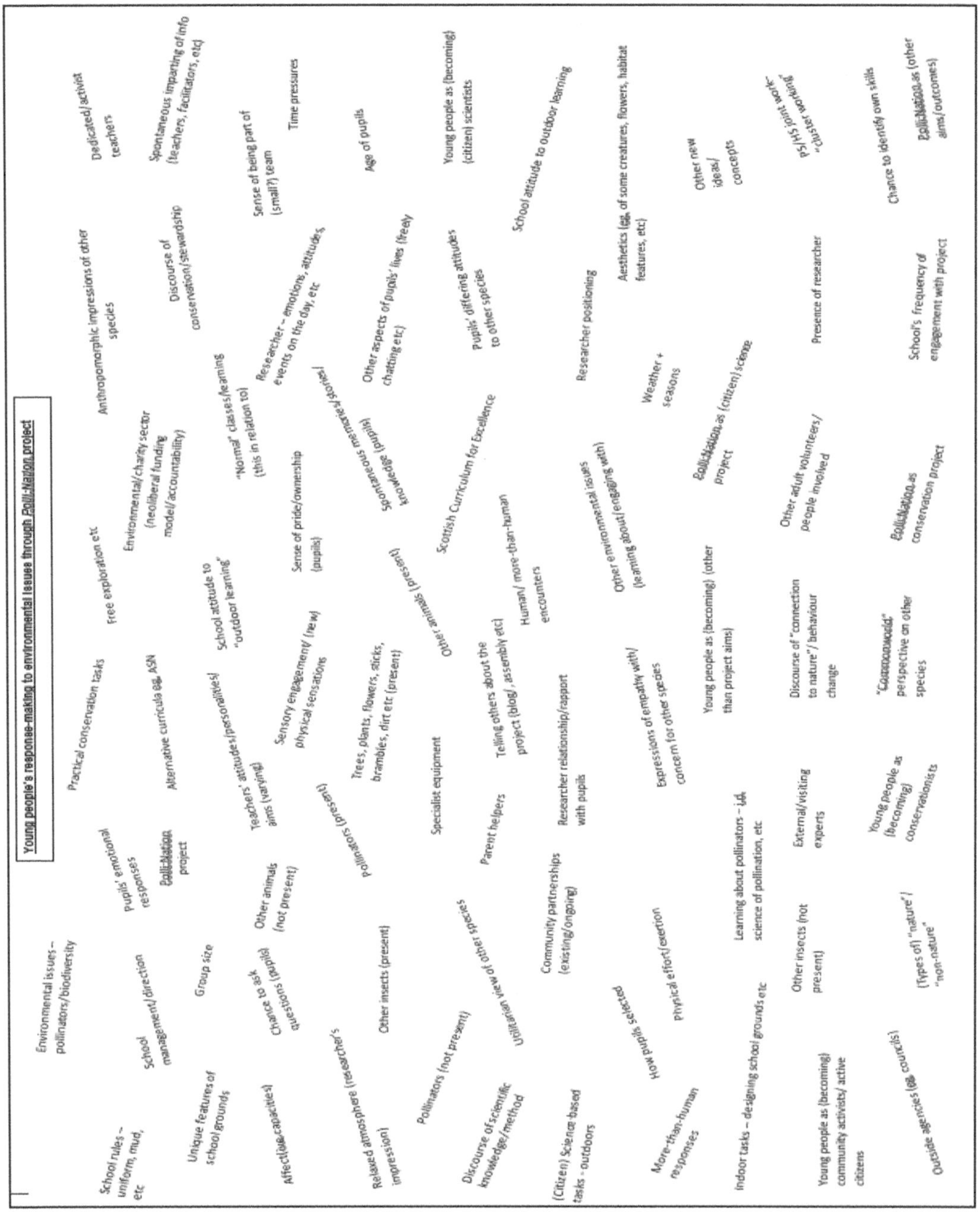

Figure 2. Situational map produced after all participant-observation sessions in the study outlined here. Relational maps are then created based on this map, by taking a particular element as a starting point, and tracing its relations to all other elements. The process is then repeated for each other element.

epistemologies', to instead consider humans as 'members of interconnected and interdependent multispecies common worlds' (Taylor and Pacini-Ketchabaw 2015, 511).

Secondly, despite the challenges with the suspension of meaning-making highlighted in the previous section, it is clear that these re-orientations ensured that the ongoing agencies and affective capacities of more-than-human elements were, to some degree, sustained through the practice of writing fieldnotes. My feeling is that this, in turn, served to reinforce my sense that other species 'co-shape' (Taylor and Pacini-Ketchabaw 2015, 512) the processes by which these ideas were formed or re-produced. At this stage, there are perhaps no obvious ways in which this fundamentally shifted my thinking – it is, as I say, a feeling. The influence of multi-species ethnography on the fieldnotes in Figure 1, however, is clear. As well as the woodlouse/slater, the extract above also contains references to the pungent smell of the stream, the rhododendron bushes and silver birches, the twigs and dry leaves on the woodland floor, the long spikey 'strings' of re-rooted brambles. Despite my sense of the impossibility of denouncing my performative privilege within the research assemblage outlined here, I am also persuaded by Bowden's (2015, 78) view that our 'human intentional action' ought still to be viewed as inextricably bound up with the 'nonhuman forces' making up a given assemblage. That is, while we cannot escape our human perspectives, acknowledging and exploring our co-constitution with other species and materials can nonetheless serve to produce new ways of knowing, and of doing research (Childers 2013).

As will be demonstrated in the following sections, the new thoughts activated by this new way of taking fieldnotes also fed into the process of creating situational maps – a method central to the wider approach of SA.

Situational analysis – introduction

SA is an emergent approach to qualitative data analysis that has grown out of grounded theory, and whose principal aims are to redress the 'positivist tendencies' of the grounded theory method in order to 'understand the dense complexities of a particular situation broadly conceived' (Clarke, Friese, and Washburn 2017, xxiv). Three books principally authored by Adele Clarke (Clarke 2005; Clarke, Friese, and Washburn 2015, 2017) provide practical and theoretical orientations for the use of SA. Practically, SA is characterised by the extensive use of maps throughout the research process. These are divided into three main types of map: Situational and relational maps, social worlds/arenas maps and positional maps. It should be noted that for the authors, all three types of map should be used during the research process in order to engage as fully as possible with the 'situation broadly conceived' (Clarke, Friese, and Washburn 2017, 361). In this article, however, I focus on situational and relational maps which, for me, stood out as being of greatest relevance to a post-qualitative methodology.

'Messy' situational maps are usually the first to be carried out in SA. These are usually created from the very start of the research process, and can therefore be based on early fieldnotes. Messy situational maps consist of simply laying out 'all the major human, non-human, discursive, historical, symbolic, cultural, political and other elements in the research situation of concern' (Clarke, Friese, and Washburn 2015, 100). The process of making these maps is simple, with the researcher plotting on the page any 'elements' of the situation that come to mind. At this early, 'messy' stage, there is no specified way to position these elements on the page. Relational maps are then created based on these existing situational maps, with the researcher drawing lines between elements on the map in order to explore the relations between these elements. The researcher takes one element of the map, and beginning with this element as a 'starting point', then traces its relationship with each other element in turn (Clarke, Friese, and Washburn 2017, 140). The process is then begun again with another element as the starting point, and repeated with each element. For Clarke, Friese, and Washburn (2017, 197), simultaneous creation of memos is also key in the process of creating situational and relational maps – that is, making notes detailing one's thoughts provoked through creation of the maps.

The stand-out link between new materialist theories and SA is the direct influence of Deleuze and Guattari's (1987) concepts of rhizomes and assemblages on situational and relational maps in particular. Clarke, Friese, and Washburn (2017, 96) see the maps as 'provid(ing) a direct means of operationalising these concepts', and as fitting with Deleuze and Guattari's (1987, 21) description of a rhizome as 'pertain(ing) to a map that must be produced, constructed, a map that is always detachable, connectable, reversible, modifiable, and has multiple entryways and exits and its own lines of flight'. In the horizontal, non-hierarchical positioning of elements on situational maps, and tracing of the relations between elements through relational maps, these can together be seen as a pictorial representation of a rhizome (Clarke, Friese, and Washburn 2017, 92). Adding to this, Clarke Friese, and Washburn (2017, 95) explain that they see both SA as a whole, and 'the situation broadly conceived' that it seeks to explore, as assemblages. Relational maps, they argue, address the key question 'how is (the assemblage) working?' (95). In summary, the authors state that since 'situational and relational maps were inspired in part by Deleuze and Guattari's (1987) rhizomic assemblages, we see SA as fairly compatible with those approaches' (366).

Broader links can also be made between new materialist theories and the theoretical orientations of SA as a whole. As well as the aim of engaging with the complexities of a given 'situation', Clarke, Friese, and Washburn (2017) point to the importance of 'taking the non-human explicitly into account' (xxv), researcher reflexivity and the importance of 'making sure we include ourselves, as researchers, in the maps we are producing' (35), and the rejection of a 'tripartite framework' that upholds the perceived separation between 'micro', 'meso' and 'macro' elements in social research (62). Additionally, Clarke (2005, 29) argues that the simple process of map-making itself offers the potential to analyse qualitative data in a manner more sensitive to new materialist theories, simply through presenting data in a new and re-assembled manner, thus 'provok(ing) us to see things afresh'.

'Elements' and 'producing order'

In the study detailed here, I began creating situational maps after my first session of participant-observation, and despite the synergies outlined above, a key tension in relation to new materialist theories quickly emerged. It was clear already that seeking to identify the distinct elements present within a research encounter carried remnants of the desire, rooted in Cartesian dualist thought, to 'produce order and regularity in the guise of categories that erase difference and privilege identity among seemingly similar things' (Jackson 2013, 742). For Jackson, it is from this deep-rooted, essentialist thinking that the practice of coding, so commonplace in qualitative research, stems. The focus on *patterns* inherent in the practice of coding, say St. Pierre and Jackson (2014, 716), is 'unthinkable' in Deleuzian ontologies that 'describe the world as unstable and becoming'. MacLure (2013b), however, argues *for* coding in spite of this ontological mismatch, since it encourages a long, slow 'immersion in, and entanglement with, the minutiae of "the data"' (170). She encourages researchers to think of coding as 'an experiment with order and disorder' that is 'not a static representation or translation of a world laid out before us… but an open-ended and ongoing practice of *making* sense' (171).

Was SA, however, representative of the sort of coding that encourages such static representations? Or was it sufficiently open-ended and experimental? Certainly, Clarke (2005), Clarke, Friese, and Washburn (2015, 2017) do not explicitly address the difficulty in using the reductive term 'elements', and use it in a fairly commonsense manner throughout their books. As Mathar (2008, 13) points out in a review of Clarke's first SA book (2005), the method 'does not ask… how these different elements are being produced and how they condense themselves into elements'. Clearly, the categorisation required to create situational maps is bound up with the researcher's own pre-conceived impressions and prior experiences. In the example map in Figure 2, this is perhaps most clearly illustrated by the element 'relaxed atmosphere'. By this, I reflected through memos, I meant the 'atmosphere' produced by some of the other elements I had included: a small group, the chance to ask spontaneous questions, the chance for spontaneous close-up encounters with other species and the opportunities for what I labelled 'free play/exploration' in-between planned activities. Labelling this as 'relaxed' was, of course, my own judgement, deep-rooted in my own experiences of school, and professional experience in outdoor learning contexts.

Further indicative of the potentially essentialising nature of SA is Clarke, Friese, and Washburn's (2017, 182) provision of orientations for the creation of 'ordered' situational maps, based on the researcher's existing 'messy' maps. Here, the researcher begins to categorise the elements they have identified into broader categories – for example, 'more-than-human elements', 'discourses' or 'key people'. Clarke (2005, 89) notes that the creation of these ordered maps 'isn't necessary', perhaps indicating her awareness of the potential tensions (Mathar 2008).

Researcher reflexivity

While Clarke, Friese, and Washburn (2017) do not directly address the issue of identifying elements, the key to their thinking is in their insistence on researcher reflexivity. For the authors, an acceptance of the importance of 'reflexively analysing what "we" do as well as what "they" do' (56) is a key way in which SA differs from Grounded Theory, which 'has long been burdened with the assumption that researchers should be tabula rasa' (35). Practically, this reflexivity involves making sure we include ourselves, as researchers, in the maps we are producing, thereby acknowledging (in accompanying memos, if not obviously in the maps themselves) how each element came to be defined as such. Following this, we might conceive of each element as an assemblage in itself – a unique entanglement of concepts, percepts and affects (Colebrook 2002) arising both through the research encounter, and through the researcher's pre-existing 'ideas, hunches, and emergent concepts' (Hammersley and Atkinson 1983, 174).

This leads us back to the debate previously discussed in relation to multi-species ethnography. For some, to create situational maps will be to jump to humanist interpretation too early in the research process, when we should instead be attempting to suspend our pull towards meaning making. For proponents of SA, however, researcher reflexivity is so important because suspension of the researcher's own thoughts, and humanist meaning-making, is *never* possible. The key, for them, is in acknowledging this, and instead using situational maps to explore and exemplify when and how our thoughts and privileges are inextricably bound up with other human and non-human elements comprising the research situation. Clarke makes her position on this clear in the FAQ on her own website (Situational Analysis, 2018). In response to the suggestion that SA 'still retains the voice of the analyst/researcher/author', she responds, 'In my reading of the postmodern, there is no turning that can erase the situatedness of research and researchers. All we can offer is our best and most reflexive take to date in terms of understanding something'. Similarly, with regard to coding, MacLure (2013b) contends that since '(a)ll language shares the fixative ambitions of coding', it is never entirely possible to abandon the human tendency towards classification.

Importantly, the ways in which I conceived of the elements making up Figure 2 was also strongly influenced by the new approach to fieldnote-writing outlined previously, in which I responded to questions drawn from new materialist theories and multi-species ethnography. This can be seen in the labelling of concepts that I had identified as being 'invented in practice' (Semetsky 2015), such as 'young people as (becoming) conservationists', and 'young people as (becoming) community activists/active citizens'. I also acknowledged the ever-present 'affect(ive capacities)' within the map, as well as numerous elements emphasising relations between humans and more-than-human elements ('human/more-than-human encounters', 'more-than-human responses', pollinators (not present)'). When added to the orientations of SA outlined previously, these subtle shifts suggest that if SA *does* encourage a greater degree of categorisation than is desirable when thinking with new materialist theories, this unique research assemblage has at least enabled this researcher to do so in ways that enabled greater consideration of ideas drawn from new materialist theories.

Returning to Clarke, Friese, and Washburn's (2017) drawing of links between SA and new materialist theories, the creation of relational maps – using the situational maps created after participant-observation sessions – was also key to conceiving of the map-making process in a manner more sensitive to new materialist thinking. The process of drawing links between elements (however human-defined they were), I felt, furthered my sense of the complex relations between these. The limitation of these maps, however, is that whilst they encourage one to visualise the complexities and entanglements within the research encounter, the maps themselves do not directly encourage one to consider what is *produced* by the assemblages they represent. Clarke, Friese, and Washburn (2017, 69), with reference to Dewey (1938), do point to the idea of 'the situation as itself having a *gestalt* that makes the whole greater than the sum of its parts', but aside from ongoing memoing, do not offer a prescribed way for the researcher to consider this. Again, however, I felt that this was enabled by the theoretically-sensitive fieldnotes outlined in the previous sections, which helped to further provoke new ways of thinking, and encourage me to consider what was produced.

Figure 2 overleaf shows the situational map I had produced based on all participant-observation sessions during the Polli:Nation study.

Later use of situational and relational maps

Whilst situational and relational maps proved useful as a means of analysing fieldnotes shortly after participant-observation sessions, it is also worth briefly outlining what was enabled by these early orientations in the later stages of the study in question. Firstly, the use of situational maps to analyse fieldnotes enabled the identification of key features and activities making up the Polli:Nation project, which then helped to inform the focus groups I carried out in twenty participating schools. The focus groups included an activity where young people were asked to choose from a series of flashcards based on characteristics of the project identified through my iterative and on-going analysis of what seemed salient within my fieldnotes. These included, for example, 'working with experts from outside of school', 'doing practical conservation tasks' and 'doing science outdoors'. In turn, this activity facilitated insightful discussions about these elements that would almost certainly not have occurred had I simply asked 'what do you find significant/memorable about the Polli:Nation project?' Later, on completion of my focus groups, I updated my existing situational map to include new elements that had emerged during the focus groups – a further acknowledgement of the ideas produced by the assemblage the map represented.

Additionally, following on from the flashcards activity described above, I created a series of relational maps starting from the key processes and characteristics that young people had been most keen to talk about. Creating these maps helped me to continue visualising those elements as part of a unique assemblage, rather than a singular process occurring in isolation.

Discussion

This article has detailed the approach taken to the writing and analysis of fieldnotes during a recent doctoral study in an environmental education context, drawing upon multi-species ethnography, new materialist theories, and SA. Rather than being a prescriptive guide to carrying out research in a new materialist or post-qualitative manner, it is instead an account of my lived experience of attempting to use these existing qualitative methods in a manner more sensitive to new materialist theories. Its chief contribution is to highlight the ways in which these methods came to be re-oriented through attending to ideas from the new materialist and post-qualitative literatures, as well as via the onward effects these had on what then amounted to my data, and on myself as researcher. The re-orientations enabled by the unique 'research assemblage' (Fox and Alldred 2018) included the following: Firstly, a new approach to taking fieldnotes that, through drawing on ideas from new materialist theories, activated new ways of thinking that went beyond SA to consider what was produced by the assemblages under examination. Secondly, a greater focus on more-than-human elements through multi-species ethnography, including a greater sense of the ways in which these co-shape young people's learning experiences. Thirdly, a way of thinking in a more entangled, relational manner through the creation of situational and relational maps. And, finally, changes to the ways in which I conceived of the elements making up situational maps, owing to the ideas from new materialist theories with which I directed my fieldnote-writing.

It is important here to highlight the transitory, unstable nature of the research assemblage created through this study, and that subsequent research assemblages will most likely bring about new understandings and further re-orientations. This may be through re-reading, and/or a changed or further developed understanding of ideas such as concepts, percepts and affects. Such an altered understanding may, in turn, influence the identification of elements making up future situational maps. This lack of permanence is in fact central to new materialist thinking, with Strom (2018, 112) pointing to the need to 'resist overcoding Deleuzian thought and subjecting it to petrification'. She points to Deleuze and Guattari's own thinking on this, which highlights the danger, having used their ideas in a given research assemblage, 'that you will reencounter organizations that restratify everything, formations that restore power to a signifier, attributions that reconstitute a subject' (1987, 9).

For some, the ongoing tensions highlighted throughout this article will be enough to render the approach discussed here 'unthinkable' within new materialist ontologies (St. Pierre and Jackson 2014, 716). Indeed, by St. Pierre's (2018) definition, the presence of an existing research design already disqualifies this study from being post-qualitative. Following Strom (2018), however, I am inclined to question this incompatibility. Strom (2018, 109) argues that whilst the 'strategic use of selective concepts' from new materialist theory within qualitative research might be criticised for 'failing to holistically employ rhizomatic language or engage in discourse that completely breaks with Western academic writing and research norms', such an approach may nonetheless result in 'micro transformations' that serve to bring these concepts towards 'mainstream' educational research. This, she suggests, might also serve to 'translate' Deleuzian concepts in a manner more accessible to wider audiences, thus working against the 'exclusionary' nature of 'high theory'. Clearly, not all will be persuaded by this argument. Nonetheless, I hope that at the very least, this article has provided a useful insight into the on-the-ground methodological struggles that come when attempting to carry out research in a manner sensitive to new materialist/post-qualitative thinking, whilst operating within the at-times restrictive norms of academic research. In this case, these amounted to the time-bound context of a collaborative, funded PhD study where there was an emphasis on 'demonstrable impact' (Hammersley 2018).

The approach set out in this article represents a series of small, initial steps towards exploring the struggles that take place when conducting research within a new materialist framing. It provides a potential starting point from which there is ample opportunity for further exploration. I suggest three further directions for this. Firstly, there is potential for further experimentation with the orientations outlined here. These may include, for example, the alternatives suggested in the section 'Language practices' and 'performative privilege', which acknowledged the possibility of, for example, not taking fieldnotes at all. Secondly, since this article has focused on situational and relational maps, there is potential for exploration of other aspects of SA, and their compatibility with new materialist theories. This might include the use of two further types of map identified by Clarke (2005), usually completed later in the analysis process – social worlds/arenas maps, and positional maps. Finally, as the focus of this article has been on fieldnotes, there is potential for further exploration of the tensions encountered with regard to the use of other qualitative methods – such as interviews and focus groups – in a manner sensitive to new materialist theories.

Disclosure statement

No potential conflict of interest was reported by the authors.

Notes

1. The first author, who carried out all research described here. The second author was responsible for setting up this doctoral study, and played a key role in the development of many of the ideas explored in this article.
2. Quotation marks are used here either to indicate the inherent tensions in words such as 'data', 'elements' and 'research design', as well as when terms are attributable to a particular author. We place problematic words in quotation marks at their first appearance, then use them without quotation marks thereafter.

ORCID

Andy Ruck http://orcid.org/0000-0003-1892-8205
Greg Mannion http://orcid.org/0000-0003-2233-9278

References

Banerjee, B., and M. Blaise. 2013. "There's Something in the Air: Becoming-with Research Practices." *Cultural Studies, Critical Methodologies* 13 (4):240–245. doi:10.1177/1532708613487867.

Barad, K. 2007. *Meeting the Universe Half Way: Quantum Physics and the Entanglement of Matter and Meaning*. Durham: Duke University Press.

Bennett, J. 2010. *Vibrant Matter: A Political Ecology of Things*. Duke University Press.

Bowden, S. 2015. "Human and Nonhuman Agency in Deleuze". In *Deleuze and the Non/Human*, edited by Stark, H. and J. Roffe. Hampshire: Palgrave Macmillan UK.

Bridges-Rhoads, S. 2015. "Writing Paralysis in (Post) Qualitative Research." *Qualitative Inquiry* 21 (8):704–710. doi: 10.1177/1077800414566690.

Bridges-Rhoads, S. 2017. "Philosophical Fieldnotes." *Qualitative Inquiry*. doi:10.1177/1077800417733498.

Brinkmann, S. 2014. "Doing Without Data." *Qualitative Inquiry* 20 (6):720–725. doi:10.1177/1077800414530254.

Brewer, J. D. 2000. *Ethnography*. Philadelphia, PA: Open University Press.

Childers, S. M. 2013. "The Materiality of Fieldwork: An Ontology of Feminist Becoming." *International Journal of Qualitative Studies in Education* 26 (5):599–609. doi:10.1080/09518398.2013.786845.

Clarke, A. E. 2005. *Situational Analysis: Grounded Theory after the Postmodern Turn*. Sage.

Clarke, A. E., C. Friese, and R. Washburn. 2015. *Situational Analysis in Practice: Mapping Research with Grounded Theory*. Left Coast Press.

Clarke, A. E., C. Friese, and R. Washburn. 2017. *Situational Analysis: grounded Theory after the Interpretive Turn*. Sage Publications.

Clarke, D. A. G., and J. Mcphie. 2014. "Becoming Animate in Education: immanent Materiality and Outdoor Learning for Sustainability." *Journal of Adventure Education & Outdoor Learning* 14 (3):198–216. doi:10.1080/14729679.2014.919866.

Clarke, D. A. G., and J. Mcphie. 2016. "From Places to Paths: Learning for Sustainability, Teacher Education and a Philosophy of Becoming." *Environmental Education Research* 22 (7):1002–1024. doi:10.1080/13504622.2015.1057554.

Colebrook, C. 2002. *Understanding Deleuze*. Allen and Unwin.

Coole, D., and S. Frost. 2010. "Introducing the new materialisms". In *New Materialisms: Ontology, Agency and Politics*, edited by Coole, D. and S. Frost. Duke University Press.

Deleuze, G., and F. Guattari. 1987. *A Thousand Plateaus: Capitalism and Schizophrenia*. Minneapolis: University of Minnesota Press.

Deleuze, G. 1990. *The Logic of Sense*. New York, NY: Columbia University Press.

Desmond, M. 2014. "Relational Ethnography." *Theory and Society* 43 (5):547–579. doi:10.1007/s11186-014-9232-5.

Dewey, J. 1938. *Logic: The Theory of Inquiry*. New York: H. Holt.

Emerson, R. M., R. I. Fretz, and L. L. Shaw. 1995. *Writing Ethnographic Fieldnotes*. University of Chicago Press.

Fenwick, T., S. Doyle, M. K. Michael, and J. Scoles. 2015. "Matters of Learning and Education: Sociomaterial Approaches in Ethnographic Research". In *MultiPluriTrans in Educational Ethnography: Approaching the Multimodality, Plurality and Translocality of Educational Realities*, edited by Bollig, S., M-S. Honig, S. Neumann and C. Seele. Bielefeld, Germany: Transcript Verlag/Columbia University Press.

Fox, N. J., and P. Alldred. 2015. "New Materialist Social Inquiry: Designs, Methods and the Research-Assemblage." *International Journal of Social Research Methodology* 18 (4):399–414. doi:10.1080/13645579.2014.921458.

Fox, N. J., and P. Alldred. 2018. "Mixed Methods, Materialism and the Micropolitics of the Research-Assemblage." *International Journal of Social Research Methodology* 21 (2):191–204. doi:10.1080/13645579.2017.1350015.

Goulson, D. 2014. *A Sting in the Tale: My Adventures with Bumblebees*. London: Jonathan Cape.

Green, M., and I. Duhn. 2015. "The Force of Gardening: Investigating Children's Learning in a Food Garden." *Australian Journal of Environmental Education* 31 (01):60–73. doi:10.1017/aee.2014.45.

Hammersley, M. 2018. "What Is Ethnography? Can It Survive? Should It?" *Ethnography and Education* 13 (1):1–17. doi:10.1080/17457823.2017.1298458.

Hammersley, M., and P. Atkinson. 1983. *Ethnology: Principles in Practice*. New York: Routledge.

Haraway, D. J. 2008. *When Species Meet*. Minneapolis: University of Minnesota Press.

Higgins, P., R. Nicol, and H. Ross. 2006. "Teachers' Approaches and Attitudes to Engaging with the Natural Heritage Through the Curriculum." PDF file. http://www.docs.hss.ed.ac.uk/education/outdoored/teachers_approaches_snh.pdf

Hultman, K., and H. Lenz Taguchi. 2010. "Challenging Anthropocentric Analysis of Visual Data: A Relational Materialist Methodological Approach to Educational Research." *International Journal of Qualitative Studies in Education* 23 (5):525–542. doi:10.1080/09518398.2010.500628.

Jackson, A. Y. 2013. "Posthumanist Ethnography of Mangling Practices." *International Journal of Qualitative Studies in Education* 26 (6):741–748. doi:10.1080/09518398.2013.788762.

Lather, P. 1993. "Fertile Obsession: Validity after Poststructuralism." *The Sociological Quarterly* 34 (4):673–693. doi:10.1111/j.1533-8525.1993.tb00112.x.

Lather, P. 2000. "Against Empathy, Voice and Authenticity." *Women, Gender and Research* 4:16–25.

Lather, P. 2016. "Top Ten + List: (Re) Thinking Ontology in (Post) Qualitative Research." *Cultural Studies, Critical Methodologies* 16 (2):125–131. doi:10.1177/1532708616634734.

Learning through Landscapes (LtL). 2014. *Polli:Nation Activity Plan 2015–18. Winchester, UK.*

Lenz Taguchi, H. 2012. "A Diffractive and Deleuzian Approach to Analysing Interview Data." *Feminist Theory* 13 (3):265–281. doi:10.1177/1464700112456001.

Mack, N. C., C. Woodsong, K. M. MacQueen, G. Guest, and F. Namey. 2005. *Qualitative Research Methods: A Data Collector's Field Guide*. Family Health International.

MacLure, M. 2013a. "Researching without Representation? Language and Materiality in Post-Qualitative Methodology." *International Journal of Qualitative Studies in Education* 26 (6):658–667. doi:10.1080/09518398.2013.788755.

MacLure, M. 2013b. "Classification or wonder? Coding as an analytic practice in qualitative research". In *Deleuze and Research Methodologies*, edited by Coleman, R. and J. Ringrose. Edinburgh University Press.

Mannion, G., A. Fenwick, and J. Lynch. 2013. "Place-Responsive Pedagogy: Learning from Teachers' Experiences of Excursions in Nature." *Environmental Education Research* 19 (6):792–809. doi:10.1080/13504622.2012.749980.

Mannion, G. 2019. "Re-Assembling Environmental and Sustainability Education: orientations from New Materialism." *Environmental Education Research* :1. doi:10.1080/13504622.2018.1536926.

Mathar, T. 2008. "Review Essay: Making a Mess with Situational Analysis?" *Forum Qualitative Sozialforschung/Forum: Qualitative Social Research* 9 (2). doi:10.17169/fqs-9.2.432.

Mazzei, L. A. 2014. "Beyond an Easy Sense: A Diffractive Analysis." *Qualitative Inquiry* 20 (6):742–746. doi:10.1177/1077800414530257.

Ogden, L. A., B. Hall, and K. Tanita. 2013. "Animals, Plants, People, and Things: A Review of Multispecies Ethnography." *Environment and Society* 4 (1):5–24. doi:10.3167/ares.2013.040102.

Pacini-Ketchabaw, V., A. Taylor, and M. Blaise. 2016. "Decentring the human in multispecies ethnographies". In *Posthuman Research Practices in Education*, edited by Taylor, C. and C. Hughes. Palgrave Macmillan.

Petersen, E. B. 2018. "'Data Found Us': A Critique of Some New Materialist Tropes in Educational Research." *Research in Education* 101 (1):5–16. doi:10.1177/0034523718792161.

Pink, S. 2009. *Doing Sensory Ethnography*. London: Sage Publications.

Richardson, L., and E. A. St. Pierre. 2005. "Writing: A method of inquiry". In *SAGE Handbook of Qualitative Research*, edited by Denzin, N.K and Y. S. Lincoln (3rd ed). Thousand Oaks, CA: Sage.

Rickinson, M., C. Lundholm, and N. Hopwood. 2009. *Environmental Learning: Insights from Research into the Student Experience*. Springer.

Semetsky, I. 2015. "Deleuze's philosophy for education". In *Encyclopedia of Educational Philosophy and Theory*, edited by Peters, M. Singapore: Springer.

Situational Analysis. 2018. "Frequently Asked Questions". Accessed October 15 2018. http://clarkessituationalanalysis.blogspot.com/

Springgay, S., and S. E. Truman. 2018. "On the Need for Methods beyond Proceduralism: Speculative Middles, (in) Tensions, and Response-Ability in Research." *Qualitative Inquiry* 24 (3):203–214. doi:10.1177/1077800417704464.

St. Pierre, E. A. 2011. "Post qualitative research: The critique and the coming after". In *Sage Handbook of Qualitative Inquiry*, edited by Denzin, N. K. and Y. S. Lincoln. 4th ed. Los Angeles, CA: Sage.

St. Pierre, E. A. 2014. "A Brief and Personal History of Post Qualitative Research: Toward 'Post Inquiry." *Journal of Curriculum Theorizing* 30 (2):2–19.

St. Pierre, E. A., and A. Y. Jackson. 2014. "Qualitative Data Analysis after Coding." *Qualitative Inquiry* 20 (6):715–719. doi:10.1177/1077800414532435.

St. Pierre, E. A. 2017. "Writing Post Qualitative Inquiry." *Qualitative Inquiry*. doi:10.1177/1077800417734567.

St. Pierre, E. A. 2018. "Post Qualitative Inquiry in an Ontology of Immanence." *Qualitative Inquiry*. doi:10.1177/1077800418772634.

Strom, K. J. 2018. "'That's Not Very Deleuzian': Thoughts on Interrupting the Exclusionary Nature of 'High Theory'." *Educational Philosophy and Theory* 50 (1):104–113. doi:10.1080/00131857.2017.1339340.

Taylor, A., and V. Pacini-Ketchabaw. 2015. "Learning with Children, Ants, and Worms in the Anthropocene: Towards a Common World Pedagogy of Multispecies Vulnerability." *Pedagogy, Culture & Society* 23 (4):507–529. doi:10.1080/14681366.2015.1039050.

Taylor, C. A. 2016. "Edu-crafting a Cacophonous Ecology: Posthumanist Research Practices for Education". In *Posthuman Research Practices in Education*, edited by Taylor, C.A and Hughes, C. Palgrave Macmillan, London.

Taylor, A. 2017. "Beyond Stewardship: Common World Pedagogies for the Anthropocene." *Environmental Education Research* 23 (10):1448–1461. doi:10.1080/13504622.2017.1325452.

Tsing, A. 2013. "More-than-Human Sociality". In *Anthropology and Nature*, edited by Hastrup, K. New York: Routledge.

Vannini, P. 2015. "Non-Representational Ethnography: New Ways of Animating Lifeworlds." *Cultural Geographies* 22 (2):317–327. doi:10.1177/1474474014555657.

Doing little justices: speculative propositions for an immanent environmental ethics

David Rousell

ABSTRACT

This article develops a series of speculative propositions for an immanent environmental ethics that is responsive to the challenges of the Anthropocene epoch. The article is framed within a new materialist approach to environmental education, and specifically works to re-imagine the notion of justice in terms of performative gestures, multiplicities, processes, and speculative thought experiments. Drawing on Whitehead's speculative philosophy in conjunction with recent new materialist thought, the article proposes the concept of 'doing little justices' as a way of enacting micropolitical interventions into everyday patterns of environmental thought, learning, sociality, and behaviour. The concept of 'little justices' is further elaborated through the analysis of vignettes that problematise issues of climate change, human exceptionalism, ecological sovereignty, and environmental justice with university students in the fields of education and the philosophy of law. The article concludes that an immanent ethics cannot be reduced to a set of predetermined values or prescriptions for environmental education, but should proceed through a speculative process of creative experimentation and negotiation in the pursuit of unforeseen openings and potentials for co-existence.

Introduction

Recent movements in posthumanist scholarship have called for reconceptualisations of environmental education in response to the onset of the Anthropocene epoch, a time typified by anthropogenic climate change, catastrophic loss of biodiversity, ubiquitous digitisation and rapid advances in biotechnology and computer science (Lloro-Bidart 2015; Malone 2016; Payne 2016; Rotas 2015; Rousell 2016). In revealing the ontological inseparability of human culture and natural phenomena in the contemporary world, the advent of the Anthropocene has had a destabilising effect on dualistic philosophies and binary logics that have upheld ontological boundaries between the human and the non-human, the organic and the inorganic, the natural and the artificial, the geologic and the social (Braidotti 2013; Colebrook 2014; Ellsworth and Kruse 2012; Haraway 2016; Yusoff 2015). In many ways the onset of the Anthropocene has acted like a 'storm on thought' in environmental education, calling into question many historical 'presumptions, logics and methods of reason' and forcing both fields to re-evaluate 'what really matters' (Payne 2016, 169). Innovative concepts and practices of environmental education are called for that can mobilise potentials for ethical thinking and action that are both attuned and responsive to the rapidly changing material conditions of everyday life (Rousell, Cutter-Mackenzie, and Foster 2017). Such geophilosophical problematics call for a renewed consideration of justice in environmental education which admits, on the one hand, that 'humans have secreted themselves into every corner of the environment', and on the other hand, 'that the environment is actually inside human bodies and minds' (Bennett 2010, 116).

In addressing the need for a renewed conception of justice in an age of climate change, this article develops a series of speculative propositions for an immanent environmental ethics that is responsive to the challenges of the Anthropocene. Drawing together a series of emerging conceptions of justice in new materialist thought, the article proposes the concept of 'doing little justices' as a way of enacting micropolitical interventions into everyday patterns of thought and behaviour. 'Doing little justices' is a concept of immanence to the extent that it proposes an ethics that is located *inside* material encounters, events, socialities, and actual occasions of experience that exceed human knowledge and understanding (Whitehead 1967). To do a 'little justice' is to perform a collective experiment within places and times of 'mutual immanence' (197). Akin to what Manning (2016) refers to as a minor gesture, a little justice can be as small as a movement, a word, an image, or an idea that brings care and attention to the fragilities, entanglements, and uncertainties of life in the Anthropocene. A little justice is therefore what Deleuze and Guattari (1986) call a minoritarian proposition, as it plays out in the minor keys that lurk in the interstices between the majoritarian political structures and discursive social agendas of 'Justice' writ large. Where majoritarian or State regimes present a 'constant and homogenous system', a minoritarian politics operates through 'subsystems ... seeds [and] crystals of becoming whose value is to trigger uncontrollable movements and deterritorialisations of the mean or majority' (106).

In developing the concept of 'doing little justices' through this article, I draw on the *States and Territories* project (2013–2016) which has collectively reimagined the learning environments of a regional Australian university in response to the Anthropocene epoch (Rousell 2017). The project involved the creation of a network of twelve environmental artworks across the university campus, each of which is connected to a specific concept and an associated digital archive on the project website (see www.statesandterritories.org). The project was framed through a process-orientated theoretical approach and methodology, with a specialised focus on the non-anthropocentric philosophies of A.N. Whitehead (1978, 1967) and Deleuze and Guattari (1987, 1994). In this article, I focus on how the concept of justice was re-imagined through a specific 'data event' from this project, which problemetised issues of climate change, human exceptionalism, ecological sovereignty, and environmental justice with university students in the fields of education and the philosophy of law. In analysing the performative unfolding of this data event, I work to assemble an alternative conceptualisation of justice that is steeped in new materialist notions of the performative, the speculative, the collective, the processual, and the minor. In order to establish a context for my theorisation of justice within this project, the following sections provide a brief account of new materialist approaches to ethical thought and inquiry in the context of environmental education, as well as three variations on the concept of 'immanence' as developed in Whitehead's (1967) later philosophical work.

An ethics of the impossible

> The ethical value of an action is what it brings out in a situation, for its transformation, how it breaks sociality open. Ethics is about how we inhabit uncertainty, together. (Massumi 2015, 11)

The mounting social and ecological crises of the Anthropocene challenge environmental educators to assemble an entirely new imperative for ethical experimentation and inquiry in response to the mounting uncertainties of contemporary life (Zylinska 2014). Bennett (2010, 116) argues that the onset of the Anthropocene has eroded any possibility for the resurrection of a 'pure nature unpolluted by humanity', while at the same time dissolving any meaningful endeavor to 'define the self as something purely human'. As Massumi suggests in the quote above, this situation calls for a reconceptualisation of ethics in terms of processual activity that responds to changing conditions, situations, and qualities of life, rather than an ethics based on transcendent moral principles or universal ideals. This sense of radical openness to the immanent creativity and uncertainty of ethical relations is shared across many posthumanist philosophies of education, which are resistant to predetermined ethical frameworks or principles that delimit the possibilities for learning and inquiry (Snaza and Weaver 2015). Yet this has also been part of the core work of certain environmental educators who have been active for many years (see, for instance, Jickling 2013; Stables 2007; Weston 1992). Harvester and Blenkinsop (2011, 123) also highlight the extensive genealogy of ecofeminist thought that has contested the patriarchal 'logic of domination' over women, animals and environments since the 1970s, while actively promoting 'non-dualistic and non-hierarchical systems of relations' as alternatives to humanist metaphors, concepts and understandings of nature and the environment. This rich history of feminist critique has emphasised that concepts actually *matter* in environmental education, and that replacing or reconfiguring dualistic concepts of nature are crucial for the advancement of social and environmental justice (see also Braidotti 2013, 2017).

The imperative to reconceptualise environmental education has become an increasingly urgent and precarious undertaking as the Anthropocene epoch comes into focus (Malone 2016), as we find ourselves already implicated within planetary scale environmental catastrophes that exceed human capacities to understand, conceptualise, manage, or foresee (Morton 2016). In responding to ecological catastrophe on a planetary scale, Jickling (2013, 167) has suggested that environmental education now needs to enable learners 'to grapple with issues critically and creatively and find ways to do things that currently appear impossible'. As a practice of grappling with the impossible, environmental education comes to involve the creation of transdisciplinary spaces for experimenting with environmental ethics and values, rather than prescribing and predetermining certain values as environmentally ethical (168). Jickling's comments bring the function and operation of environmental education into close proximity with new materialist thought, as an onto-epistemological orientation that emphasises the situational, relational, and speculatively pragmatic nature of ethical thinking and action. Such arguments shift the role and purview of environmental education towards the creation of spaces for experimenting with a speculative 'ethics of the impossible' (168). As Bryant (2011, 27) further argues:

> The question of the ethical is not the question of how crisis can be ameliorated by recourse to pre-existing principles for the simple reason that the ethical is encountered at just that moment where "principles" governing a composition no longer hold. Rather, the question of the ethical is that of how situations must be re-composed in response to this moment of crisis.

In other words, we encounter the ethical through the specificity of *an event*, as a moment of encounter that forces us to think and act differently in response to the unknown, the unthinkable, and the impossible.

Old and new materialisms

It is within these moments of encounter with the impossible that new concepts become necessary to think and make the world differently, and thus to produce new kinds of events as situations of mutual encounter and co-existence. What is critically at stake in new materialist revisions of environmental education is the contested ontological status of 'environment' and 'education' as *concepts* that are inseparable from the material-semiotic conditions under which they become thinkable. When the concepts of 'environment' and 'education' are harnessed to anthropocentric, dualistic and humanist systems of thought then the practices they engender are similarly constrained; often to the point where these practices promote injustice through the unquestioned proliferation of dominant logics, inhibiting value-systems and patriarchal norms (Morton 2016; Snaza and Weaver 2015).

Alternatively, new materialist approaches work to push concepts of 'environment' and 'education' to the limits of what they might include and become, with the visionary aim of opening up spaces for new modes of ethical thinking and practice to emerge. Weston (2015), for instance, has been working 'the dark edges' of environmental education by expanding the concept of environment to include the Cosmic darkness of the universe beyond the boundaries of the Earth, and the Chthonic darkness that lies in the qualitative, inner powers of the Earth's landscapes and living creatures (71). Weston's dark ecological approach radically alters the scale of environmental education's ethical purview by inviting the unknowable and unthinkable elements of existence to impact on environmental ethics and pedagogical practice. In working the dark edges of environmental education and thought, Weston's (2015) approach also fosters a deep acknowledgement and consideration of Indigenous cosmologies and practices associated with animism, vitalism, totemism and magic. This exemplifies the pressing relevance of Indigenous conceptions of the environment as an emergent and actively constituted field of entangled relations, rather than a passive container or backdrop for human occupation and use (Cajete 2006; Todd 2015). Geo-ontological analyses of Indigenous art, philosophy, and culture by feminist scholars such as Grosz (2008) and Povinelli (2016) also highlight an emerging sense of compatibility between the 'traditional' materialisms of Indigenous peoples and the 'new' materialisms proposed by today's speculative theorists, artists, and scientific practitioners.

This convergence of 'old and new' materialisms recognises the primacy of embodied experience as a constitutive process, a movement that also establishes many resonances between Indigenous knowledge systems and recent development in the life sciences (Rousell and Williams 2018). For instance, postgenomic research in epigenetics indicates that social and physical environments affect the biological constitution of living bodies not only over the course of a single lifetime, but across generations and even across species (Frost 2016; Meloni 2016). Recent findings in embodied cognitive science further suggest that environmental and social experiences are intricately connected with biological processes, such as sensory-motor activity, directional motility, biochemical gradients, and pre-cognitive affective responses (Protevi 2013).

These dynamic notions of the 'environment' as an interpenetrating series of geo-bio-cultural milieus have significant implications for how we understand the experiential, ethical, political and pedagogical dimensions of environmental education, and indeed force us to question the very nature of 'learning' as an environmentally distributed and transgenerational process. Put simply, the environment becomes the very medium and milieu *through* which all creatures come to learn, grow, and develop in dynamic relation to one another (Frost 2016). As noted by Frost, however, we need to be careful not to unwittingly adopt an environmental reductionism that posits all forms of life, thought, and experience as 'epiphenomena' of a totally conditioning environment that operates through a linear regime of time and causality. Importantly, Frost (145) stresses the transgenerational plasticity of biological and social patterns and variations across multiple spaces and times, such that organisms are always both contemporaneous *and* non-contemporaneous with their immediate habitats and environments.

> As a product of the many generations preceding it, an organism's habitat-responsive development and growth is noncontemporaneous with its current habitat. Those histories, as they manifest in cellular structures, protein activities, and organismic processes, set the organism apart from its habitat in the sense of providing the possibilities for and the constraints on the biological responses that the habitat can evoke or induce… it is this corporeal history, this noncontemporaneity, that gives a living organism its "itness" even as it composes and recomposes itself continuously in response to and through engagement with its habitat. (145)

In recalibrating our understanding of the organism–environment relation along a 'differential temporal horizon' (Frost 2016, 152), the new life sciences offer important insights into the radically distributed and multi-temporal nature of environmental experience, learning, thought and behaviour.

Three variations on the concept of immanence

As discussed in the previous sections, there are a number of important ethical implications currently emerging in the spaces between critical posthumanism, new materialism, environmental education, Indigenous philosophies, and the life sciences. For the purposes of this article, I am interested in how the philosophical concept of *immanence* might help us to develop these lines of ethical inquiry in relation to the overwhelming environmental challenges of our times. In the process-relational philosophies of Whitehead and Deleuze, immanence is attributed (in different ways) to *becoming* as the ontogenetic process through which reality comes into being.[1] For Deleuze (1994), there is the repetition of an immanent 'difference in itself' which produces a *continuity of becoming* across all scales, planes, and modes of existence. For Whitehead (1978), there is a *becoming of continuity* that is actively produced through processual events within an immanent unity of nature as an 'extensive continuum'. Although these accounts of ontogenesis differ in subtle and significant ways, both Whitehead and Deleuze propose that immanence is expressed in *the differential modes and manners* through which things and events comes into being (Robinson 2009). For both of these thinkers, the question of the ethical comes to rest in the affective dilation or diminishing of a body's capacities to feel, sense, think, and act in the world, as well as the aesthetic qualities and forms of experience that are rendered through this process. Each of these philosophers could be said to contribute a different *axiology* to the concept of immanence, as the ethical and aesthetic dimensions of the concept come to fall in configurations that differ in subtle and yet important ways.

For the purposes of this article, I would like to continue a little further with Whitehead's pluralistic account of immanence as a concept that takes on multiple variations and operational functions in his later book, *Adventures of Ideas* (1967). This connects the concept of immanence with the 'differential temporal horizon' introduced by Frost (2016) in the previous section, as Whitehead posits different expressions and instantiations of immanence in the relations between contemporaneous and non-contemporaneous events. While Whitehead (1967) upholds a monist immanence of nature as an extensive continuum that pervades and sustains all things and their relations, he also affirms a plurality of variations on the concept of immanence within this continuum. These variations can be largely attributed to his physico-mathematical theorisation of multiple time systems, as he distinguishes between three temporal variations of immanence: the immanence of the past in the present as an 'inherited immanence' (p. 188); the immanence of the future in the present, as what might be termed an 'anticipatory immanence' (p. 192); and the 'mutual immanence' that is attributed to the contemporaneous relations between two of more occasions within a given time system (p. 197). In this article, these distinctions prove useful to the extent that they allow us to attend to the complex ethical implications of the concept of immanence as expressed through the differential relations between multiple times, spaces, events, bodies, environments and societies.

The notion of 'inherited immanence' is useful in thinking through the ethical implications of epigenetics and the agency of a conditioning environment as raised by Frost (2016) in the previous section. Not only do we inherit and express genetic codings and characteristics from the past, but we also inherit feelings, memories, thoughts, experiences, interactions, and significantly, environmental and social patterns of organisation that are intricately interwoven with our everyday lives. On the one hand we inherit our own 'corporeal history' (145), or 'life-thread of occasions' as a 'special strand of unity within the general unity of nature' (Whitehead 1967, 187). With each passing moment, we inherit the immanence of our own lifeworlds, bodies, affects, and mental states as they existed a tenth of a second ago, a day ago, a year ago, a decade ago, and so on. And yet in a certain speculative and metaphysical sense, we also inherit the immanent totality of the entire settled world, and indeed, *the whole Universe as it exists at any given moment.* Hence, a key ethical implication to be garnered from the notion of 'inherited immanence' is that multiple occasions of the past always directly condition the present occasions in which ethical encounters occur, and through which ethical activities are carried out. Our present capacity to act is always a product of everything that we (and the whole world) have ever experienced in the past.

This brings us to the concept of 'anticipatory immanence' which proposes an objectively real presence of the future within the present. Whitehead (1967) offers a number of quotidian examples to support this proposition, including the existence of legal contracts, social understandings, anxieties, and railway schedules. Without the future as a real ingredient in the present, Whitehead argues, the entire edifice of the contemporary world would fall apart. Hence, 'the future is immanent in the present… as a general fact belonging to the nature of things' (194). It therefore seems safe to suggest that an 'anticipatory immanence' expresses the relation between the future and the present, such that 'the present bears in its own realised constitution relationships to a future beyond itself' (191). The present both contemplates and anticipates a future that is yet to come through the 'real potentiality' of actual occasions *as they are in the process of becoming.* It is within this real potentiality of each occasion that particular ethical implications arise, as the anticipatory force of potentiality generates a modicum of causal independence or 'elbow room' which grants ethical freedom to contemporary events in the passing present.

This brings us to the concept of 'mutual immanence' which refers to the relations between two or more occasions of experience which are contemporaneous within the present. Whitehead (1967) writes that, 'Any set of actual occasions are united in the mutual immanence of occasions, each in the other' (197). In other words, there is a mutual immanence that inheres to any event of encounter between contemporaneous things in the world, and this mutuality of contemporary relation entails both an enabling unification (each *in* the others) and a constraint (entanglement *with* others). We are both enabled and constrained by the mutual immanence of our contemporary relations with others, including the relations between those elements of our own bodies, affects, and mental states that constitute our existence as 'living persons'. For the purposes of this article, my suggestion is that the affordances and constraints of these mutual relations implies an 'immanent ethics' that inheres to each event of encounter as it occurs in the passing present. This makes every moment an opportunity to sense-, feel-, think- and act-with others in ways that are ethically attuned to the mutual immanence of co-existence within a shared environmental manifold.

Reimagining the concept of justice

The concept of 'doing little justices' emerges from this mixed milieu of new and old materialist thought, sprouting in the cracks and intersections between process-relational philosophy, environmental education, participatory art, and the new life sciences. Thinking about ethics through Deleuze and Whitehead's philosophies helps us to conceive of justice as a 'minor concept' that inheres to an immediate event of encounter, as conditioned by the immanence of the past and the future in the present. To do a little justice is to become attuned to *this* situation of mutual immanence, to acknowledge in some small way its entangled route of inheritance as well as its real potentiality, and to act in a way that creatively anticipates the necessity of a future that only becomes possible through mutual co-existence.

This re-imagining of the concept of justice was undertaken as part of my doctoral project, *States and Territories*, and involved a series of collaborative artworks and pedagogical experiments with university students and academics in the Philosophy of Law and Primary Education. Central to the *States and Territories* project has been the permanent installation of interactive

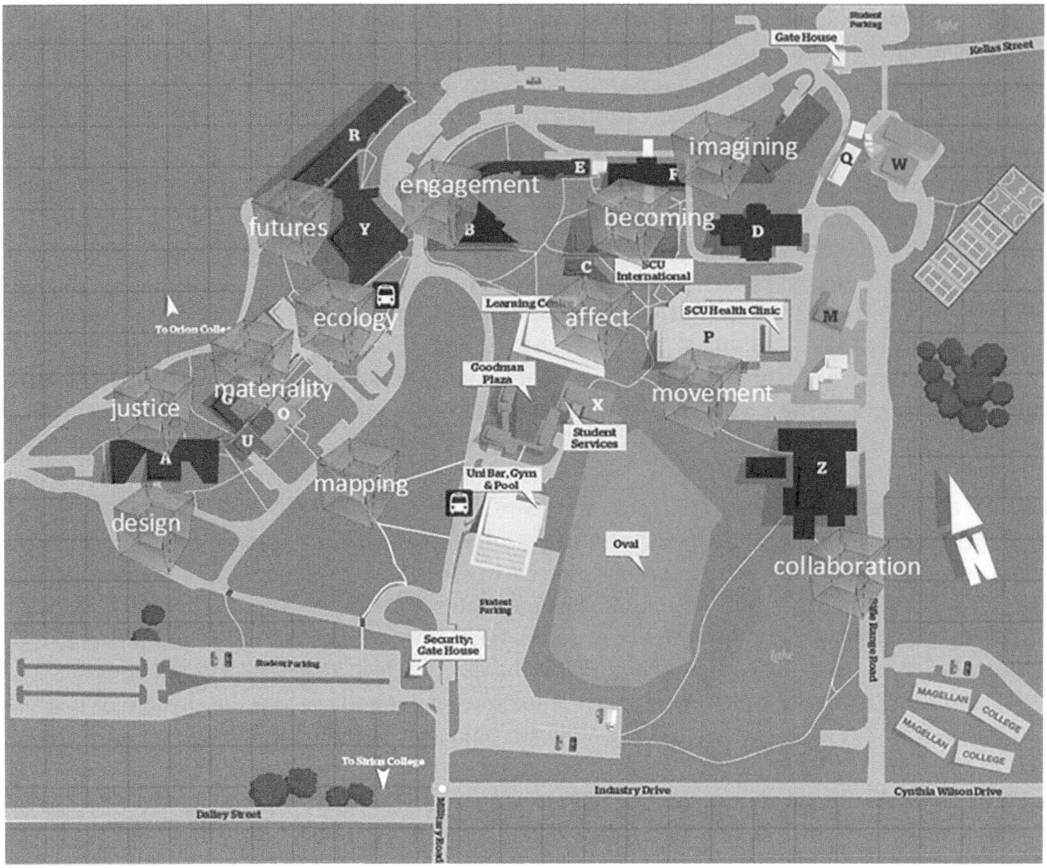

Figure 1. Network of cube locations and concepts distributed across SCU's Lismore campus.

artworks and multimedia interfaces across the learning environments of Southern Cross University's (SCU's) campus in Lismore, NSW, Australia. Each installation takes the form of a glass cube that maps the surrounding learning environment using photographic images captured from the object's own point of view within the landscape. These 12 cubes have operated as conceptual pivot points for a series of data events that were collectively produced with students and academics working in different units of study across the university, including units in the visual arts, media studies, cultural studies, education, law and justice, engineering, ecology, chemistry and environmental science. Each of these 12 data events has explored a different concept through curricular experimentations with new materialist ideas and practices. As shown in Figure 1, 'justice' is one of 12 concepts that was explored in collaboration with university students and academics as part of the larger project, including experiments with the concepts of 'becoming' and 'affect' in cultural studies; the concept of 'materiality' with environmental science students; and the concept of 'imagining' with students in the arts and media studies. These collaborative experiments resulted in the production of a wide variety of artefacts, including videos, images, artworks, animations, sound pieces, and texts which have subsequently been archived on the project website: http://www.statesandterritories.org/archives

The cube associated with justice is located within a sloping section of landscaped garden in between the School of Law and Justice and the School of Engineering (see Figure 2). The cube is surrounded by a corridor of moss-covered trees, flowering plants, and an unusual assortment of lichens and fungi that have proliferated along the ground in this particular area. This environment changes dynamically over the course of the year, as various trees and plants flower and

Figure 2. The justice cube is located in a stand of trees adjacent to the School of Law and Justice.

drop their leaves and the patterns of coloured light and shadow shift with variations in moisture, sunlight, and weather conditions. This cube is also situated in a rather dynamic social milieu, as the School of Law and Justice has become an international centre for research in the areas of Wild Law and Ecological Jurisprudence. It is within these emerging areas of study that speculative, posthumanist, and new materialist theories are beginning to exert an impact on legal philosophies, practices, pedagogies, and policy frameworks (see, for instance, Pelizzon and Ricketts 2015).

The data event associated with the Justice cube began with a series of conversations with academics in the School of Law and Justice, and the discovery that we were actually reading quite a few of the same books (for instance, Latour 2013). Our shared interest in thinking justice beyond the human was then extended to a group of postgraduate students studying the philosophy of law, who participated in a series of seminars exploring the concept of 'doing little justices' that I had been working on. The notion of 'doing little justices' had emerged through my studies and experiments with new materialist theories of ecology, aesthetics, and pedagogy, and had taken shape in a series of four propositions:

1. Justice as a performative gesture by which ethical encounters are negotiated (Manning 2016);
2. Justice as a multiplicity that works 'across differences' and is 'internally differentiated, but still grounded and accountable' (Braidotti 2013, 49);
3. Justice as a process that is embodied and embedded in the actualisation of micropolitical events (Massumi 2015); and
4. Justice as a speculative experimentation that produces new forms of togetherness amongst species, practices, technologies, and modes of existence (Haraway 2016; Stengers 2005).

In my initial work with the philosophy of law students, I also connected these four propositions with environmental and socially engaged artists and artworks, including the works of Suzanne Lacy, Meirle Ukeles, and Andy Goldsworthy. We discussed Lacy's and Ukeles' 'minor' interventions into social and environmental justice through their community-based installations and performances in public spaces and institutions, as well as Goldsworthy's ephemeral and subtle aesthetic interventions into wild and remote environments. We thought about how these actions constituted 'little justices' which were connected to an ethics of events, encounters, and occasions of experience rather than transcendent principles or moral ideals. We also talked about simple events in our own lives that might be thought of as little justices, such as sitting outside around a fire, or exploring a part of the creek we had never seen, or engaging in the very conversations that were having. Coming out of these discussions was the idea that doing a little justice could be as simple as becoming aware of the environment as a multiplicity, and learning to sense, inhabit, articulate, and respond to this temporally and spatially distributed environment in ways that were artful, sensitive, and productive of difference.

This initial test-run of the concept of 'little justices' formed the basis for a second series of seminars and collaborative activities, this time with a group of 29 primary education students undertaking a Bachelor of Education unit called *Foundations: Human society and its environment*. This is the core unit associated with the teaching of geography and social studies within the Australian National Curriculum, and which deals (in small part) with topics associated with sustainability, climate change, and geopolitics. For this series of seminars and activities, I focused more closely on 'game changers' that the Anthropocene brings to bear on questions of justice in the primary school classroom. These game changers included:

- the detonation of the first atomic bomb in 1945, and the embedding of radionuclides within the Earth's geologic strata
- the Great Acceleration of human enterprise since 1950, including the tripling of human populations within a single lifetime and the loss of 40% of the world's animal life
- the default legal categorisation of non-human animals as 'possessions' rather than 'persons'
- the Whanganui river in New Zealand being granted the rights of a 'person' in 2012, as connected with its traditional human (Maori), animal, and plant inhabitants
- the current impacts of climate change on the habitats of human and non-human animals alike, including the submergence of Pacific islands and glacier melts
- the rapid acceleration of biotechnology, artificial intelligence, robotics, genetics, and nanotechnology

After discussing the ways that these game changers might impact on our understandings of justice, the students were invited to create a series of videos that expressed their attempts to grapple with these issues. Specifically, students were encouraged to explore the rights of humans in relation to animals, mountains, and rivers; the impacts of climate change on notions of social and environmental justice; and the question of what an education would be like that did not insist on human superiority and dominance. The students worked in groups with iPads and used the various environments of the university campus as the settings for their video productions. They were also made aware that these videos would be publically archived within the *States and Territories* website, and would be made available for future students to encounter and engage with through the Justices cube. While the range of responses to this activity was diverse, there was an unexpected consensus amongst all 29 students that the rights of a 'person' should be granted to non-human entities such as animals, mountains, and rivers. This question of the rights of a 'person' (whether in legal, ethical, psychological or ontological terms) links to Whitehead's (1978, 1967) expanded definition of a 'living person' to include all organisms which, more or less, exhibit a 'personality' as an individual within a more widely distributed ecological society. In the following brief vignettes, I offer an account and analysis of two of the more fascinating video responses produced by the primary education students. Edited versions of the videos can also be found in the Justices archive on the *States and Territories* website: http://www.statesandterritories.org/justice-archive

First vignette: 'We can fuckin' clone them and eat the clones'

In this video, a group of four students conduct a lively, speculative and at times hilarious discussion in which they entertain the possibilities of non-human animals having political agency. They begin by asking 'should an animal have the right to suffrage?', which leads them to consider the possibility that animals might democratically 'elect a leader within their own kingdom' through 'natural selection'. The students further question the capacity for animals to communicate across species, and wonder how a 'bat could talk to a dolphin'. This offers an interesting example, as bats and dolphins have both evolved to use sonar as a primary communication and navigation function. This is a rare example of convergent evolution, in which species co-evolve symbiotic biological functions without having any direct contact with one another. While this discussion flirts with the fraught ethical implications of a 'social Darwinism' as applied to human and non-human animals, it also gestures towards a political ecology in which life processes and social processes are intricately intertwined and dynamically modulated. Further connections can be made to recent developments in the ecological sciences, in which new theories of eco-evolutionary plasticity are being posed which account for variations and exchanges of genetic material within ecological systems (West-Eberhard 2003).

The students go on to suggest that the rights of humans are no more important than those of other animal species, but that humans are obligated with the responsibility of protecting the rights of other animals because they have the capacity to do so. They go on to apply this ethical proposition equally across species: 'I guess once any animal becomes the top on the food chain, then it becomes their responsibility to take care of the rest of it'. This offers an interesting question regarding the increased response-abilities of apex predators such as wolves, which have been shown to foster ecological rehabilitation when re-introduced into Yellowstone National Park (National Geographic Society 2017) as well as the nuclear fallout zone at Chernobyl (WNET 2016). While these examples may at first glance reinforce a hierarchical image of organisms arranged along the ecological food chain, from another perspective they gesture towards the immanent ethics of an ecological system that is dynamically modulated and re-assembled by the activities of wolves as 'living persons'. Rather than simply filling an empty position in the ecosystem that is preinscribed for an apex predator to manage, the wolves introduce an entire concatenation of biological, social, and mental processes that effect a qualitative transformation of the system's working as a whole. The reinvigoration of the ecosystem through the dynamic activities of the wolves can be considered a prime example of 'doing little justices', even though this justice is partially predicated on the wolves' capacity to consume other animals.

The students further speculate that human technologies have become so advanced that they have disconnected humans from the rest of the food chain. 'We have ways to get around it; we have spaceships, we have genetic modification, we have crops that we can grow in space … we have genetic material from cattle and sheep, we can fuckin' clone them and eat the clones'. Notwithstanding the students' palpable sense of hilarity during this discussion, there is something disturbing and strange in the idea that animals might achieve the 'right' to be cloned, while ethical protocols continue to prevent the cloning of humans. Would the cloning of an animal raise or diminish its status as a 'person'? Would eating a clone make the consumption of animals seems less like eating a 'person'? What rights do techno-scientific inventions create for the category of the 'human' in relation to other animals? As a result of their certainty regarding the techno-scientific superiority of humans, the students conclude that 'animals need the environment more than we do, but the environment doesn't need us at all'. For these students, humanity's technological capabilities contribute to their increased ethical obligations and responsibilities to maintain the ecological systems that support other animals. At the same time, they recognise a profound alienation as the rapid evolution of techno-scientific invention detaches human lives from a sense of mutual immanence with the natural world (see also Simondon 1958/2017). Furthermore, this group of students contends that human rights such as freedom of speech and freedom of movement should be attributed to non-human animals as 'the right to roam free ... without trying to adapt animals better for our own needs'. Hence, as consistent with all of the groups who participated in this series of experiments, this group of students argues for far more inclusive rights for animals than are commonly accorded under normative practices in Australia or under international law.

Second vignette: 'Qualities of life'

In the second video example, three students begin by conducting a more laconic discussion of the rights that might be accorded to geographical entities such as mountains and rivers. 'A lot of the rights are similar', they say. 'Access to food and water, rights to reproduction, free borders for animals and water. The right to socialise with whomever you please ... qualities of life and freedom'. While the students do not elaborate on the speculative implications of these rights, they do suggest that mountains and rivers could exhibit a desire for sociality, as well as the need to maintain open and porous boundaries for the movements of lifeforms and geochemical resources. They suggest that the Murray River should have the right to flow across political boundaries between States without 'persecution', and allude to the river's right to maintain its own catchments without excessive drainage by farmers. It is interesting that they use the proper name of the Murray river to designate its status as a living person. The students even refer to the river as 'Murray', in the same congenial sense that they might refer to a personal friend or family member. They go on to argue that animals, mountains, and rivers should have equal rights to 'qualities of life', and should not be 'enslaved for their materials, or for what they're producing'. Furthermore, the students identify climate change as a phenomenon that is 'escalating and building up and getting mixed with other problems ... decreasing the qualities of life' by contributing directly to 'diseases, lack of food, lack of free space really'.

These considerations lead the students to question the mass slaughter of animals as a result of the rapid growth in human population and enterprise. 'Would we like it if *we* killed *them*, I mean, if *they* killed *us*?' one student asks. This slippage in the students' questioning highlights the habitual normativity and overwhelming statistical frequency of humans killing other animals, as opposed to the relative infrequency of humans being killed by other animals. Perhaps more disconcertingly, the slip also highlights the zone of indiscernibility between 'us' and 'them': if humans are animals, then doesn't killing and eating animals entail a perverse form of cannibalism? 'Would we eat meat if we didn't have that kind of mentality?' another student asks. 'Maybe we would, but we would eat less. Only for survival you would. Which is what it was traditionally. Hunting down what you could for you and your family to survive'. It is interesting to note that the students describe the act of eating meat as the function of a particular 'mentality', rather than a biological function necessary for the acquisition of energy and nutrients. Morton (2016, 69) further argues that the Mesopotamian shift from a nomadic way of life to a sedentary, territorialised, and agrarian civilisation inaugurated the blind execution of what he calls an 'agri-logistic mentality'. For Morton, Enlightenment claims to scientific rationality and universal reason are linked to the rise of agri-logistics as an 'algorithmic' mental virus that is largely subconscious, 'and therefore liable to be repeated and prolonged like a zombie stumbling forward' (58). Such a subconscious mental algorithm makes the mass killing of animals both a habituated and a 'logical' necessity, while the over-consumption of animals remains a primary factor in the production of greenhouse gases, soil degradation, water depletion, and the relentless destruction of oxygen-producing vegetation (McNeill and Engelke 2014).

From an immanent ethics to the creation of rights

The two vignettes above provide a brief series of glimpses into the modes of thinking provoked by the concept of little justices. In shifting the discourse from legal jurisprudence to an applied ethics of teacher education, the question of human and non-human rights becomes central to the inquiry. This shift in register brings an immanent ethics into productive tension with more normative conceptions of rights as they apply to the lives and activities of 'persons'. Yet this does not necessarily entail a reduction of rights to transcendent humanist ideals. As Patton (2012, 15) argues from a Deleuzian perspective, an immanent ethics leads to *the creation of new rights* rather than the extension or application of axiomatic rights that somehow transcend events and occasions of experience. For Deleuze, rights are created through jurisprudence as the philosophical contemplation and expression of an ethics that is immanent within events. This becomes evident through the creation of rights with respect to biogenetic technologies, such as cloning, that produce new ontological problems and material conditions under which new rights must be invented. Such an appeal to the speculative creation of rights breaks with humanist beliefs that 'individuals possess certain rights by virtue of some rights-bearing feature of human nature, such as rationality, sentience, or the capacity to act in pursuit of chosen ends' (Patton 2012, 21). Rather, there is a continuous becoming, distribution, and re-assembling of rights in relation to the mutual construction of an ethical problem. The creation of rights thus occurs at the level of a subterranean micropolitics, in which consensus and dissensus come into being as coeval powers of contagion, negotiation, fostering and invention. Similarly there appears a contiguous relation between the right to exist and the right not to exist; the right to agree and the right not to agree; the right to know and the right not to know.

These productive movements from an immanent ethics to the creation of rights permeate the vignettes above, as well as the broader collection of video artefacts created by this cohort of students. Perhaps most surprising was the concurrence amongst all 29 students that the rights of a 'person' should be extended to non-human entities such as animals, mountains, and rivers. Moreover, students were willing to make relatively radical leaps in their thinking through the creation of rights for animals to roam freely and elect chosen leaders through natural selection, or for rivers to maintain an uninterrupted flow across political borders. Whilst this could be seen as an anthropomorphic extension of human rights to animals, it was arguably the capacity for the students to re-imagine issues of justice under immanent terms that provoked them to engage so enthusiastically with the creation of rights.

Unknowable outcomes and antecedents

By feeling, thinking and embodying issues of justice through a speculative and affective sense of mutual immanence, students came to generate a series of aberrant and provocative ideas that connect with some of the critical concerns of environmental education and new materialist scholarship. In the case of the second vignette, for instance, there appeared to be a genuine reappraisal of the mass killing of animals as a function of blind, algorithmic execution inherited from millennia of agri-logistic practices. However, the extent to which such realisations will impact on the philosophies and behaviours of students when they become classroom teachers can never be known. But can the actual 'outcomes' or 'results' of our research ever be known? This question becomes especially problematic when we consider the immanence of future events as they 'unknowingly' participate in the constitution of bodies, places, events, and milieus in the passing present.

The problematic nature of anticipatory immanence in our research also begs the question of 'unknowable' antecedents and inherited immanence. What corporeal histories could have contributed to the students' consensual agreement that all animals, mountains, and rivers should be accorded the legal status of a person? Perhaps their responses were conditioned by their engagement with different curriculum content and ideas, such as the Anthropocene thesis and the posthuman game changers associated with climate change, the Great Acceleration, and biogenetic technologies. Maybe they were affected by asking themselves provocative questions that disrupted normative beliefs about human superiority, dominance and exceptionalism. Perhaps their engagement with the immersive art and media of the cube network and project website had a significant effect, motivating them to articulate and archive their ideas for a future public. It's also possible that their responses were coloured by the socio-cultural milieu of Northern NSW, which is colloquially known as a haven for 'green', 'alternative', 'spiritual', and 'hippie' communities. Maybe it was simply the concept of 'doing little justices' that captured their environmental imagination? The aim of all of this speculation is simply to suggest that the outcomes and antecedents of a research event can never be known because they do not follow a linear causality. Rather, they follow an emergent causal logic of the event itself as a novel form of collective experience, and an immanent ethics that only becomes sensible in the co-occurrence of events across multiple spaces and time systems.

(In)Conclusion

By developing and prototyping the concept of 'doing little justices' as a theoretical and pedagogical figure, I have endeavoured to sketch out the contours of an environmental education that is responsive to the Anthropocene, and an immanent ethics that is embedded in the performative, speculative, collective, and minor fluctuations of a micropolitics. In lieu of claims regarding the outcomes, results, or causal antecedents of this research, I would like to propose a few pedagogical propositions that connect the events described in this article with the broader concerns of environmental education in the Anthropocene. Firstly, the concept of 'little justices' has demonstrated a pedagogical potential to produce the conditions under which immanent ethical understandings might emerge, even if only for a short time. This indicates that the concept may have pedagogical import for environmental educators who are interested in thinking-with students (animals, rivers, mountains, etc.), and exploring the speculative possibilities of life, politics, sociality, and experience beyond the human. The concept should also be regarded as 'portable' like a piece of modular software (Bryant 2014), in the sense that 'doing little justices' can be plugged into other assemblings of disciplines, studies, concerns, practices and ideas without losing its consistency as a conceptual proposition. Secondly, I would suggest that the process of embodying the speculative possibilities of non-human rights and affective capacities constitutes, in itself, the enactment of a 'little justice' that re-arranges the real potentiality for new patterns of thought and behaviour to emerge. This proposition gestures towards an immanent ethics that is *in* and *for* itself (you do justice by thinking it, and think justice by doing it). And lastly, and perhaps in summary, 'doing little justices' cannot be reduced to a set of predetermined values or prescriptions for environmental education. The concept needs to proceed through a speculative process of ethical experimentation and the creation of rights in the pursuit of unforeseen openings and potentials for mutual immanence. It is in this way that *not knowing* can become an asset, rather than a hindrance, in the ongoing pursuit of an immanent ethics for environmental education in the Anthropocene.

Note

1. Both Deleuze and Whitehead emphasise the singular-plural nature of reality, but lean differently towards a 'pure' monism and a 'pluralistic' monism respectively. This difference is noticeable in their concepts of immanence, as Deleuze (2001) poses a 'pure' plane of immanence that is indivisible and immanent only to itself, while Whitehead (1967) poses multiple variations and qualities of immanence in relation to nature, or the Universe, as an immanent totality. Whitehead's system of thought is ultimately a monist philosophy, but an unusual monism in which the notion of immanence is multiplied, organic, creaturely, relational, experiential and processual (Whitehead 1978, 19). Whitehead therefore leans more towards a pluralist monism than Deleuze or Spinoza. While Spinoza posits affective modes and their attributes as the manifold expressions of a single substance, Whitehead retains only the modes themselves as 'sheer actualities' or 'becomings' without any underlying substance. For Whitehead the basic units of reality are 'actual occasions' that are self-causing and immanently creative, rather than inheriting a final causation derived from a single underlying substance (such as God, Being, nature or matter). Hence, Whitehead attributes the emergence of complex systems (such as bodies, societies and ecosystems) to the patterning of interconnected events within the 'extensive continuum' of nature as an immanent totality, or 'common world'.

Disclosure statement

No potential conflict of interest was reported by the authors.

References

Bennett, Jane. 2010. *Vibrant Matter: A Political Ecology of Things*. Durham, NC: Duke University Press.

Braidotti, Rosi. 2013. *The Posthuman*. Cambridge, UK: Polity Press.

Braidotti, Rosi. 2017. "Critical Posthuman Knowledges." *South Atlantic Quarterly* 116 (1), 83–96.

Bryant, Levi. 2011. "The ethics of the event: Deleuze and Ethics Without Αρχή." In *Deleuze and Ethics*, edited by N. Jun and D. W. Smith, 21–43. Scotland: Edinburgh University Press.

Bryant, Levi. 2014. *Onto-Cartography: An Ontology of Machines and Media*. Edinburgh, UK: Edinburgh University Press.

Cajete, Greg. 2006. "Western Science and the Loss of Natural Creativity." In *Unlearning The Language of Conquest*, edited by D. T. Jacobs, 247–59. Austin: University of Texas Press.

Colebrook, Claire. 2014. *Death of the PostHuman: Essays on Extinction*, vol. 1. Ann Arbor, MI: Open Humanities Press.

Deleuze, Gilles, and Felix Guattari. 1986. *Kafka: A Minor Literature*. Translated by D. Polan. Minneapolis: University of Minnesota Press.

Deleuze, Gilles, and Felix Guattari. 1987. *A Thousand Plateaus: Capitalism and Schizophrenia*. Translated by B. Massumi. Minneapolis: University of Minnesota Press.

Deleuze, Gilles, and Felix Guattari. 1994. *What is Philosophy*? Translated by H. Tomlinson and G. Burchell. New York, NY: Columbia University Press.

Deleuze, Gilles. 2001. *Pure Immanence: Essays on a Life*. Translated by A. Boyman. Brooklyn, NY: Zone Books.

Deleuze, Gilles. 1994. *Difference and Repetition*. Translated by P. Patton. New York, NY: Columbia University Press.

Ellsworth, Elizabeth, and Jaime Kruse. 2012. *Making the Geologic Now: Responses to the Material Conditions of Everyday Life*. Brooklyn, NY: Punctum Books.

Frost, Samantha. 2016. *Biocultural Creatures: Toward a New Theory of the Human*. Durham, NC: Duke University Press.

Grosz, Elizabeth. 2008. *Chaos, Territory, Art: Deleuze and the Framing of the Earth*. New York, NY: Columbia University Press.

Haraway, Donna. 2016. *Staying With the Trouble: Making Kin in the Cthulucene*. Durham, NC: Duke University Press.

Harvester, Lara, and Blenkinsop, Sean. 2011. "Environmental Education and Ecofeminist Pedagogy: Bridging the Environmental and the Social." *Canadian Journal of Environmental Education* 15: 120–34.

Jickling, Bob. 2013. "Normalising Catastrophe: An Educational Response." *Environmental Education Research* 19 (2): 161–76.

Latour, Bruno. 2013. *An Inquiry Into Modes of Existence: An Anthropology of the Moderns*. Translated by C. Porter. Cambridge, MA: Harvard University Press.

Lloro-Bidart, Teresa. 2015. "A Political Ecology of Education in/for the Anthropocene." *Environment and Society* 6(1): 128.

Malone, Karen. 2016. Reconsidering Children's Encounters with Nature and Place Using Posthumanism. *Australian Journal of Environmental Education* 32(01): 42–56.

Manning, Erin. 2016. *The Minor Gesture*. Durham, NC: Duke University Press.

Massumi, Brian. 2015. *The Politics of Affect*. Cambridge, UK: Polity Press.

McNeill, J.R. and Peter Engelke. 2014. *The Great Acceleration: An Environmental History of the Anthropocene Since 1945*. Cambridge, MA: Harvard University Press.

Meloni, Maurizio. 2016. *Political Biology: Science and Social Value in Human Heredity from Eugenics to Epigenetics*. Basingstoke, UK: Palgrave Macmillan.

Morton, Timothy. 2016. *Dark Ecology: For a Logic of Future Coexistence*. New York, NY: Columbia University Press.

National Geographic Society. 2017. *Wolves of Yellowstone: How does wolf reintroduction benefit Earth's natural systems?* Retrieved 24 January, 2017 from http://www.nationalgeographic.org/media/wolves-yellowstone/.

Patton, Paul. 2012. "Immanence, Transcendence, and the Creation of Rights." In *Deleuze and Law*, edited by Laurent de Sutter and Kyle McGee, 15–31. Edinburgh: Edinburgh University Press.

Payne, Phillip. 2016. "What next? Post-critical materialisms in environmental education." *The Journal of Environmental Education* 47(2): 169–78.

Pelizzon, Alessandro, and Aidan Ricketts. 2015. "Beyond Anthropocentrism and Back Again: From Ontological to Normative Anthropocentrism." *Australasian Journal of Natural Resources Law and Policy* 18 (2): 105–25.

Povinelli, Elizabeth. 2016. *Geontologies: A Requiem to Late Liberalism*. Durham, NC: Duke University Press.

Protevi, John. 2013. *Life, War, Earth: Deleuze and the Sciences*. Minneapolis: University of Minnesota Press.

Robinson, Keith. 2009. "Introduction: Deleuze, Whitehead, Bergson- Rhizomatic connections." In *Deleuze, Whitehead, Bergson- Rhizomatic Connections*, edited by Keith Robinson, 1–27. Basingstoke, UK: Palgrave Macmillan.

Rotas, Nikki. 2015. "Ecologies of Praxis: Teaching and Learning Against the Obvious." In *Posthumanism and Educational Research*, edited by Nathan Snaza and John Weaver, 91–104. New York, NY: Routledge.

Rousell, David. 2017. "Mapping the Data Event: A Posthumanist Approach to Art|Education|Research in a Regional University." In *Arts, Research, Education: Connections and Directions*, edited by Linda Knight and Alexandra Cutcher, 203–20. New York: Springer International Publishing.

Rousell, David. 2016. "Dwelling in the Anthropocene: Re-imagining University Learning Environments in Response to Social and Ecological Change". *The Australian Journal of Environmental Education* 32(2): 1–20.

Rousell, David, and Dilafruz Williams. 2018. "Ecological Aesthetics: New Spaces, Directions, and Potentials". In *Research Handbook on Childhoodnature*, edited by Amy Cutter-Mackenzie, Karen Malone, and Elizabeth Barrat-Hacking. The Netherlands: Springer.

Rousell, David, Cutter-Mackenzie, Amy, and Foster, Jasmyne. 2017. "Children of an Earth to Come: Speculative Fiction, Geophilosophy and Climate Change Education Research". *Educational Studies* 53(6): 654–669.

Snaza, Nathan, and John Weaver. 2015. "Introduction: Education and the Posthumanist Turn." In *Posthumanism and Educational Research*, edited by Nathan Snaza and John Weaver, 1–16. New York, NY: Routledge.

Stables, Andrew. 2007. "Is Nature Immaterial? The Possibilities for Environmental Education Without an Environment." *Canadian Journal of Environmental Education* 12(1): 55–67.

Stengers, Isabelle. (2005). "Introductory Notes on an Ecology of Practices". *Cultural Studies Review* 11(1): 183–96.

Todd, Zoe. (2015). "Indigenizing the Anthropocene". In *Art in the Anthropocene: Encounters Among Aesthetics, Politics, Environments and Epistemologies*, edited by Heather Davis and Etienne Turpin, 241–54. London: Open Humanities Press.

West-Eberhard, Mary. 2003. *Developmental Plasticity and Evolution*. Oxford, UK: Oxford University Press.

Weston, Anthony. 1992. "Before Environmental Ethics." *Environmental Ethics* 14(4): 321–38.

Weston, Anthony. 2015. "Working the Dark Edges." *Canadian Journal of Environmental Education* 19: 70–79.

Whitehead, Alfred North. 1967. *Adventures of Ideas*. New York: The Free Press.

Whitehead, Alfred North. 1978. *Process and reality*. New York: The Free Press.

WNET. 2016. *"Radioactive Wolves."* Retrieved 26 January, 2017 from http://www.pbs.org/wnet/nature/radioactive-wolves-introduction/7108/.

Yusoff, Kathryn. 2015. "Geologic Subjects: Nonhuman Origins, Geomorphic Aesthetics and the Art of Becoming Inhuman." *Cultural Geographies* 22(3): 383–407.

Zylinska, Joanna. 2014. *A Minimal Ethics for the Anthropocene*. Ann Arbour, MI: Open Humanities Press.

Painting trees in the wind: socio-material ambiguity and sustainability politics in early childhood education with refugee children in Denmark

Nanna Jordt Jørgensen and Asger Martiny-Bruun

ABSTRACT

In this article, we explore new materialist 'common world' approaches to early childhood environmental and sustainability education and the ambition of these approaches to challenge social–material and nature--culture dichotomies often taken for granted in Western education systems. Firstly, we point out the socio-cultural inequalities inherent in taken-for-granted approaches to children and nature in Denmark, and argue that attention to children's relationships with the more-than-human should not overshadow the consideration of social justice agendas; rather, these should be seen as fundamentally linked. Secondly, while we support a greater attention towards children's experiences of living with other beings in entangled, enmeshed common worlds, we nevertheless argue against completely erasing or overlooking the extent to which these experiences are also intertwined with experiences of being separated from the world. Acknowledging the fundamental ambiguity of simultaneous immersion in, and detachment from, the world, we propose, is of key importance in terms of becoming able to take action for social and ecological sustainability.

Introduction

Yara, a four-year old Syrian girl, sits stiffly on the periphery of a small group of children in a park. They all face a tree, and in front of them are large sheets of paper, and glasses filled with coloured India ink. Yara has been in Denmark for about six months and is enrolled in a day care institution. Today she is taking part in an artwork session with the aim of learning about nature and sustainability. Initially she was very shy and hesitant about leaving the premises of the day care institution, but the atmosphere around her changes over the course of the session. As the wind blows through her hair and the branches of the tree move slowly above her, she takes a firm grip on a branch dipped in black India ink. Her painting starts to fill out the paper, her body moves and the sound of the wind mingles with a song in Arabic that Yara sings while drawing.

What is going on between the child, the tree, the wind, the ink, the paper and the song? Why do we get the impression that something important is at stake in this moment, something that transgresses the ordinary atmosphere in the day care institution as well as the omnipresent public debate in Denmark that problematizes immigrant minorities and their integration? What does the situation tell us about the ambiguities of socio-material relations; about the complex ways in which we touch and are being touched by human and nonhuman others, and about our sensory experiences of oscillation between immersion in and detachment from the world?

This article arises from a motivation to contribute to research on sustainability education in early childhood. Firstly, we wish to point out the socio-cultural inequalities inherent in taken-for-granted approaches to children and nature in Denmark. During our research on the encounters between refugee children, their families and Danish day care institutions, our attention has been drawn to the ways in which these institutions have attempted to domesticate socio-cultural and natural 'otherness' (cf. Hage 2017) in ways that create polarized positions. Secondly, we will explore some of the tensions and processes at work in sensuous and aesthetic experiences, as in the art session described above, where the sensory experiences dissolve the distance between Yara, the wind, the ink, the paper, the moving branches of the tree and the other children. We suggest that providing spaces where children can experience such alternative ways of relating to their socio-material environment may potentially transgress the non-sustainable modes of inhabitance otherwise dominant in the Western part of the world. Providing such space may potentially provide children with starting points for more socially and ecologically sustainable lives.

In recent years, scholars of early childhood, such as Affrica Taylor, Fikele Nxumalo and Veronica Pacini-Ketchabaw, have suggested that 'reconceptualising the nature of childhood' (Taylor 2013), or rethinking dominant understandings of children's relationships with the more-than-human world, constitute a key opportunity to challenge 'the nature/culture divide that has contributed to the challenges of the Anthropocene' (Nxumalo and Pacini-Ketchabaw 2017, 2). In this article, following the anthropologist Ghassan Hage, we propose that the nature–culture divide positions nonhuman others, but also human others, as inferior and in need of domestication – or integration (Hage 2017). Yet, rather than attacking dualism per se, and with inspiration from the philosophical writings of Maurice Merleau-Ponty, we argue that the experience of separation underlying dualistic thinking, as well as the experience of immersion related to notions of entanglement, are fundamental aspects of being in the world. By exploring and analyzing empirical case material from aesthetic sustainability educational activities, we propose that pedagogical attention and openness to ambiguity may offer an entry point for what Hage has called: 'the recovery of the multiplicities of modes of inhabitance that capitalist modernity has excluded and marginalized' (Hage 2017, 116).

In the following, we describe the empirical material that our analysis draws on as well as our methodological approach. Thereafter, we line up the theoretical discussions about new materialist perspectives on early childhood sustainability education that we wish to enter into a dialogue with. In the first part of our analysis, we then explore the social othering processes involved in taken-for-granted approaches to children and nature in Danish day care institutions. Hage's work has inspired us to look for practices which may present alternatives to these dominating approaches, or 'cracks' in the dominating logic, to paraphrase Hage (2012, 294). Hence, in the latter part of the article, we explore aesthetic sustainability education activities, which open up other ways of inhabiting the world, acknowledging its fundamental ambiguity.

Methodology

This article is based on an anthropological research project carried out in 2016 and 2017, which focused on the reception and participation in early childhood education of refugee children newly arrived in Denmark.[1] The empirical material stems from ethnographic fieldwork in four day care institutions located in four different municipalities in Denmark. Over the course of one year, we visited the institutions regularly, carried out participant observation of daily activities, interviewed professional staff and visited the families of children with a refugee background. Some of the families taking part in the study had been granted asylum while others were still in the process of seeking asylum. Having been granted asylum, refugee families are placed in a municipality and enrolled in a so-called integration program focusing on language training and job internships. As part of the program, younger children are offered a place in a municipal day care institution while older children are enrolled in municipal schools. In principle, day care is not mandatory for children in Denmark. However, the current law on day care states that all children must carry out a language test at the age of three, and it is compulsory for parents of children who are found to be in need of 'language stimulation' to send their children to day care (or ensure equivalent language stimulation). Furthermore, following the government strategy 'One Denmark without parallel communities – no ghettoes in 2030' (Regeringen 2018, our translation), day care has become mandatory from the age of 1 for all children residing in an area classified as a ghetto (Børne- og Socialministeriet 2018a)

Our research has been informed by phenomenological thinking (e.g. Merleau-Ponty 1968, 2012; Ingold 2000, 2011; Jackson 1996, 2013). Our methodology rests on participant observation aiming for a reflexive consideration of sensations (Hastrup 2010; Pink 2009), hereby moving our research attention beyond words and verbal interactions. Following the sociologist Jennifer Mason, we see sensations as experiences and feelings which are embodied, but also emanating from and flowing 'in things that happen and things that come into contact' (Mason 2018, 9). As such, sensory ethnography (Pink 2009) may be seen as offering an entry into embodied being in the world, as well as into the socio-material coexistence ('intercorpeoreity' in Merleau-Ponty's terms, see Merleau-Ponty 1968, 141) of different beings. Our analytical approach may be characterized as abductive (Coffey and Atkinson 1996). Rather than clarifying theory and concepts before the empirical study, as in deductive approaches, or inductively letting phenomenological exploration lead to conceptualizations that are subsequently related to existing theories, in this analysis we have established a continuous dialogue between empirical findings and theoretical conceptualizations.

Our analysis of the place of immigrant children in taken-for-granted approaches to nature and children draws broadly on empirical material generated through participant observation and interviews in the four day care institutions investigated. The case discussed in the latter part of the article is a description of an artwork session in one of the day care institutions, based on field notes from participant observation.[2] The institution is situated in a social housing area with a large number of immigrant families, and thus represents a pedagogical space with particular attention to 'integration'. In our analysis, this detailed empirical example constitutes an opening for rethinking early childhood environmental and sustainability education.

Studies of ESE in early childhood education and the new materialist turn

In 2009, Julie Davis pointed to the lack of research on early childhood education for sustainability as a 'hole' in environment and sustainability education research (Davis 2009). However, in the decade following Davis' article, numerous publications addressing early childhood and sustainability education have been published, and in 2015, two reviews followed up on Davis' work and pointed to new developments. On the one hand, Hedefalk, Almqvist, and Östman (2015) suggested that recent research on early childhood and sustainability focused primarily on children's democratic participation, their outdoor experiences, their knowledge about the environment, and, to a lesser degree, their actions for change. Somerville and Williams (2015), on the other hand, pointed to new theoretical developments and in particular to the rise of research informed by new materialist and post-humanist theory, such as, for instance, Affrica Taylor's work. In an article in Environment Education Research from 2017, Taylor calls for 'a serious rethinking of the business-as-usual of environmental education, particularly the education of children' (Taylor 2017, 1).

While the term 'new materialism' covers a range of perspectives with diverse theoretical sources, they share an ambition towards challenging dichotomies often taken for granted in Western education systems (nature-culture, mind-matter, etc.) and work towards new ways of conceptualizing humans and the world (see for instance Coole and Frost 2010)[3]. In this article, we are particularly interested in the 'common world' approaches that have been rather influential in the research debate on sustainability education in early childhood education (see e.g. Somerville and Williams 2015). We have chosen the writings of Affrica Taylor as an example of this perspective, which emphasises the entangled nature–cultures of children's worlds. In her book on *Reconceptualizing the Nature of Childhood* (2013), Taylor, with inspiration from Donna Haraway, suggests overcoming dichotomies between childhood and adulthood, nature and culture by queering childhood as well as nature. Haraway addresses the current socio-ecological crisis with a call for 'staying with the trouble', which challenges blind faith in technical fixes on the one hand, and a complete loss of hope on the other (Haraway 2016, 3). Staying with the trouble, according to Haraway, is about being present 'as mortal critters entwined in myriad unfinished configurations of places, times, matters, meaning' (Haraway 2016, 1). In line with this idea, Taylor's 'common worlds pedagogies' takes as its starting point that children's worlds are already and always collectives, 'messy, queerly reconfigured, mixed up' (Taylor 2013, 115) common nature–cultures. Rather than learning *about* nature or *in* nature, her pedagogical approach is based on the ambition to learn '*with* those others with whom we are already entangled' (Taylor 2013, 120). Along similar lines, recently published articles within the research areas of environmental education research and childhood studies challenge nature/culture binaries. They point to children's messy, troubling and entangled relations to, for instance, animals (e.g. Nxumalo and Pacini-Ketchabaw 2017; Pacini-Ketchabaw and Nxumalo 2015), places (mountains, forests, cities) (Nxumalo 2017; Pacini-Ketchabaw 2013; Duhn, Malone, and Tesar 2017), weather (Rooney 2018) and materials (Rautio 2013).

Taylor takes issue with the 'rational quest to know about the world from a distance' (Taylor 2017, 9–10) that characterises environmental education approaches focusing on children's knowledge about the environment, as well as with the occupation with children's participation and actions for change. The vision of children as contemporary or future environmental stewards and agents of change, Taylor claims, reflects a problematic human centeredness, resting on a 'belief in a radical separation and superiority of "rational" humans from all other life forms' (2017, 6). In line with Kopnina (2012), Taylor criticizes education for sustainability, arguing that it is too occupied with social justice agendas, hereby reiterating 'human-exceptionalism through its renewed emphasis upon the transformative powers of collective (and individual) human agency' (Taylor 2017, 4).

To a large degree, we sympathise with Taylor's critique and call for rethinking taken-for-granted approaches to environment and sustainability education and their foundation on dichotomous conceptualisations that tend to legitimise relations of domination and control. At the same time, we wish to discuss and add nuances to two of Taylor's claims. Firstly, we argue, human relations with the nonhuman cannot be understood independently of social relations. Hence, greater attention to children's relations with the more-than-human need not overshadow consideration of socio-cultural inequalities and social justice agendas; rather these should be seen as fundamentally linked. Secondly, while we agree with Taylor and others that children, like adults, live with other beings in entangled, enmeshed, messy common worlds, we argue against completely erasing or overlooking the experiences of being separated from the world which have been seen as a characteristic of modern Western thought (see Taylor 2017, 10).

Nature and other 'others' in Danish early childhood education

We begin by exploring how human relations to nature and nonhuman others are intertwined with cultural and social mechanisms of differentiation. In Denmark, nature and outdoor play are important topics in early childhood pedagogies. Nature and natural phenomena form part of the current learning curriculum, and are consequently the focus of structured educational activities aiming for different kinds of learning objectives.[4] Furthermore, nature and free outdoor play have important roles as cultural, taken-for-granted practices in early childhood pedagogies in Denmark. During our research, we found that social educators[5] often perceive migrant children's relations to nature and their ability to engage in free outdoor play as an indicator of how well the children and their families are integrated into Danish society and the everyday life of the institutions. When describing their encounters with families with a refugee background, the social educators explained how they spent time clarifying to the parents the local 'normality' of outdoor life and free play. The fact that children were outdoors several hours every day, spending time at the playground or taking part in excursions, and furthermore that the younger children slept outside in the afternoon in all kinds of weather, gave occasion to surprise and sometimes concern among the parents. The social educators would argue with families who decided to keep their children at home during bad weather, or questioned the learning outcomes of outdoor activities. 'It is the way we do things in Denmark', the social educators would tell newly arrived refugee parents as well as immigrants who had been in the country for a longer time. Some families became the cause of concern because they could not afford or would not prioritize clothing appropriate for cold or wet weather, or because they protested about the dirty clothes that their children brought home after having played outside. Social educators would also observe whether children had 'learned' to play outside and whether they were involved in curiosity-based interactions with, for instance, insects. The refugee children were supposed to learn to play in the natural environment, however the social educators also stressed that their play should not be too 'wild', as in climbing tall trees or digging large holes in the ground. Hence, the outdoor activities of immigrant children are linked both to understandings of children's relationship to nature, but also to politicized questions of belonging and integration.

Taylor has proposed (2013, xiii) that the link between childhood and nature, seen as dichotomous to adulthood and culture, has become 'firmly cemented within the Western popular imaginary' since Rousseau. In the Nordic countries, the relation between childhood and nature appears particularly strong as it connects to understandings of childhood, as well as of national identity. In Nordic educational thinking, as argued by Wagner and Einarsdottir (2008), childhood and thereby children's 'natural', free play has been considered of value in its own right and free play is often understood as outdoor play (see also Kristjansson 2006). A number of studies have pointed to the cultural significance of children's outdoor play in Nordic early childhood education (see e.g. Halldén 2009; Coninck-Smith and Bygholm 2011; Melhuus 2013; Gulløv 2012), a significance related to the strong position of nature, landscape and outdoor life in the Nordic national traditions (see, e.g. Sandell and Öhman 2010; Olwig 2005; Taylor 2013, 54; Ween and Abram 2012).

While the coupling of childhood and nature builds on romantic notions of nature as benign, mild, beautiful and innocent, Western imaginaries also include notions of nature as wild, uncivilized and threatening (see Mcphie and Clarke 2018, for an outline of various other western conceptualizations of nature). The imagined connection between children and nature refers to both these notions of nature – the naturally good and innocent child *and* the bad, wild, natural child - parallel to notions of the noble savage (cf. James, Jenks, and Prout 1998; Gulløv 2012). In contrast to the romanticized notion of the child's closeness to nature, the natural wildness of children is considered problematic, and thereby in need of civilization. According to the anthropologist Eva Gulløv, both understandings of children are present in the organization of Danish early childhood institutions. On the one hand, children's 'natural' play is encouraged; on the other hand, early childhood institutions also aim to teach children to be in control of their bodies and to become ready for school. 'Nature' as presented to children in day care institutions is therefore nature regulated and domesticated, to a large degree free of dangers and risks (Gulløv 2012, 72). As suggested by the research of the anthropologist Laura Gilliam, migrant children take up a specific position in relation to notions of nature, as they are often considered 'wild' children in need of civilization, especially the boys (Gilliam 2017).

These understandings of the relationship between children and nature are situated in a Danish context, where the public debate is heavily focused on integration and on immigrants and immigrant children as a problem category. The anthropologist Mikkel Rytter has described the vision of society embedded in the emic concept of integration as resembling a bowl of soup, 'with newcomers being expected to blend in and dissolve as fast as possible' (Rytter 2018, 4). Thus, in politics and everyday life in Denmark, integration refers to the goal, a utopian horizon of absolute integration (i.e. assimilation), as well as to the process of getting there. According to Rytter, 'the emic notion of integration rests on widespread images and ideas about culture, race and belonging that seem to expand the already existing gap between Danish majorities and immigrant (Muslim) minorities' (Rytter 2018, 14). In other words, integration paradoxically creates the otherness that it attempts to dissolve. This argument resonates with research on day care institutions suggesting that these institutions tend to reinforce rather than break down social inequalities between ethnic minority and ethnic majority children – in spite of good intentions (see, e.g. Larsen and Jørgensen 2017; Palludan 2007). Ethnographic studies show how through social processes children experience the reproduction of social classes and distinctions based on ethnicity and religion (see, e.g. Gilliam and Gulløv 2017). The social and political goal of integrating minority children into an imagined whole thus appears to be the implicit background for many pedagogical approaches (cf. Bundgaard and Gulløv 2008), including the particular attention paid by social educators to the ability of migrant children to engage in outdoor activities in a 'normal' and 'free' way. In this context, pedagogical activities that focus on environmental education and nature run the risk of reproducing socially unsustainable practices of inequality and experiences of marginalisation.

Searching for the cracks

In his book *Is racism an environmental threat?* (2017), Ghassan Hage points out that the position of 'the Muslim' in Western imaginary is often classified as analogue to the 'threatening undomesticated other of nature' (Hage 2017, 33). Hage argues that the governmental response is integration, following a logic of domestication. He introduces the concept of 'generalized domestication', defined as 'a mode of inhabiting the world through dominating it for the purpose of making it yield value' (Hage 2017, 87), yet not following a purely instrumental logic. Domestication is simultaneously and paradoxically a quest to 'make things partake in the making of one's home' (Hage 2017, 91) and a mode of domination and control. An important element of generalized domestication, according to Hage, is the polarization process that it involves. Polarization is the process of turning difference into polarity – between, for instance, human and nonhuman, man and woman, black and white, child and adult or Muslim and non-Muslim (2017, 110–111). Hage posits a close connection between the dominating ways in which white westerners practice their relationship to human and nonhuman others; a connection that appears to shed light on the polarisation processes involved in practices in Danish day care institutions that attempt to integrate immigrant families and include the natural environment in young children's lives.

Still, according to Hage, the answer to this kind of thinking should not be anti-dualism, but rather to oppose the *dominance* of this way of inhabiting the world (Hage 2017, 116). 'Examining both the possibilities of other modes of existence and, as important, the possibility of negotiation and entanglement of a multiplicity of modes of existence, serves to highlight the poverty of our mono-relational present', Hage asserts (2017, 124–125). This may be an important starting point for creating socially and ecologically sustainable lives. Sustainability, in other words, may be seen as a question of finding viable alternatives to the practices of exploitation and racism rooted in colonialism (cf. Hage 2017, 132–133; see also Lotz-Sisitka 2017). Hage suggests that anthropology offers a certain mode of critical thought which in the first place shows us the possibility of other ways of living, and secondly exposes us 'to the possibility of being other than we are' (2012, 289). 'Our otherness is always dwelling within us: there is more to us than we think, so to speak', Hage asserts (2012, 289). This kind of critique has inspired what Hage terms 'alter politics': politics which not only aims at resisting existing power structures, but aims for an alternative political order (2012, 292). Experiencing the otherness within ourselves may happen in an encounter with the radical alterity of human life in other parts of the world (e.g. human others with radically different experiences of 'nature'). But it may also happen within our own societies in what Hage terms 'cracks' or 'ungovernable spaces' that presents alternatives to the dominating logic (Hage 2012, 294). In the following, we explore what such a crack may look like in the context of Danish early childhood education.

Painting trees with Jenny

A day care institution situated in a social housing area with many immigrant families had a particular focus on art activities, involving an artist, Jenny, who visited the institution every now and then to do art activities with the children. During one period, Jenny's activities addressed 'nature and sustainability', responding in this way to the curriculum theme of nature and natural phenomena. The activity focused on trees, but it was characteristic for Jenny's work that the theme and approach were open to what emerged within the pedagogical framework she had established, as illustrated in the field notes describing the activity.

> Jenny has gathered a group of children around the table in the art workshop of the kindergarten, an open room with cupboards and shelves filled up with paint, paper and other materials. Jenny informs the children that they are going outside to look at trees, and the children point to the trees on the playground. Jenny asks them if they climb the trees sometimes, but the children shake their heads. Ulla, a social educator who is also in the room, informs Jenny that in the kindergarten the children are not allowed to climb trees. Jenny looks surprised, 'You aren't?', she asks. 'No, no', the children answer, and one of the older boys points out the risk of falling down, if you climb. Jenny nods thoughtfully and continues a dialogue with the children: 'What do trees look like?', 'Why are they important?' and so on. The discussion is interrupted by a girl, Ayla, who has spotted a tractor outside and shouts out excitedly. Jenny looks at the tractor and smiles at Ayla. 'Yes, that *is* important', she says.
>
> Jenny packs up large sheets of paper, bottles of India ink and wooden boards. Ulla helps the children to get dressed; it is freezing cold outside, but the sun is out. The small group leaves the premises of the kindergarten and some of the older children run towards a tall tree in the green area in between the social housing blocks. The Syrian girl, Yara, walks with Ulla. She appears timid.
>
> Jenny asks the children to find sticks on the ground to draw with and shows them how they can soak the stick in India ink and draw on paper. Yara sits down, her feet in the ground in front of her. She dips a wooden stick into a glass filled with black India ink and starts painting, or scratching on the white paper in front of her. She looks elsewhere, yawns. Then she pauses, noticing the wind against her face. She taps the stick on the paper and starts laughing, soaking the branch in ink again, and enthusiastically continues painting, using her entire body. Above her, the branches of the tree move slowly in the wind. The scratching sound of sticks on paper fills the air and mingles with the sound of the wind that makes the paper sheets flutter. Yara starts singing in Arabic.
>
> The tractor that Ayla saw from the art workshop passes in front of the children and Ayla shouts out again: 'Tractor, tractor!'. Jenny nods, mirroring the excitement of the child with her body and asks her 'could you paint it, the tractor?' Ayla starts drawing on a new piece of paper and points, 'tractor'. Then she puts the stick with ink into her mouth. Ulla, who is standing next to her, utters a worried 'Oh', and turns to Jenny. 'What about the ink? In her mouth?' Jenny laughs and says that it is not dangerous.
>
> Jacob, one of the older boys, approaches the large tree. Jenny follows him and they touch the tree together. Jenny points to a patch of green moss and says 'try touching it, it is really soft'. Jacob touches the

moss and nods. 'And this is all green', Jenny says, touching a different spot on the tree, 'but here, yes, it is also soft. You get all "zen"'. She suggests that they lie down to look at the tree from below. The big branches are moving slowly in the wind. Jacob stands up again and Jenny says, 'I will find the green ink, then you can paint some green on your tree'.

Back in the workshop, Jenny encourages the children to take a look at their paintings from a distance and discuss the colours, the sizes of the branches and so on.

(Field note, April 2017)

The field note offers a glimpse into a pedagogical space which opens up for a variety of relationships to the surroundings and other people. This space does not guide children towards any specific conceptualization of nature; in fact, Jenny later said that she is not interested in teaching children the 'correct' names of 'natural phenomena' encountered.[6] Rather, the art activity acknowledges different kinds of experiences of and relationships with trees (and tractors). Our attention is drawn to the visual characteristics of trees, to the ways in which you might interact with them (e.g. by climbing, if you are allowed to), to the sound of the wind in the branches, to the way they feel (the softness of the moss) and to the atmosphere of immersion and the 'zen' feeling they offer. The branches of the trees become tools for painting, incorporated into the body space of the children and connecting them to ink and paper.

During our research project on children with a refugee background, we observed and participated in numerous situations and activities in which social educators more or less consciously worked on the 'integration' of refugee children into Danish society. These activities often had a strong focus on cognitive language learning, or on passing on taken-for-granted cultural norms, such as the ones about children's outdoor play described in the first part of this article. However, we also came across a few pedagogical activities that created a quite different space. The activities that stood out as different were all grounded in sensuous, aesthetic relationships. Jenny's art activity is one of these, and we propose that the activity may be seen as constituting what Hage terms a 'crack' – a space which shows us a glimpse of an alternative to dominant ways of inhabiting the world in the contexts of Danish day care institutions. One of the reasons is that Jenny's sessions allow for ambiguity.

The acknowledgement of existential ambiguity

The philosophical work of Merleau-Ponty and his understanding of the fundamental ambiguity of being in the world may be seen as supporting Hage's argument that eradicating dualisms from the surface of the earth is an impossible quest. In his later years, Merleau-Ponty, sometimes considered to be a type of thinker who accords a privileged position to consciousness and subjectivity, developed concepts which have been taken up by new materialist writers (see, e.g. Coole 2010). In particular his last, unfinished work, *The visible and the invisible* (1968), offers a new understanding of human preconscious being in the world (e.g. Merleau-Ponty 1968, 141) through the introduction of the concepts of flesh (*chair* in French) and chiasm. Flesh is not a direct synonym of matter; rather it is an 'intercorporeal field' of coexistence (Coole 2010, 106) in which matter and meaning are entwined, and human existence is woven corporeally into the material and social world (Merleau-Ponty 1968; cf. Carman 2008, 126). Flesh describes what we have in common with the world, and is thus more ontologically fundamental than our conscious experience (Carman 2008, 181; Coole 2010, 102). Merleau-Ponty's figure of the chiasm describes the interlaced relationships between humans and the world that account for the paradox of the human body being at once sentient and sensible – that is, capable of sensing the world and being sensed by it (Merleau-Ponty 1968; Toadvine 2009, 182–183). Hence, our relationship to the world is characterised by the constant ambiguity of being simultaneously immersed in and detached from the world, of 'transcendence and immanence' (Merleau-Ponty 2012, 347). 'This ambiguity is not an imperfection of consciousness or of existence', Merleau-Ponty maintains, 'it is their very definition' (Merleau-Ponty 2012, 347). In other words, the experience of being separated from the world is as existentially fundamental as the experience of being part of it (cf. Jørgensen 2015).

Immersion and detachment

In terms of pedagogical practice – as well as of research methodology – one way to approach the ambiguity of being in the world is through attention to sentient experience. Sensuous experience plays a key role in Jenny's pedagogical approach. She makes space for what Taylor terms 'children's intimate, immediate and embodied impulses to touch and become with others' (Taylor 2017, 9) by encouraging the children to see, touch and smell the surroundings, and even to taste the ink. The connection between artistic creation and outdoor experiences of landscape and weather brings into focus the chiasmic experience of sensing and being sensed by the world. The understanding of a chiasmic relationship between the human body and the world leads to an understanding of sensations as not just subjective, first person experiences, but rather as situated in relationships. In order to perceive things, we must 'live' them, Merleau-Ponty suggests:

> Thus, a thing is not actually given in perception, it is inwardly taken up by us, reconstituted and lived by us insofar as it is linked to a world whose fundamental structures we carry with ourselves and of which this thing is just one of several possible concretions. (Merleau-Ponty 2012, 341).

Along these lines, Jenifer Mason (inspired by Merleau-Ponty, amongst others) argues that sensations emanate from and in other people, in environments, situations and things (Mason 2018, 45). Sentient experience, in other words, could be seen as existing in an ambiguous space in between the human and the nonhuman. By working with, and allowing for, experiences of relations with trees through touch, sound, vision, taste and smell, and through sensing them, painting them and using their branches as tools, Jenny plays with the ambiguity and openness of our relationship to the world.

On the one hand, the activity offers an experience of embodied immersion in the world. On the other hand, back in the workshop, Jenny encourages a more detached position in relation to the world, when she asks the children to step back from their paintings and describe them as objects, thereby playing on the oscillation between immersion in an embodied activity and detached reflection. A main point in Merleau-Ponty's philosophy is the view that reflexive thinking (the ability to see oneself as an object) is part of embodied being in the world. Rejecting both Cartesian rationalism and Husserl's phenomenological reduction (*epoché*) (which removes thinking from phenomenological experience), Merleau-Ponty thus connects reflexivity to embodied existence. In his writing about the chiasm, Merleau-Ponty describes the body as simultaneously sensing and 'sensible for itself' (Merleau-Ponty 1968, 135), and this is, according to Merleau-Ponty, the opening for reflexive thought. Using the example of how we can touch one of our hands with the other, Merleau-Ponty suggests that, in this act, we are simultaneously sensed and being sensed; however, there is still a 'shift' or 'spread' (Merleau-Ponty 1968, 148) between the two experiences. This shift opens up for a 'new type of being' (Merleau-Ponty 1968, 149), which in the words of Coole 'opens up a crack for contingency and reflexivity' (Coole 2007, 147). Intellectual reflection upon human existence, in other words, is anchored in embodied, sensuous processes (cf. Jørgensen 2015). Or, as expressed by Tim Ingold: '[o]nly because we are thus immersed in the world can we imagine ourselves as existing separately from it' (Ingold 2000, 169). The anthropologist Michael Jackson has conceptualized the dynamics and relationships between direct, unreflected engagement with the world and more reflective modes of being as 'a perpetually unsettling *oscillation*' (Jackson 2013, 198, italics added).

If we pay attention to Yara's engagement in the art activity, we notice that it is not one of complete immersion; rather we see an oscillation between being immersed in relationships with the world and adopting an observing distance to them. Working pedagogically with this oscillation acknowledges the fundamental ambivalence of human existence, which, it seems to us, is often down-played in the emphasis on entanglement characteristic of new materialist 'common world' approaches.

Beyond social polarizations?

Jenny's activities not only open up an aesthetic and reflective relationship between the children and the more-than-human beings that were a part of the activity; Jenny also tries to sidestep the focus on integration and language learning that is a part of many other activities by trying to avoid the identity categories of minority and majority. She tells us that,

> when I am with the children, I do not consider whether they are from Turkey or Denmark or wherever. Or how old they are. Or what kind of challenges or diagnosis they have. It is not a political decision; my approach is just: "You are you – get involved, let's go to this party".

So, by focusing on the *common* sensory 'party' – or, as we have put it, playing with immersion in and detachment from the world – Jenny attempts to work against the polarisation inherent in identity and integration politics that underlie many other pedagogical activities with minority children, as mentioned earlier.[7]

The art activities do not have language learning as an explicit objective and appear to avoid the polarization created by language politics in day care institutions. However, language is present in other ways in the art activities, as grounded in embodied, sensuous experiences and in the oscillation between immersed activity and detached reflexion. When Yara sings in Arabic while painting, we get the impression that the words and the song arise from her sensuous experiences of the wind in her hair, the flow of painting, the scratching stick, the ink on the paper. According to Merleau-Ponty, language is not radically different from other experiences, but 'deepens and transforms' our experiences of the world through refinement, expansion and variations of meaning already found in situations, things and events (cf. Carman 2008, 18). Language is 'grounded in the spontaneous expressive capacities of embodied speakers' (Carman 2008, 28), and is thus not a separate layer of 'culture' encasing the natural person (Merleau-Ponty 2012, 195). In his later work, Merleau-Ponty describes language as expressions which build on the chiasmic relation to the world. We speak about the world, but the world, in a way, also speaks through us: '[I]n a sense [...] language is everything, since it is the voice of no one, since it is the very voice of the things, the waves, and the forests.' (Merleau-Ponty 1968, 155). Yara's song, just as her painting, seems to emerge in a space in between the child and her surroundings (cf. Rautio 2013, 396). After the session, Jenny highlights Yara's experience as the success of the day:

> When we started here [in the workshop], she was very shy and introvert. She is often like that [...], she doesn't have a lot of language. [...] And then suddenly, out there... Sitting there, feet on the ground, starting to sing. She had her own little party. [...] She opened up, she felt freedom.

The acknowledgement of ambiguities, playing with the oscillation between immersion and detachment, the accentuation of sensuous experience, the embodied approach to language: these elements provide the art activity with an atmosphere indicating the potential of aesthetic sustainability pedagogies to move beyond social divisions as well as polarizations between the human and the nonhuman.

Ambiguity as the politics of sustainability

In this article, we have explored the culturally rooted understandings of childhood and nature that form part of everyday life in Danish day care institutions through the lens of Hage's concept of generalized domestication. According to Hage, nature–culture dichotomies are created in processes of domestication which mirror the ways in which (Western) humans domesticate, and hereby establish, a polarization towards human others. Therefore, nature–culture and culture–culture binaries are necessarily linked. In the Danish setting, we propose that paying attention to the ways in which the naturalization of children's play in (regulated) nature often work to emphasize the otherness of immigrant families is essential, are we to avoid reproduction of social inequalities in day care institutions.

Within environmental and sustainability education research, new materialist thinking has offered a timely input to studies of early childhood education concerning nature and environment by challenging nature–culture dichotomies and bringing attention to the messy entanglements of the human and the nonhuman. These critiques position binary dichotomies as a key figure of Western imagery. Merleau-Ponty's concepts of flesh and chiasm have helped us understand that ambivalent and ambiguous experiences of being immersed in the world, while placing a reflective distance to it, are not only paradoxes created by the failures of enlightenment thinking, but also reflections of a more basic ambivalence inherent in human existence. Acknowledging this fundamental ambiguity, we propose, is of key importance for becoming able to take action for social and ecological sustainability. This requires the experience of our entanglement and connectivity with other beings in the world, but also the reflective understanding of our own responsibility for taking care of those others.

A situation from an art session encountered in a Danish day care institution has provided us with an example of a pedagogical space in which ambiguous relationships to the world are acknowledged and encouraged. The art session, we have suggested, may be seen as a 'crack' in dominant modes of inhabiting the world in this particular socio-cultural context, entailing a potential for moving beyond human–human and human–nonhuman polarization processes and creating more open and inclusive spaces of being for children. The approach of the artist transgresses the focus on language learning and integration dominant in pedagogical approaches to immigrant children in the Danish educational and larger societal contexts by opening up a wider space for participation through a variety of embodied and aesthetic engagements with the natural environment and art materials. The artist bypasses dominating ways of understanding immigrant children's outdoor play and relations to 'nature' in terms of the children's lacking competences or their parents' lacking cultural understanding and support. Rather, during the art activity the outdoor becomes a space for embodied immersion, as well as for reflective, aesthetic distance, explored in different ways by different children.

We do not suggest that this should be the only way of working with sustainability education in day care institutions. Our point is that the art activity questions several key understandings dominating pedagogical work with immigrant children and nature in Denmark, hence offering an opening for opposing the dominance of these modes and a potential for exploring alternative modes. As such, and in spite of the artist's own declaration, the art session could be seen as an act of radical early childhood sustainability politics.

Disclosure statement

No potential conflict of interest was reported by the authors.

Notes

1. The research was carried out in collaboration with Susanne Bregnbaek, Annelise Arent and Hanne Ellegaard under the auspices of University College Copenhagen in Denmark.
2. In this situation, the researcher shifted between the positions of participating in the art activities with the children, observing, and conducting a dialogue with the adults.
3. New materialist approaches have been developed in reaction to the strong dominance of what Coole and Frost term 'the cultural turn' in social theorizing of the last decades of the twentieth century. They argue, in

their widely cited introduction to the book 'New Materialisms', that the analytical privileging of language and discourse has to a large degree neglected material processes (Coole and Frost 2010, 3). While many new materialist thinkers draw on earlier materialist philosophies, Coole and Frost insist that there is something new in the ways in which these older materialist traditions are pushed 'in novel, sometimes experimental, directions towards fresh applications' (Coole and Frost 2010, 5). The 'horrors of the Anthropocene' (Haraway 2016, 3) have played an important part in the push for rethinking materialism.

4. A revised version of the learning curriculum passed in 2018 renames the topic 'nature, outdoor life and science'. Sustainability has only recently become a term used more widely in the context of early childhood education, and the new curriculum, for the first time, includes references to sustainability (Børne- og Socialministeriet 2018b).
5. 'Social educator' is the official Ministry of Education translation of the Danish term *paedagog*; a member of staff trained in pedagogical theory and practice in relation to younger children.
6. By highlighting this approach, we do not mean to suggest that teaching children words and concepts for natural phenomena is wrong; rather, we propose, this is not the only way to approach education for sustainability, and Jenny's approach with its focus on sensuous and aesthetic experience constitutes one alternative.
7. It should be noted that an approach that downplays the differences in children's backgrounds may also have disadvantages, if it assumes that all children have equal possibilities for participating in the activities of Danish day care institutions which in many cases reflect white Danish middle class norms and priorities (see, e.g. Bundgaard and Gulløv 2008).

References

Børne- og Socialministeriet. 2018a. *Bekendtgørelse Af Lov Om Dag-, Fritids- Og Klubtilbud m.v. Til Børn Og Unge (Dagtilbudsloven)*.

Børne- og Socialministeriet. 2018b. *Den Styrkede Paedagogiske Laereplan: Rammer Og Indhold*.

Bundgaard, H., and E. Gulløv. 2008. *Forskel Og Faellesskab: Minoritetsbørn i Daginstitution*. København: Hans Reitzel.

Carman, T. 2008. *Merleau-Ponty*. London and New York: Routledge.

Coffey, A., and P. Atkinson. 1996. Making Sense of Qualitative Data: Complementary Research Strategies. London, Thousand Oaks, CA and New Delhi: Sage Publications

Coninck-Smith, N. D., and J. Bygholm. 2011. *Barndom Og Arkitektur: Rum Til Danske Børn Gennem 300 År*. Århus: Klim.

Coole, D. 2007. *Merleau-Ponty and Modern Politics after anti-Humanism*. Lanham: Rowman & Littlefield Publishers.

Coole, D. 2010. "The Inertia of Matter and the Generativity of Flesh." In *New Materialisms: Ontology, Agency, and Politics*, edited by Diana Coole and Samantha Frost, 92–114. Durham: Duke University Press.

Coole, D., and S. Frost. 2010. "Introducing New Materialims." In *New Materialisms: Ontology, Agency, and Politics*, edited by Diana Coole and Samantha Frost, 1–43. Durham: Duke University Press.

Davis, J. 2009. "Revealing the Research 'Hole' of Early Childhood Education for Sustainability: A Preliminary Survey of the Literature." *Environmental Education Research* 15 (2): 227–241. doi:10.1080/13504620802710607.

Duhn, I., K. Malone, and M. Tesar. 2017. "Troubling the Intersections of Urban/Nature/Childhood in Environmental Education." *Environmental Education Research* 23 (10): 1357–1368. doi:10.1080/13504622.2017.1390884.

Gilliam, L. 2017. "The Impossible Bilingual Boys: Civilising Efforts and Oppositional Forms in a Multi-Ethnic Class." In *Children of the Welfare State: Civilising Practices in Schools, Childcare and Families*, edited by Laura Gilliam and Eva Gulløv, 138–164. London: Pluto Press.

Gilliam, L., and E. Gulløv. 2017. *Children of the Welfare State : Civilising Practices in Schools, Childcare and Families*. London: Pluto Press.

Gulløv, E. 2012. "Den Tidlige Civilisering: En Flertydig Bestraebelse." In *Civiliserende Institutioner: Om Idealer Og Distinktioner i Opdragelse*, edited by Laura Gilliam and Eva Gulløv. Aarhus: Aarhus Universitetsforlag.

Hage, G. 2012. "Critical Anthropological Thought and the Radical Political Imaginary Today." *Critique of Anthropology* 32 (3): 285–308. doi:10.1177/0308275X12449105.

Hage, G. 2017. *Is Racism an Environmental Threat?* Cambridge: Wiley.

Halldén, G. 2009. *Naturen Som Symbol För Den Goda Barndomen*. Stockholm: Carlssons.

Haraway, D. J. 2016. *Staying with the Trouble : Making Kin in the Chthulucene. Experimental Futures*. Durham, NC: Duke University Press.

Hastrup, K. 2010. "Feltarbejde." In *Kvalitative Metoder Og Tilgange: En Grundbog*, edited by Lene Tanggaard and Sven Brinkmann, 55–80. København: Hans Reitzels Forlag.

Hedefalk, M., Almqvist, J., and Östman, L. (2015). Education for sustainable development in early childhood education: a review of the research literature. *Environmental Education Research*, 21(7), 975–990.

Ingold, T. 2000. *The Perception of the Environment: Essays on Livelihood, Dwelling and Skill*. New York: Psychology Press.

Ingold, T. 2011. *Being Alive: Essays on Movement, Knowledge and Description*. London: Routledge.

Jackson, M. 1996. "Introduction: Phenomenology, Radical Empiricism, and Anthropological Critique." In *Things as They Are: New Directions in Phenomenological Anthropology*, Vol. 278. Indiana: Indiana University Press.

Jackson, M. 2013. *The Wherewithal of Life: Ethics, Migration, and the Question of Well-Being*. Oakland: University of California Press.

James, A., C. Jenks, and A. Prout. 1998. *Theorizing Childhood*. Camridge: Polity Press in association with Blackwell Publishers Ltd.

Jørgensen, N. J. 2015. "From Herders to Hustlers: Young People's Engagements with Sand, Sustainability and Education in Northern Kenya." Ph.thesis, Aarhus Universitet.

Kopnina, H. 2012. "Education for Sustainable Development (ESD): The Turn Away from 'Environment' in Environmental Education?" *Environmental Education Research* 18 (5): 699–717. doi:10.1080/13504622.2012.658028.

Kristjansson, B. 2006. "The Making of Nordic Childhoods." In *Nordic Childhoods and Early Education: Philosophy, Research, Policy and Practice in Denmark, Finland, Iceland, Norway, and Sweden*, edited by Johanna Einarsdottir and Judith T. Wagner. Greenwhich: IAP.

Larsen, V., and N. J. Jørgensen. 2017. "Review of Danish Research on Social Inclusion in Early Childhood Education." København.

Lotz-Sisitka, H. 2017. "Decolonisation as Future Frame for Environmental and Sustainability Education: Embracing the Commons with Absence and Emergence." In *Envisioning Futures for Environmental and Sustainability Education*, edited by Peter B. Corcoran, Joseph P. Weakland, and Arjen E. J. Wals, 45–62. Wageningen: Wageningen Academic Pub.

Mason, J. 2018. *Affinities: Potent Connections in Personal Life*. Cambridge: Polity Press.

Mcphie, J., and D. A. G. Clarke. 2018. "Nature Matters: Diffracting a Keystone Concept of Environmental Education Research – Just for Kicks." *Environmental Education Research* 1–18. doi:10.1080/13504622.2018.1531387.

Melhuus, E. C. 2013. "Utebarnehager – et Sted for Demokratisk Praksis?" Nordisk barnehageforskning 66 (1): 1–18.

Merleau-Ponty, M. 1968. *The Visible and the Invisible: Followed by Working Notes*. Evanston: Northwestern University Press.

Merleau-Ponty, M. 2012. *Phenomenology of Perception*. London and New York: Routledge.

Nxumalo, F. 2017. "Geotheorizing Mountain – Child Relations within Anthropogenic Inheritances." *Children's Geographies* 15 (5): 558–569. doi:10.1080/14733285.2017.1291909.

Nxumalo, F., and V. Pacini-Ketchabaw. 2017. "Staying with the Trouble' in Child–Insect–Educator Common Worlds." *Environmental Education Research* 4622: 1–13.

Olwig, K. R. 2005. "Introduction: The Nature of Cultural Heritage, and the Culture of Natural Heritage – Northern Perspectives on a Contested Patrimony." *International Journal of Heritage Studies* 11 (1): 3–7. doi:10.1080/13527250500036742.

Pacini-Ketchabaw, V. 2013. "Frictions in Forest Pedagogies: Common Worlds in Settler Colonial Spaces." *Global Studies of Childhood* 3 (4): 355–365. doi:10.2304/gsch.2013.3.4.355.

Pacini-Ketchabaw, V., and F. Nxumalo. 2015. "Unruly Raccoons and Troubled Educators: Nature/Culture Divides in a Childcare Centre." *Environmental Humanities* 7 (1): 151–168. doi:10.1215/22011919-3616380.

Palludan, C. 2007. "Two Tones: The Core of Inequality in Kindergarten?" *International Journal of Early Childhood* 39 (1): 75–91. doi:10.1007/BF03165949.

Pink, S. 2009. *Doing Sensory Ethnography*. London: SAGE.

Rautio, P. 2013. "Children Who Carry Stones in Their Pockets: On Autotelic Material Practices in Everyday Life." *Children's Geographies* 11 (4): 394–408. doi:10.1080/14733285.2013.812278.

Regeringen, 2018. "Ét Danmark Uden Parallelsamfund. Ingen Ghettoer i 2030."

Rooney, T. 2018. "Weather Worlding: Learning with the Elements in Early Childhood." *Environmental Education Research* 24 (1): 1–12. doi:10.1080/13504622.2016.1217398.

Rytter, M. 2018. "Writing against Integration: Danish Imaginaries of Culture, Race and Belonging." *Ethnos* 1–20. doi: 10.1080/00141844.2018.1458745.

Sandell, K., and J. Öhman. 2010. "Educational Potentials of Encounters with Nature: Reflections from a Swedish Outdoor Perspective." *Environmental Education Research* 16 (1): 113–132. doi:10.1080/13504620903504065.

Somerville, M., and C. Williams. 2015. "Sustainability Education in Early Childhood: An Updated Review of Research in the Field." *Contemporary Issues in Early Childhood* 16 (2): 102–117. doi:10.1177/1463949115585658.

Taylor, A. 2013. *Reconfiguring the Natures of Childhood*. New York: Routledge.

Taylor, A. 2017. "Beyond Stewardship: Common World Pedagogies for the Anthropocene." *Environmental Education Research* 23 (10): 1448–1461.

Toadvine, T. 2009. *Merleau-Ponty's Philosophy of Nature*. Evanston: Northwestern University Press.

Wagner, J. T., and J. Einarsdottir. 2008. "The Good Childhood: Nordic Ideals and Educational Practice." *International Journal of Educational Research* 47 (5): 265–269. doi:10.1016/j.ijer.2008.12.005.

Ween, G., and S. Abram. 2012. "The Norwegian Trekking Association: Trekking as Constituting the Nation." *Landscape Research* 37 (2): 155–171. doi:10.1080/01426397.2011.651112.

Challenging amnesias: re-collecting feminist new materialism/ecofeminism/climate/education

Annette Gough and Hilary Whitehouse

ABSTRACT

This paper discusses the apparent amnesia with regard to insights manifested in ecofeminist thought and applies a re-collective analysis to thinking on the implications of an ecofeminist new materialism for contemporary environmental education research, and curriculum practice. We engage with a conversation between feminist new materialism and the tropes of ecofeminism at this very unusual time in human history, making visible such interactions. Drawing attention to this and other apparent amnesias and, arguing from a genealogical perspective, we argue the scholarly and conceptual disruption caused by rapidly changing environmental (hence social and cultural) conditions can be fruitfully understood and analysed through a reconceived new materialist ecofeminism. This is especially important given the unequal impact of the climate emergency on women and the continued absence of a truly coherent focus on women's interests – another amnesia – at this moment when the climate dominates all human and other than human life. In exploring the relationships between feminist new materialism, ecofeminism and the more-than-human, we theoretically + materially consider the conceptual challenges of confronting the climate emergency as viewed through the lens of articulating feminisms; and we promote possibilities for further conceptual and practical environmental education research.

Introduction

Amnesia is a term with its etymological origins in *amnēsía*, a variant of *amnēstía* meaning oblivion and obliviousness. The term is related to the Greek terms *mnemnon,* (mindful), *mneme,* (memory) and *mnasthai* (to remember). In social analysis, amnesia is a forgetfulness, a failure to cause to remember, or recall; or more seriously, the loss of memory. We use the term cautiously in order to indicate that there are conceptual and practical intersections in environmental education research and practice where forgetting to make the connections is not, or will not prove, useful for the third decade of the twenty-first century. In arguing for an un-forgetting (an undoing of forgetfulness) we draw on the act of recollection, or re-collection, an act of regathering from the past to manifest useful theoretical + material concepts for understanding the deep challenges of the Anthropocene, particularly the climate emergency. The noun recollection has its etymological roots in the Latin *recollectus,* past participle of *recolligere,* meaning to take up again, to regain, to collect once more, from *re* (again) + *colligere* (gather). Recollection is defined as the ability to remember things, to possess a memory of past events (and concepts). The verb, recollect, refers to the act of remembering, to recover or recall knowledge and to bring back to the mind or memory. As Karen Barad (2012, 48) argues, "Matter feels, converses, suffers, desires, yearns and remembers".

Ecofeminism + feminist new materialism

The first amnesia we draw attention to concerns the intersections between ecofeminism and feminist new materialism (Gaard 2015, 2017; Gough and Whitehouse 2018; Merrick 2017; Sandilands 2017; Stevens, Tait, and Varney 2017b). Richard Twine (2010, 402) has argued that "the emergence of a feminist new materialism ought to usher in a renewed conversation between feminism and ecofeminism due to shared interests" and such a conversation would be a very productive space for environmental education research. It is our analysis that historically ecofeminism and the new materialist feminisms have much to offer each other, and that "a more robust and inclusive feminism" (Gaard 2017, 117) is possible, one that replaces a

> culture of domination with a world of participatory economics grounded in communalism and social democracy, a world without discrimination based on race or gender, a world where recognition of mutuality and interdependency would be the dominant ethos, a global ecological vision of how the planet can survive and how everyone on it can have access to peace and well-being. (Hooks 2000, 10)

However, as discussed later in this article, this analysis needs to acknowledge the many erasures of ecofeminist arguments through the genealogical shift from feminist philosophy to Europatriarchal thought that underpins new materialism (Gaard 2011, 2015, 2017; Sandilands 2017).

Historically, the term *ecofeminisme* was coined by Francoise d'Eaubonne in her book, *Le feminisme ou la mort* (1974). Ecofeminism was intended "to represent women's potential for bringing about an ecological revolution to ensure human survival on the planet. Such an ecological revolution would entail new gender relations between women and men and between humans and nature" (Merchant 1990, 100). According to Greta Gaard (2011, 28), a significant expansion of ecofeminist analysis "emerged from the intersections of feminist research and the various movements for social justice and environmental health, explorations that uncovered the linked oppressions of gender, ecology, race, species, and nation". The importance of ecofeminist thinking was its direct focus on 'nature' (and "Nature") (see Mcphie and Clarke 2018), how women were historically positioned, and explicitly embracing the world beyond the human. Rosemary Ruether (1975) argued that that the liberation of women and the project of addressing the (then) emerging ecological crisis were intricately linked through societal practices of domination. Her theoretical solution was to envision a "radical re-shaping" (204) of socio-economic and environmental relations. Although, as Karen Warren et al. (1993) pointed out, there are many feminisms and many ecofeminisms, a general conception of ecofeminism suffered from a backlash in the late 1990s, being criticised as essentialist, elitist and ethnocentrist, and ecofeminism was "effectively discarded" (Gaard 2011, 26). However, studies at the intersection of feminism and environmentalism have seen a resurgence of the label ecofeminism during this century (see for example, Gaard 2011, 2015; MacGregor 2017; Phillips and Rumens 2016a, 2016b; Stevens, Tait, and Varney 2017a; Thompson 2006), as distinct from other labels such as "ecological feminism" (Warren 1994; Warren and Cheney 1991), "feminist environmentalism" (Seager 1993), "feminist ecologies" (Stevens, Tait, and Varney 2017b), or "global feminist environmental justice" (Sturgeon 2009).

While clarifying labels, we also note that there is possible confusion between material feminism and feminist new materialism. Stacy Alaimo and Susan Hekman (2008, 18) distinguish what they call "material feminism" emerging "primarily from corporeal feminism, environmental feminism, and science studies" and "new materialism" which is "from, or is synonymous with Marxist feminism". While many material feminist theorists have been influenced by Marxist theory, "their definition of materiality is not, or is not exclusively, Marxist". Alaimo and Hekman see "materialist feminism" as focusing on race, sexuality, imperialism and colonialism, and anthropocentrism, but not necessarily encompassing "the materiality of human corporeality or, certainly, of nonhuman nature" (18). As new materialism developed over the past decade these fine distinctions have become blurred. We do not see this as necessarily problematic for two reasons.

The first is that strictly delineated 'schools of thought' are possibly more relatable to the modernist project rather than to the post, postmodernist turn in social thinking. The privilege of arguing over categories may come to belong to a time before we properly understood we were going to be overwhelmed by the chemical, geophysical and eco-biological realities of climate change (which we call the climate emergency in this article). This leads us to the second reason, that, in practice, scholars are running out of the luxury of time in which to debate the fixities of categories. The climate emergency is forcing a realisation there is no one conceptual model that is going to see us through. In a globalised situation where it is becoming evident that we need an 'all hands on deck' approach, scholars can draw on the strengths of relevant theoretical perspectives in order to generate means for more fully comprehending our present moment. This is why we argue against certain amnesias concerning past active theories and argue for a recollection of the strengths of historical perspectives, including those insights generated through ecofeminism.

From philosophical and theoretical perspectives, feminist new materialism attempts to 'denature' bodies (human, other, organic and inorganic) as a means to, "establish a radical break with both universalism and dualism as the co-constitutive-ness of cultural discourse and materiality" (Lenz Taguchi 2013, 707). Importantly, new materialism throws up provocative challenges for interdisciplinary feminist scholarship in that matter (including the body) is conceived as being in possession of "a distinctive kind of agency, one that is neither a direct nor an incidental outgrowth of human intentionality but rather one with its own impetus and trajectory" (Frost 2011, 70). New materialist feminism challenges some of the politically useful foundations of ecofeminism, the purposes of which were to undo the normalised power of gendered essentialisms that kept women and 'Nature' oppressed. In considering the relationship between new material feminism and ecofeminism, the challenge is to denaturalize nature *and* deculturalize culture, as "matter is not inert, neither does it form an empty stage for, or background space to, human activity. Instead, matter is conceptualised as agentic and all sorts of bodies, not just human bodies, are recognised as having agency" (Taylor and Ivinson 2013, 666).

Another amnesia that is relevant to this discussion is what was being articulated by ecofeminists in the 1990s is not fully acknowledged (Gaard 2017, 118), and that,

> feminist engagement with theories of posthumanism (e.g.Barad 2003) and the emergence of 'new materialist feminists' (e.g. Hird 2004) do not address the relationship between feminism and ecofeminism: many new materialists do not acknowledge ecofeminist scholarship, despite its foundational contributions to new materialist feminisms and the continuing intersections of these two theoretical perspectives.

New materialist feminists are embracing posthumanism (Gaard 2017, 124) but not acknowledging its associations with ecofeminism, and also "disavow[ing] the history of biopolitics in feminism" and forgetting, ignoring, or turning away "from questions of matter, biology, and nature" (Sandilands 2017, 232). Yet posthumanism is relevant, as Karen Barad (2007, 32) argues, for "posthumanism marks a refusal to take the distinction between 'human' and 'nonhuman' for granted, and to found analysis on this presumably fixed and inherent set of categories". However, rather than use the term posthumanism, we follow Elspeth Probyn (2016) and adopt the term more-than-human.

Probyn argues that, while the terms "posthuman" and "more-than-human" "are generative in that they seek to shake up any assumptions that we might have had about what conjoins and what separates us, not to mention what that profoundly confusing 'us' might be,". Probyn (2016) prefers "more-than-human" because it is "ontologically and materially relational, and opens up new epistemologies as it narrows the diverse and shifting relations between and among humans, and the many different aspects of that are so much more-than-human" (p. 110).

The "renewed conversation" (Twine 2010, 402) between feminism and ecofeminism is not yet happening to any great extent, but it is important because without it there is an amnesia in new material feminism towards what had been discussed and achieved by ecofeminist theorising. Both of these forms of feminism share a concern with challenging what Bronwyn Davies (2018, 114) calls "the takenfor-granted ascendance of all things human ... and the co-implication of humans with non-human matter" and "the apparent indifference ... to the ways human thinking-doing affects what will matter in the world". Before Barad (2007) offered her new concepts and new ways of thinking in new materialism, Carolyn Merchant (2003, 201) recognised that epistemology and ontology are always relational and interacting:

> The material world is mentally constructed and interpreted in terms of language, but it exists as biophysical entities. Only biophysical beings, such as humans, have the tongues and brains to create and perform speech acts – to invent the logos, or word. The real physical world and the constructed mental world thus exist in dialectical relation to each other. Reality and narrative, I argue, interact with each other.

So why is it that new materialist feminists have tended to ignore the genealogical connections with ecofeminism? Historical amnesia seems to be a growing issue with new generation academics who perhaps do not have the time to immerse themselves in historical literature from theirs and related fields, however, we believe it is important to look at where we have come from, where we are now and where we can go from here in seeking understanding and changes: as Alasdair MacIntyre (1984, 216) wrote, "I can only answer the question 'What am I to do?' if I can answer the prior question 'Of what story or stories do I find myself a part?'". In an environmental education context, the stories of which we are a part include ecofeminism, and these stories must not be forgotten. More than this, ecofeminism describes "a socially constructed association among women (sex), femininity (gender), and nature that was contextual and fluid, not ahistorical and static" (Merchant 1992, 31). The research programs of which we are a part can flow from these associations, as reflected in the two recent special issues of *The Journal of Environmental Education* on gender and environmental education (Volumes 48(1) and 49(4)). The articles demonstrated commitments to different gendered methodologies – ranging from the personal to the political and drawing on a range of theoretical resources including ecofeminism, poststructuralist, materialist and other feminisms, intersectionality, Foucauldian analysis, assemblages, masculinity and environmental justice – and how the different perspectives of such approaches invigorate environmental education scholarship and show what is possible when gendered methodologies are adopted. However, gendered methodologies still remain on the margins of most environmental education research programs, as the editors of these special issues note (Gough, Russell, and Whitehouse 2017).

What ecofeminism told us in the past was that systematic oppression of a multiply categorised, naturalised, otherness, was indelibly linked with systems of patriarchal, hegemonic and colonising thinking that serves to oppress the agency of bodies actively constituted into a category 'nature' and a category of 'women' to their mutual disadvantage. This remains a profound insight. A genealogical examination of the ecofeminist writings of the 1980s and 1990s reveals how such hierarchy of favouritism – dominion over nature, domination over women - has underpinned the accelerating disaster that befalls us all in the present. We must recollect that older ecofeminism contains many insights relevant to new materialist feminism. Yes, the critiques of ecofeminist essentialism are valid (see Gaard 2011). The essentialist dustings were always empirically suspect anyway, as there certainly is no essential, empirically delineated nature to be found, and women were always agentic, and seeking agency, whether this was recognised or not. However, the ecofeminist analysis of the workings of the system against the fabric of life on earth and the struggles of women against this dual domination remains extant and meaningful in material ways. And, on a historical note, it is not wise to abandon powerful insights solely on the basis of critique.

Barad (2012, 49) is quite emphatic on this point, saying:

> Critique is all too often not a deconstructive practice, that is, a practice of reading for the constitutive exclusions of those ideas we cannot do without, but a destructive practice meant to dismiss, to turn aside, to put someone or something down—another scholar, another feminist, a discipline, an approach, et cetera. So, this is a practice of negativity that I think is about subtraction, distancing and othering.

And Barad (2012, 50) herself is always keen for any proposal (theoretical, philosophical, conceptual, practical) to be read "in the sense of it being suggestive, creative and visionary".

It is our view that 20_{th} century ecofeminism was certainly visionary. For example, according to Karen Warren (2000, 1), "Nature is a feminist issue, might be the slogan of ecofeminism". The purpose of earlier ecofeminism was to recognise and investigate "the interconnection among all unjustified human domination" (Warren 2000, 2) with a focus on women, because the focus on women (rather than on geography, poverty, race, class, indigeneity etc.) "reveals important interconnected systems of human domination" (2). And, significantly, when human groups are harmed by ecological destruction (inadvertent or deliberate), it is generally women and their children who suffer more material harm than men – a social reality that has been well understood for decades. For example, Warren (2000, 4) quotes from a 1995 United Nations document titled *The World of Women* that highlights how rural women were, at the time, "excluded from traditional rural development programmes that … provide training - and from the credit and other institutional support needed for rural development". Overcoming systematised and institutional disadvantage in relation to social development and environmental care experienced by the majority of the world's women remains a work in progress, as evidenced by Goal 5 on Gender Equality of the United Nations (2016) Sustainable Development Goals, and that gender is recognised as "a prerequisite for the achievement of other Goals, including Goal 4", which is concerned with Education (United Nations 2019, 8).

Warren (2000) is but one example of how ecofeminist analysis pre-configured the new materialism. To quote Barad (2012, 54) again; "Agency is not … a property of persons or things; rather, agency is an enactment, a matter of possibilities for reconfiguring entanglements … entailed in reconfiguring material-discursive apparatuses of bodily production, including the boundary articulations and *exclusions* that are marked by those practices". Ecofeminism was always concerned with exclusions, of women's bodies and of "natural", "Othered" bodies from the domains of privilege, and from privilege's rapacious pursuits.

Climate matters [for] eco-feminist new materialism

Barad (2007, 3) has argued that what is 'new' about new materialism is that we come to know that "matter and meaning are not separate entities" (2007, 3). Jamie Mcphie and David Clarke (2018, 5) point out that concepts "matter in very literal ways" and that, "conceiving the world in terms of material – and materiality – allows researchers to think with different conceptions, different stories that matter different consequences. For research is a story and it does matter consequences" (6). As thinking counts, recollecting the 'eco' into new materialist feminism allows for a larger story to be told, the amnesia forgotten. The term *eco* is an adjective derived from the Greek *oiko* meaning house, home, settlement, place of being. In contemporary use, *eco* is combined with nouns and adjectives to indicate an additional consideration of the ecological, the space beyond the wholly human. Eco + feminism produced a sophisticated and complex analysis of human systems of oppression and domination in the later 20_{th} century in order to reconfigure those relations and relationships towards a larger vision of justice and inequality (Warren 2000). Gaard (2011, 31) sees that what is needed in the 21_{st} century is "an *intersectional*" ecological-feminist approach. One that will frame conditions "in such a way that people can recognize common cause across the boundaries of race, class, gender, sexuality, species, age, ability, nation—and affords a basis for engaged theory, education, and activism" (Gaard 2011, 44). Both ecofeminism and new material feminisms similarly shift the emphasis from humans at the centre and "propose that all manner of bodies, objects and things have agency within a confederation of meaning-making" (Taylor in press).

For Stacy Alaimo (2010), the way forward is through a deeper understanding of "trans-corporeality", which she describes as a way of theorizing the relationship between humanity and the world at large as not being clearly delineated and separate, but as fluid. Alaimo (2008, 238) has pointed out that, "human corporeality in all its material fleshiness is inseparable from 'nature' or 'environment'", and that body/nature is also intertwined with 'culture'. However, as Greta Gaard (2015, 25) points out, this is another example of amnesia as

> transcorporeality, the physical fact of our co-constituted embodiment with other flows of life, matter, and energy… rests on four decades of feminist science studies and ecofeminist perspectives on the human-environment connection, developing knowledge in the study of gender, race, class, age, and public health.

Considering inherent agency of material bodies is a means for better apprehending the active workings of power (Barad 2003) and for questions regarding causation, resistance and effect (Frost 2011). Blanche Verlie (2017, 2019) has discussed this in terms of climate as entanglement and becoming-climate. Mcphie and Clarke (2018) argue strongly for immanent thinking. All environmental education research is best served when we are able to draw inclusively on all available ideas across ecofeminism and new material feminisms (Gough and Whitehouse 2018). Any form of eco-amnesia in theory and practice is a significant conceptual oversight, particularly when we are facing a climate emergency, which is quite possibly the 'Mother' of all materialities and made all the more frightening because how each and all of us will experience the effects will be a matter of uncertainty. In her complex, historical analysis, Merchant (2016) explores concepts of chance, randomness unruliness and unpredictability and argues that coming to terms with our collective pollution of the atmosphere-ocean system means coming to terms with a world in which "chance and randomness will constitute our new reality" (6).

Viewed temporally, the whole condition of risk 'we' (every material body, large and small) collectively face, is the matter of carbon pollution changing the chemistry of our oceans and atmosphere. George Marshall (2014, 2) calls this "the ultimate challenge to our ability to make sense of the world around us". The urgency to make sense is generated through sophisticated scientific modelling, empirical evidence and our lived experiences of what Isabelle Stengers (2015, 2) says is "the brevity of time to avoid brutal climate change". Driving our dread is the fear that "human action risks being ineffective in a system which is already exhibiting a powerful momentum in an apocalyptic direction" (Knox 2015, 104). As Lawrence Torcello (2017, np) succinctly states, "time is running out and physics is managing the clock". Some recent environmental education research is reconsidering philosophical, theoretical and practical approaches to climate change education and communication (Börebäck and Schweiler 2018; Stapleton 2019; Verlie 2017, 2019; Verlie and CCR 15 2018), exploring the problematics of material entanglements to challenge the hierarchical hegemonies implicit within the climate emergency. For example, Sarah Riggs Stapleton (2019) argues for a climate justice approach to climate change education, through "storying" climate change (Crate 2017) and providing opportunities "to provide needed context, nuance, and personal connection to the problem" (736). Judy Horacek (see Figure 1) encapsulates such storying.

Blanche Verlie and CCR15 (2018, 1) take another approach and "explore cultivating climate action if we decentre the human from our climate pedagogies". However, gender is still not generally brought to the forefront, which is particularly dismaying given women and their children are already, *materially*, caught in the grip of rapid environmental change, to wit: "The women of Bangladesh are among the first to face the impacts of climate change, and their suffering is

Figure 1. Storying climate change 'Cartoon by Judy Horacek ©2019, first published in *The Age*, reprinted with permission'.

disproportionate. In the cyclone disaster of 1991, for example, 90% of the 140,000 people who died in the country were women." (Environmental Justice Foundation 2017, online).

It is the case that the *material* effects of the climate emergency remain profoundly gendered. Anna Ines Abelenda (2014, 125) remarks that the international order is yet to "seek to bring about just and healthy economies, which form the basis for the realisation of the fulfilment of the full gamut of rights and freedoms, including economic, social and women's rights, rather than merely attempt to make the existing order appear less egregious in its exploitation". And Jeremy Deaton (2018, np) states that, given "the climate crisis is tightly bound with other urgent issues … societies cannot effectively deal with the carbon crisis without addressing, racism, xenophobia and economic inequality … [and] without tackling pervasive sexism". Such analyses have always formed part of the ecofeminist project. Vandana Shiva (1989, 1991, 1997, 2005) has consistently argued that the nexus between a dominant patriarchy and a dominant hyper-capitalism (and now, a hyper-digital revolution with all its attendant pollution and toxicity problems) is the underlying cause for the rise of global, profit orientated corporations, the biodiversity crisis, the homogenisation of human cultures and the attendant acceleration of the climate crisis.

For Gaard (2015, 21), the historical focus of the discussion has shifted from being *about* women, particularly as "*victims* of environmental degradation in need of rescue" to more thoughtfully reconsidered "gender as a system structuring power relations" (Gaard 2015, 22). For example, Atieno Mboya Samandari (2016, 25) uses ecofeminist analysis to argue that an inferiorised, ocean/atmosphere "has been used as a seemingly endless waste receptacle for by-products of industrialization which serve the male dominated status quo". Unequal power relations magnify the matter of physical risk. The Global Gender and Climate Alliance (GGCA) (2013) has documented the real and present dangers to women, quoting a World Bank survey finding that 103 out of 140 countries impose legal differences on people solely on the basis of gender. Two thirds of the world's illiterate adults are women, and normative oppressive relations limit women from acquiring the information and skills necessary to escape or avoid hazards (e.g.swimming or climbing trees to escape rising water levels). Even dress codes imposed on women restricts their mobility in times of immediate disaster, as does their responsibility for the safety of young children (Mann 2015, 1).

Hannah Knox (2015, 92) describes climate as "the new geosocial leviathan", and a major challenge for climate education as "how to encourage people to internalize objective accounts of climate in order that they could see themselves as part of the very dynamic that visualizations of predictive climate models were describing" (95). In a nutshell, Knox's argument is that climate *is* us, which is similar to Verlie's (2017, 569) argument that seeing "[c]limate as entanglement accounts for how climate science works, without conceptualizing humans and the climate as independent entities that preexist their intra-action". If we come to understand the climate as "not primarily a thing but a set of relationships" (Knox 2015, 103) then it follows that the gendered nature of those relationships also comes to the fore. As Megan Capriccio (2017) writes on the *One Million Women* website: "Are you also passionate about equality across the gender spectrum? Do you generally think that all living things should be treated with a sense of respect and dignity? Then jump on in, you're an ecofeminist!"

Gaard (2015, 23) is very clear when she writes, "make no mistake: women are indeed the ones most severely affected by climate change and natural disasters, but their vulnerability is not innate; rather it is a result of inequities produced through gendered social roles, discrimination, and poverty". The critiqued essentialisms of 20_{th} century ecofeminism sieved through newer understandings of material relationships and interrelationality reveal a more nuanced approach as to the value of an eco/logical, new feminist materialism. For whatever its faults, the tenets of ecofeminism were always about care and caring, agency and action. And herein lies the challenge for environmental education scholars and practitioners – we need to take on that leadership and draw attention to the silences around gender and the climate emergency in our research and practice. We have to imagine the possible, predictably unpredictable futures that hinge on our current actions in a real and material world; interpreting the prevalent graphs, models, visualisations and media reports into tangible realms of experience and agency, while knowing matter and meaning are inextricable in our educational practices.

The climate justice approach to climate change education discussed by Stapleton (2019) is a good beginning – but it needs a more robust gender dimension. As the recent report to the United Nations General Assembly on the Implementation of education for sustainable development in the framework of the 2030 Agenda for Sustainable Development (2019, 9) noted,

> At present, Sustainable Development Goal 5, "Achieve gender equality and empower all women and girls" is far from being a reality. The key message of UNESCO taken forward through its actions, programmes and strategy is consequently clear: "women and men must enjoy equal opportunities, choices, capabilities, power and knowledge as equal citizens. Equipping girls and boys, women and men with the knowledge, values, attitudes and skills to tackle gender disparities is a precondition to building a sustainable future for all".

For Claire Brault (2017, 27) "imagining climate futures needs feminist insight" and she cautions that we all have to be aware and critical of certain "hegemonic assumptions and epistemological procedures that animate many of today's debates" (5).

For the more optimistic practitioners and researchers, Naomi Klein (2016, online) argues that, while "climate change acts as an accelerant to many of our social ills – inequality, wars, racism … it can also be an accelerant for the opposite".

Moving forward

We have previously written about an apparent amnesia around women and bringing feminist perspectives into environmental education research (see, for example, Gough (1994, 1999, 2013), Gough and Whitehouse (2003, 2018)), about posthumanism, morethan-human, biopolitics and environmental education (Gough 2015, 2017, 2019; Gough and Gough 2016, 2017) and about climate change education (Stevenson, Nicholls, and Whitehouse 2017, Whitehouse 2017, Whitehouse, Larri, and Colliver 2019). In this article, we bring together an argument that amnesia can be overcome through recollection and reinvigoration; that the past and the present can be drawn together to provide meaningful impetus for research. And, we have argued elsewhere, "environmental education researchers should draw more inclusively on international sociological and cultural thinking across ecofeminism and material discursive analyses" (Gough and Whitehouse 2018, 344). Unfortunately, ecofeminist and related feminisms have not had much influence on standardised school curriculum, even as our socioenvironmental crises would logically make the need for such teaching more urgent.

The current guiding document for schooling in Australia is the *Melbourne Declaration on Educational Goals for Young Australians* (MCEETYA 2008). In the Preamble to this Declaration it states (MCEETYA 2008, 5),

> Complex environmental, social and economic pressures such as climate change that extend beyond national borders pose unprecedented challenges, requiring countries to work together in new ways. To meet these challenges, Australians must be able to engage with scientific concepts and principles and approach problem solving in new and creative ways.

These challenges are translated in Goal 2 as "All young Australians become successful learners, confident and creative individuals, and active and informed citizens … work for the common good, in particular sustaining and improving natural and social environments" (MCEETYA 2008, 8–9). However, the only action related to the environment included in the Declaration document is "a focus on environmental sustainability will be integrated across the curriculum" (MCEETYA 2008, 14), which was part of the recommendations or framing what has become the Australian Curriculum, with a cross curriculum priority of Sustainability (ACARA 2019). In addition, while the Preamble intimates that science understanding is important in meeting complex challenges such as climate change the Science curriculum does little to educate young Australians to meet these challenges. Climate change is only referred to three times in the Science curriculum, all at Year 10 level (ACARA 2019) where the content involves:

- considering the role of science in identifying and explaining the causes of climate change
- considering how computer modelling has improved knowledge and predictability of phenomena such as climate change and atmospheric pollution
- considering the scientific knowledge used in discussions relating to climate change

None of these content statements provide opportunities to engage with "ecological, economic, and political sustainability by developing theory that responds to the racialized … and gendered climate crises of the Anthropocene" (Gaard 2017, 126). (The term 'Anthropocene' does not yet gain a mention in the national Australian Curriculum.) Given these deliberate silences (Whitehouse, Larri, and Colliver 2019), the research challenge for, and within, all formal educational institutions is to transform the normative definitions and practices of environmental or sustainability education to address the material crises of a destabilised climate. And similarly, community education in the form of lifelong learning has a strong place in "the promotion of sustainable ways of living; challenging traditional views on development and the exclusively technical approach to reversing climate change; awareness raising on the need to have a gender perspective; and disseminating local, indigenous and traditional knowledge" (Agostino 2011, online).

It is our conclusion that more deeply conceptualised forms of feminism are needed now more than ever. *The Global Gender Gap Report 2017* (World Economic Forum 2017) benchmarked gender parity in 144 countries and came to the conclusion that, at present rates of social change, global gender equality was still over two hundred years away. This is not a good sign for a humankind facing a planetary accounting in the next decade when the absence of gender parity is analysed as being largely responsible for the climate crisis. Sandra Harding (2015, 79) admonishes that:

> The failure to address women's issues directly in development contexts not only damages women's chances for flourishing and for equality; it also renders it impossible to achieve the eradication of poverty and advance of other, noneconomic kinds of flourishing that supposedly have been the goals of development projects for more than six decades.

The concept of climate justice is gaining international traction through NGOs such as *The Climate Reality Project* (see https://www.climaterealityproject.org/) and through the United Nations Framework Convention on Climate Change (UNFCCC). At COP 23 in Bonn, Germany, November 2017, a Gender and Climate Change Gender Action Plan (GAP) was proposed noting that "a gender-responsive climate policy continues to require further strengthening in all activities concerning adaptation, mitigation and related means of implementation … as well as decision making on the implementation of climate policies" (UNFCCC 2017, 1). The GAP aims to increase the number of women who are active in climate decision-making, and train up policy-makers on bringing gender equality into climate funding programmes, and engage grassroots women's organisations for local and global climate action. How this plays out in practice is another question, but the aims at least are clear. Hilda Heine (2017, online), the President of the Marshall Islands, pointed out that women who face "a multitude of structural barriers … aren't making enough of the decisions, and the decisions aren't yet doing enough for women". A climate justice approach to climate change education and research has the potential to bring together the personal and the political and to foreground "how climate, climate knowledge and climate knowers comerge through intra-action" (entanglement) (Verlie 2017, 569). Recent United Nations (2019) and UNESCO (2019) documents related to education for sustainable development do give much greater prominence to gender. For example, UNESCO. (2019, Annex page 3) is currently exploring education for sustainable development beyond 2019 and makes particular mention of gender together with climate change:

> ESD is an instrument to achieve all the SDGs, and each of the SDGs comes with specific gendered challenges. ESD takes on a cross-disciplinary and systemic approach that enables the question of gender equality to be linked to the various issues of sustainable development. There is, for example, a gendered facet of vulnerability to hazards induced by climate change … It should therefore become a priority to provide women with access to ESD. In this regard, ESD actively promotes gender equality, and creates conditions and strategies that empower women.

Gender justice can be further promulgated in environmental education research and practice drawing on new materialist, new feminist materialism, ecofeminist and more-than-human concepts and approaches, as all have their conceptual strengths. We need to make haste towards full realisation of the conditions and extent of the calamity of the climate emergency. This is not a time in which to condone amnesia. This is a time for recollection and reconnection– a time to remember our materialities, our corporealities, our natures, and our intersecting and desiring feminisms. Diffusing boundaries in order to achieve a great whole we have to get our heads around a present future and clearly there is much work to be done, and many spaces for new thinking, new research and resistance actions are opening up for environmental education research and praxis in a rather frightening hurry. Cecilia Åsberg (2017, 198) argues we must truly consider "what kinds of ethics and critiques, arts and sciences, politics and methods, can account for the changes in spatial and temporal scales introduced by climate change, species extinction, the life sciences, or the emergence of the politics of Life itself ".

It is reassuring to see examples of environmental education research emerging that are "making connections" (Gaard (2017, 126) found in the two recent special issues of *The Journal of Environmental Education* (volumes 48(1) and 49(4)), and in the writings of environmental education researchers such as Karen Malone (2016a, 2016b) and Teresa Lloro-Bidart (2017), apart from our own work. What is important in this recent writing, and which needs to be continued in developing a more robust feminism, is the dismantling of the nature-culture binary, the disruption of anthropocentric views, the new ways of encountering more-than-human worlds, and assertion of more political standpoints. As part of this we need to pursue more-than-human perspectives in environmental education research, including "learning how to open the presences of otherness and how to form relationship of mutuality with others we can never fully know" (Gruenewald 2003, 40) because "[t]he more-than-human, if it is to be meaningful as a perspective, makes us confront again and again the relatedness of all entities" (Probyn 2016, 163). This is not what is found in the Foundation to Year 10 Australian Curriculum (ACARA 2019) but it should be there (and in every other curriculum) if we are to achieve gender and climate justice. Given the existential risks we face can we afford to wait for curriculum to catch up? Such is our continuing challenge as environmental educators and researchers.

Acknowledgements

We would like to sincerely thank the anonymous reviewers together with David Clarke, Jamie Mcphie and Alan Reid for their constructive suggestions for strengthening our arguments. We are very grateful for this feedback.

Disclosure statement

No potential conflict of interest was reported by the authors.

ORCID

Annette Gough http://orcid.org/0000-0002-4189-216X

References

Abelenda, Anna Ines. 2014. "A Feminist Perspective on the Post-2015 Development Agenda Background: The Notion of Development at the Crossroads." *Equal Rights Review* 13: 117–128.

Agostino, Ana. 2011. "Gender Equality, Climate Change and Education for Sustainability." *Adult Education and Development*, 76. www.dvvinternational.de/en/adult-education-and-development/editions/aed-762011/

Alaimo, Stacy. 2008. "Trans-Corporeal Feminisms and the Ethical Space of Nature." In *Material Feminisms*, edited by Stacy Alaimo and Susan J. Hekman, 237–264. Bloomington, IN: Indiana University Press.

Alaimo, Stacy. 2010. *Bodily Natures: Science, Environment, and the Material Self.* Bloomington: Indiana University Press.

Alaimo, Stacy, and Susan J. Hekman. 2008. "Introduction: Emerging Models of Materiality in Feminist Theory." In *Material Feminisms*, edited by Stacy Alaimo and Susan J. Hekman, 1–19. Bloomington: Indiana University Press.

Åsberg, Cecilia. 2017. "Feminist Posthumanities in the Anthropocene: Forays into the Postnatural." *Journal of Posthuman Studies* 1 (2): 185–204. doi:10.5325/jpoststud.1.2.0185.

Australian Curriculum and Assessment and Reporting Authority (ACARA). 2019. "The Australian Curriculum." https://www.australiancurriculum.edu.au/f-10-curriculum/cross-curriculumpriorities/sustainability/

Barad, Karen. 2003. "Posthumanist Performativity: Toward an Understanding of How Matter Comes to Matter." *SIGNS: Journal of Women in Culture and Society* 28 (3): 801–831. doi:10.1086/345321.

Barad, Karen. 2007. *Meeting the Universe Halfway: Quantum Physics and the Entanglement of Matter and Meaning.* London, UK: Duke University Press.

Barad, Karen. 2012. "Matter Feels, Converses, Suffers, Desires, Yearns and Remembers" Interview with Karen Barad." In *New Materialism: Interviews and Cartographies*, edited by Rick Dolphijn and Iris van der Tuin, 48–70. Ann Arbor, MI: Open Humanities Press. doi:10.3998/ohp.11515701.0001.001.

Börebäck, Maria K., and Elias Schweiler. 2018. "Elaborating Environmental Communication within 'Posthuman' Theory." *Journal for the Philosophical Study of Education* 3: 1–30.

Brault, Claire. 2017. "Feminist Imaginations in a Heated Climate: Parody, Idiocy and Climatological Possibilities." *Catalyst: Feminism, Theory, Technoscience* 3 (2): 1–33. DOI: 10.28968/cftt.v3i2.123.g276. doi:10.28968/cftt.v3i2.28847.

Capriccio, Megan. 2017. "What is Ecofeminism?" *One Million Women*, February 13. https://www.1millionwomen.com.au/blog/what-ecofeminism/

Crate, Susie. 2017. "Storying Climate Change." *Anthropology News* 58 (2): e64–e69. doi:10.1111/AN.410.

d'Eaubonne, Francoise. 1974. *Le Feminisme ou La Mort.* Paris, France: Pierre Horay.

Davies, Bronwyn. 2018. "Ethics and the New Materialism: A Brief Genealogy of the 'Post Philosophies in the Social Sciences." *Discourse: Studies in the Cultural Politics of Education* 39 (1): 113–127.

Deaton, Jeremy. 2018. "Why Climate Deniers Target Women. Women Who Work on Climate Change Regularly Endure Sexist Attacks." *Nexus Media*, February 2. https://nexusmedianews.com/why-climate-deniers-target-womenf64776ee7c5b

Environmental Justice Foundation. 2017. "Climate Displacement in Bangladesh." https://ejfoundation.org/reports/climate-displacement-in-bangladesh

Frost, Samantha. 2011. "The Implications of New Materialism for Feminist Epistemology." In *Feminist Epistemology and the Philosophy of Science: Power in Knowledge*, edited by Heidi E. Grasswick, 69–83. Dordrecht: Springer.

Gaard, Greta. 2011. "Ecofeminism Revisited: Rejecting Essentialism and Re-Placing Species in a Material Feminist Environmentalism." *Feminist Formations* 23 (2): 26–53.

Gaard, Greta. 2015. "Ecofeminism and Climate Change." *Women's Studies International Forum* 49: 20–33. doi:10.1016/j.wsif.2015.02.004.

Gaard, Greta. 2017. "Posthumanism, Ecofeminism, and Inter-Species Relations." *In Routledge Handbook of Gender and Environment*, edited by Sherilyn MacGregor, 115–129. Abingdon, UK: Routledge.

Global Gender and Climate Alliance (GGCA). 2013. "Overview of Linkages between Gender and Climate Change." https://issuu.com/undp/docs/pb1_asiapacific_capacity_final

Gough, Annette. 1999. "Recognising Women in Environmental Education Pedagogy and Research: Toward an Ecofeminist Poststructuralist Perspective." *Environmental Education Research* 5 (2): 143–161. doi:10.1080/1350462990050202.

Gough, Annette. 2013. "Researching Differently: Generating a Gender Agenda for Research in Environmental Education." In *International Handbook of Research on Environmental Education*, edited by Robert B. Stevenson, Michael Brody, Justin Dillon, and Arjen Wals, 375–383. New York: Routledge. doi:10.4324/9780203813331.ch35.

Gough, Annette. 2015. "Resisting Becoming a Glomus Body within Posthuman Theorizing: *Mondialisation* and Embodied Agency in Educational Research." In *Posthumanism and Educational Research*, edited by Nathan Snaza and John Weaver, 254–275. New York: Routledge.

Gough, Annette. 2017. "Searching for a Crack to Let Environment Light in: Ecological Biopolitics and Education for Sustainable Development Discourses." *Cultural Studies of Science Education* 12 (4): 889–905. doi:10.1007/s11422-017-9839-8.

Gough, Annette. 2019. "Symbiopolitics, Sustainability, and Science Studies: How to Engage with Alien Oceans." *Cultural Studies ↔ Critical Methodologies*: doi: 10.1177/1532708619883314.
Gough, Annette, and Noel Gough. 2016. "The Denaturation of Environmental Education: Exploring the Role of Ecotechnologies." *Australian Journal of Environmental Education* 32 (1): 30–41. doi:10.1017/aee.2015.34.
Gough, Annette, and Noel Gough. 2017. "Beyond Cyborg Subjectivities: Becomingposthuman Educational Researchers." *Educational Philosophy and Theory* 49 (11): 1112–1124. doi:10.1080/00131857.2016.1174099.
Gough, Annette, Constance Russell, and Hilary Whitehouse. 2017. "Moving Gender from Margin to Center in Environmental Education." *The Journal of Environmental Education* 48 (1): 5–4.
Gough, Annette, and Hilary Whitehouse. 2003. "The 'Nature' of Environmental Education from a Feminist Poststructuralist Viewpoint." *Canadian Journal of Environmental Education* 8: 31–43.
Gough, Annette, and Hilary Whitehouse. 2018. "New Vintages and New Bottles: The 'Nature' of Environmental Education from New Material Feminist and Ecofeminist Viewpoints." *The Journal of Environmental Education* 49 (4): 336–349. doi:10.1080/00958964.2017.1409186.
Gough, Annette. 1994. "Fathoming the Fathers in Environmental Education: A Feminist Poststructuralist Analysis." Unpublished doctoral diss., Geelong, Victoria: Deakin University.
Gruenewald, David. 2003. "At Home with the Other: Reclaiming the Ecological Roots of Development and Literacy." *The Journal of Environmental Education* 35 (1): 33–43. doi:10.1080/00958960309600593.
Harding, Sandra. 2015. *Objectivity & Diversity: Another Logic of Scientific Research*. Chicago, IL: The University of Chicago Press.
Hird, M. J. 2004. Feminist Matters: New Materialist Considerations of Sexual Difference. *Feminist Theory* 5 (2): 223–232.
Heine, Hilda. 2017. "Global Climate Action Must Be Gender Equal." *The Guardian*, November 15. https://www.theguardian.com/environment/2017/nov/15/globalclimate-action-must-be-gender-equal
Hooks, bell. 2000. *Feminism is for Everybody: Passionate Politics*. Cambridge, MA: South End Press.
Klein, Naomi. 2016. "Let Them Drown: The Violence of Othering in a Warming World." *London Review of Books* 38 (11): 11–14.
Knox, Hannah. 2015. "Thinking like a Climate." *Distinktion: Journal of Social Theory* 16 (1): 91–109.
Lenz Taguchi, Hillevi. 2013. "Images of Thinking in Feminist Materialisms: Ontological Divergences and the Production of Researcher Subjectivities." *International Journal of Qualitative Studies in Education* 26 (6): 706–716.
Lloro-Bidart, Teresa. 2017. "A Feminist Posthumanist Political Ecology of Education for Theorizing Human-Animal Relations/Relationships." *Environmental Education Research* 23 (1): 111–130. doi:10.1080/13504622.2015.1135419.
MacGregor, Sherilyn, ed. 2017. *Routledge Handbook of Gender and Environment*. Abingdon, UK: Routledge.
MacIntyre, Alasdair. 1984. *After Virtue*. Notre Dame, IN: Notre Dame University Press.
Malone, Karen. 2016a. "Reconsidering Children's Encounters with Nature and Place Using Posthumanism." *Australian Journal of Environmental Education* 32 (1): 42–56. doi:10.1017/aee.2015.48.
Malone, Karen. 2016b. "Posthumanist Approaches to Theorizing Children's HumanNature Relations." In *Space, Place, and Environment*, edited by Tracey Skelton, Karen Nairn, and Peter Kraftl, 185–206. Singapore: Springer.
Mann, Tracy. 2015. "COP21 Insights: Climate Wise Women." 31 December. http://gender-climate.org/resource/cop21-insights-climate-wise-women/
Marshall, George. 2014. *Don't Even Think about It: Why Our Brains Are Wired to Ignore Climate Change*. New York: Bloomsbury.
Mcphie, Jamie, and David A. G. Clarke. 2018. "Nature Matters: Diffracting a Keystone Concept of Environmental Education Research – Just for Kicks." *Environmental Education Research*: 1–18. doi:10.1080/13504622.2018.1531387.
Merchant, Carolyn. 1990. "Ecofeminism and Feminist Theory." In *Reweaving the World: The Emergence of Ecofeminism*, edited by Irene Diamond and Gloria Feman Orenstein, 100–105. San Francisco, CA: Sierra Club Books.
Merchant, Carolyn. 1992. *Radical Ecology: The Search for a Livable World*. New York: Routledge.
Merchant, Carolyn. 2003. *Reinventing Eden: The Fate of Nature in Western Culture*. New York: Routledge.
Merchant, Carolyn. 2016. *Autonomous Nature: Problems of Prediction and Control from Ancient Times to the Scientific Revolution*. New York: Routledge.
Merrick, Helen. 2017. "Naturecultures and Feminist Materialism." In *Routledge Handbook of Gender and Environment*, edited by Sherilyn MacGregor, 101–114. Abingdon, UK: Routledge.
Ministerial Council on Education Employment Training and Youth Affairs (MCEETYA). 2008. "Melbourne Declaration on Educational Goals for Young Australians." http://www.curriculum.edu.au/verve/_resources/National_Declaration_on_the_Educational_Goals_for_Young_Australians.pdf
Phillips, Mary, and Nick Rumens, eds. 2016a. *Contemporary Perspectives on Ecofeminism*. Abingdon, UK: Routledge.
Phillips, Mary, and Nick Rumens. 2016b. "Introducing Contemporary Ecofeminism." In *Contemporary Perspectives on Ecofeminism*, edited by Mary Phillips, and Nick Rumens. 1–16. Abingdon, UK: Routledge.
Probyn, Elspeth. 2016. *Eating the Ocean*. Durham, NC: Duke University Press.
Ruether, Rosemary. 1975. *New Woman/New Earth: Sexist Ideologies and Human Liberation*. New York: The Seabury Press.

Samandari, AtienoMboya. 2016. "An Ecofeminist Perspective on the Climate Change Regime." Paper Presented to the 9th Annual Feminist Legal Theory Conference, University of Baltimore School of Law. Baltimore, MD, March 2016. https://law.ubalt.edu/centers/caf/2016_conference/Samandari_Atieno.pdf

Sandilands, Catriona. 2017. "Feminism and Biopolitics: A Cyborg account." In *Routledge Handbook of Gender and Environment*, edited by Sherilyn MacGregor, 229–238. Abingdon, UK: Routledge.

Seager, Joni. 1993. *Earth Follies: Coming to Feminist Terms with the Global Environmental Crisis*. New York: Routledge.

Shiva, Vandana. 1989. *Staying Alive: Women, Ecology and Development*. London, UK: Zed Books.

Shiva, Vandana. 1991. "Biodiversity, Biotechnology and Profits." In *Biodiversity: Social & Ecological Perspectives*, edited by Vandana Shiva, 43–58. London, UK: Zed Books/World Rainforest Movement.

Shiva, Vandana. 1997. *Biopiracy: The Plunder of Nature and Knowledge*. Boston, MA: South End Press.

Shiva, Vandana. 2005. *Earth Democracy: Justice, Sustainability, and Peace*. Boston, MA: South End Press.

Stapleton, Sarah R. 2019. "A Case for Climate Justice Education: American Youth Connecting to Intragenerational Climate Injustice in Bangladesh." *Environmental Education Research* 25 (5): 732–750. doi:10.1080/13504622.2018.1472220.

Stengers, Isabelle. 2015. *In Catastrophic Times. Resisting the Coming Barbarism*. Translated by Andrew Goffey. London, UK: Open Humanities Press. http://openhumanitiespress.org/books/download/Stengers_2015_InCatastrophic-Times.pdf

Stevenson, Robert B., Jennifer Nicholls, and Hilary Whitehouse. 2017. "What is Climate Change Education?" *Curriculum Perspectives* 37 (1): 67–71. doi:10.1007/s41297-017-0015-9.

Stevens, Lara, and Peta Tait, and Denise Varney, eds. 2017a. *Feminist Ecologies: Changing Environments in the Anthropocene*. Cham, Switzerland: Palgrave Macmillan.

Stevens, Lara, Peta Tait, and Denise Varney. 2017b. "Introduction: 'Street-Fighters and Philosophers': Traversing Ecofeminisms." In *Feminist Ecologies: Changing Environments in the Anthropocene*, edited by Lara Stevens, Peta Tait, and Denise Varney, 1–22. Cham, Switzerland: Palgrave Macmillan.

Sturgeon, Noël. 2009. *Environmentalism in Popular Culture: Gender, Race, Sexuality, and the Politics of the Natural*. Tucson, AZ: University of Arizona Press.

Taylor, Carol A. in press. "Diffracting the Curriculum: Putting 'New' Material Feminist Theory to Work to Reconfigure Knowledge-Making Practices in Undergraduate Higher Education." In *Turning Feminist Theory into Practice: Enacting Material Change in Education*, edited by Kate Scantlebury, Carol A. Taylor, and Anna Lund. Dordrecht: Sense Publishers.

Taylor, Carol A., and Gabrielle Ivinson. 2013. "Material Feminisms: New Directions for Education." *Gender and Education* 25 (6): 665–670. doi:10.1080/09540253.2013.834617.

Thompson, Charis. 2006. "Back to Nature? Resurrecting Ecofeminism after Poststructuralist and Third-Wave Feminisms." *Isis* 97 (3): 505–512. doi:10.1086/508080.

Torcello, Lawrence. 2017. "Yes, I Am a Climate Alarmist. Global Warming Is a Crime Against Humanity." *The Guardian*, April 30. https://www.theguardian.com/commentisfree/2017/apr/29/climate-alarmistglobal-warming-crime-humanity

Twine, Richard. 2010. "Intersectional Disgust? Animals and (Eco)Feminism." *Feminism & Psychology* 20 (3): 397–406. doi:10.1177/0959353510368284.

UNESCO. 2019. "SDG 4 – Education 2030. Part II. Education for Sustainable Development beyond 2019." Executive Board 206EX/6.11. https://unesdoc.unesco.org/ark:/48223/pf0000366797

United Nations Framework Convention on Climate Change (UNFCCC). 2017. *Gender and Climate Change – Establishment of a Gender Action Plan*. https://unfccc.int/resource/docs/2017/sbi/eng/l29.pdf

United Nations. 2016. "Sustainable Development Goals: 17 Goals to Transform Our World." https://www.un.org/sustainabledevelopment/sustainable-development-goals/

United Nations. 2019. "Implementation of Education for Sustainable Development in the Framework of the 2030 Agenda for Sustainable Development." https://undocs.org/A/74/258

Verlie, Blanche, and CCR 15. 2018. "From Action to Intra-Action? Agency, Identity and 'Goals' in a Relational Approach to Climate Change Education." *Environmental Education Research*: 1–15. doi:10.1080/13504622.2018.1497147.

Verlie, Blanche. 2017. "Rethinking Climate Education: Climate as Entanglement." *Educational Studies* 53 (6): 560–572. doi:10.1080/00131946.2017.1357555.

Verlie, Blanche. 2019. "Bearing Worlds: Learning to Live-with Climate Change." *Environmental Education Research* 25 (5): 751–766. doi:10.1080/13504622.2019.1637823.

Warren, Karen J., ed. 1994. *Ecological Feminism*. New York: Routledge.

Warren, Karen J. 2000. *Ecofeminist Philosophy: A Western Perspective on What It Is and Why It Matters*. Boulder, CO: Rowman and Littlefield Publishers.

Warren, Karen J., and Jim Cheney. 1991. "Ecological Feminism and Ecosystem Ecology." *Hypatia* 6 (1): 179–197. doi:10.1111/j.1527-2001.1991.tb00216.x.

Warren, Karen J. 1993. "Introduction to Ecofeminism." In *Environmental Philosophy: From Animal Rights to Radical Ecology*, edited by Michael E. Zimmerman, J. Baird Callicott, George Sessions, Karen J. Warren, and John Clark. 253–267. New York: Prentice-Hall. doi:10.1086/ahr/49.2.327.

Whitehouse, Hilary. 2017. "Point and Counterpoint: Climate Change Education." *Curriculum Perspectives* 37 (1): 63–65. doi:10.1007/s41297-017-0011-0.

Whitehouse, Hilary, Larraine Joy Larri, and Angela Colliver. 2019. "Ever Wondered What Our Curriculum Teaches Kids about Climate Change? The Answer Is "Not Much". *The Conversation*, September 19. https://theconversation.com/ever-wondered-what-our-curriculum-teaches-kidsabout-climate-change-the-answer-is-not-much-123272

World Economic Forum. 2017. *The Global Gender Gap Report 2017*. https://www.weforum.org/reports/the-global-gender-gap-report-2017

Anthropocentrism's *fluid binary*

Ramsey Affifi

ABSTRACT

I consider myself among a band of heretics seeking to deanthropocentrise environmental education. And yet, I increasingly struggle with blanket condemnations and recommendations. I do not know if the binary is as real or useful as I once thought. In this paper, I unearth some of the ways in which alleged anthropocentrisms can be nonanthropocentric, and vice versa. They seem much more fluid to me now. My purpose is not pedantic: I think environmental educators need to be more careful in their diagnoses and prescriptions. As we grope toward sustainability, we need pedagogies that help students imagine and engage with what our various claims and conceptualisations actually *do* when believed (if indeed they do anything at all); pedagogies that develop suppleness in our capacity to modify beliefs in alignment with intentions, and that help us modify these intentions in turn. I believe confronting the instability of dualistic thinking reveals paradoxes inherent in human reason, ushering a dose of humble bafflement essential for navigating the (non)Anthropocene.

Introduction

Rather than asserting that anthropocentric or nonanthropocentric positions exist and debate which is more normatively acceptable for environmental education, we might instead begin to ask in what ways a given thought, belief or practice is anthropocentric *and* nonanthropocentric, and to what effect. My purpose here is to explore anthropocentrism's fluid binary through various ethical, epistemological, ontological, causal, phenomenological, and performative or pragmatic lenses. But first, a brief historical portrayal pocked with omissions, then on to the task at hand.

Anthropocentrism has been critiqued widely in environmental philosophy and ethics (Kahn 2007), which has in turn informed (and in some cases has been informed by) environmental (e.g. Bell and Russell 2000; Kahn 2010; Kopnina, Sitka-Sage, et al. 2018) and sustainability education (Kopnina 2012). Well known to readers of this journal, a debate between anthropcentrism and nonanthropocentrism pit 'deep' against 'shallow' ecology, with the former intending environmental change through nurturing acknowledgement of the 'intrinsic' value of nonhuman beings and processes, and the latter settling instead on an 'instrumental' environmental ethic that has humans protect beings and resources necessary for their own needs (Naess 1973). In this context, shallow ecology is said to be 'anthropocentric' because it places humans as the sole bearers of morally significant concern and value (e.g. Passmore 1974). Deep ecology's 'biospheric egalitarianism,' sometimes conceived as giving humans no more ethical considerability than other species, was by contrast seen to be nonanthropocentric. In another influential vein, anthropocentrism was also critiqued by ecofeminists (e.g. Plumwood 1993), who saw anthropocentrism/biocentrism as one among many connected dualisms that worked to create and maintain hierarchies and power structures. In this sense, the split was seen as genealogically related to, and complicit in, other violent splits including male/female, reason/emotion, mind/matter, and white/other.

While recent debates have primarily assumed anthropocentrism is about where and how we place *value* (e.g. Payne 2010; Spannring 2017; Kopnina, Washington, et al. 2018), it is important to acknowledge that deanthropocentrising is not exclusively about value in either environmental philosophy or environmental education. Often the focus has been whether a claim is conceptually or ontologically anthropocentric (e.g. Rautio 2013; Mylius 2018). A certain notion or way of framing communication, freedom, causal potency, rationality, consciousness, or some other trait(s) can be seen as anthropocentric, with relevance for ethics (a few steps) removed. In some cases, ethical claims -even those apparently biocentric or ecocentric- are accused of being silently motivated by human centredness (e.g. Drenthen 2011). For instance, while deep ecology is often claimed to not be anthropocentric, some wonder whether its emphasis on self-realisation really goes beyond placing humans at the centre. For example, Grey (1993) identifies a tension between the deep ecologist's problem and solution: pollution and destruction are foregrounded as non-natural processes associated with the exceptional human, but the remedy is for people to realise humans are not exceptional and are fully a part of nature. Plumwood (1993) asserts this is because deep ecology retains a "discontinuity thesis" rather than acknowledging that humans are both a part of and separate from nature. There is also the metaethical question of whether it is anthropocentric for humans to create any value system at all, even if the system of values is itself non-anthropocentric in its allegiances (see Attfield 2011 for some discussion of this).

Is it desirable to imagine a view that has no residues of humanness? Is it possible? These are among the questions that "new materialisms" engage. As a thought experiment, let's take an extreme example. Can you imagine some wily nonanthropocentrist defending the existence of nuclear waste or the Great Pacific Garbage Patch on the grounds that such material is 'of' the world as much as anything else? Such an antagonist might claim there is no such thing as anything truly 'artificial' (because humans are never isolated causal subjects and are always co-actors in material assemblages), and that the attempt to judge one thing as good and another as bad (say ocean with or without plastic) marks the imposition of humanly derived moral criteria onto the world. We might be asked to 'go with the flow' and accept in sublime resignation the power of the self organising (or disorganising) universe, ceaselessly creating and destroying itself in a massive process we should not (and in any case, cannot) control.

Fleeting awe or fear-wrenching humility notwithstanding, dystopic sublimity is probably not the sort of nonanthropocentrism we are going for (acclerationists and dark mountainists may disagree). Rather, this abysmal asymptote gapes back at us, warning of other black holes that may also lie waiting at the edges of our eager deanthropocentrising projects. Ideas such as a Deleuzian ethics of immanent becoming; a flat ontology that rejects hierarchies in the realm of being (e.g. Bennett 2009); a commitment to removing not just normative anthropocentrism but conceptual, causal, and metaphysical ones too; a tendency to favour consideration of impersonal, postphenomenological 'affects' (e.g. Massumi, Deleuze, and Guattari 1980); and a radical doubting of any conceptual split between humans (and their artefacts), and 'nature' symbiotically creep towards the logical conclusion that there is no ethical justification against seas of plastic or skies of dioxin. Posthumanists seem to prefer theoretical dissonance to moral nihilism, because when it comes to the ethical implications of their ontologies, I too often see the urge to backpeddle from the edge and introduce cloaked anthropocentrisms (valuing life over nonlife, health over nonhealth, diversity over simplicity, values over nonvalues etc.). If the reader is committed to posthumanism or deanthropocentrising, I invite them to consider why they reject the sea of plastic, in what ways that rejection is founded on some value, and whether this value could possibly be void of anthropocentric residue. Rather than imply we should continue the wholesale eviction of the 'human', perhaps we ought to think more carefully about how anthropocentrism and nonanthropocentrism combine, co-occur and co-evolve in thought, feeling and action. This paper makes steps in that direction.

Hit the breaks we must, so where should we do it? One of the principle challenges of the posthuman project is to work out the extent to which we are able to let go of the human frame, *and* to what extent we need to hold onto it. The problem is somewhat subtle, however, as it is not at all clear where one frame begins and the other ends. I have been told by cynical friends that humans are catalysts to produce more plastic, itself an upcoming niche and opportunity for evolutionary development, perhaps not different from how oxygen, a toxic bi-product of cyanobacteria, became a crucial component of aerobic metabolism (Margulis and Sagan 2000). However callous such a view is towards the plight of those facing current and future harm, one nevertheless detects an anthropocentrism even in their appeal to the values of change, growth and prospect. What is considered nonanthropocentric in one way is always anthropocentric in another, and vice versa. This unavoidability is to be accepted, dealt with, and eventually cherished. For example, the claim that valuing an ocean free of plastic is anthropocentric itself values a humanly-constructed argument that plastic is natural (itself derived from a very culturally specific definition of 'nature') over the effects this plastic might have on the biological life of the sea and those who depend on it. Unless the term 'anthropocentric' (and its apparent alternatives) is considered with more nuance, and in particular with an eye on what these concepts actually *do*, environmental educators advocating 'worldview' change are bound to continue debating at an overgeneralised and counterproductive level of abstraction.

Prefiguring the logic of this paper's plotline, Bonnett (2004), like Nietzsche (1966) before him, noted that attempts to live according to nature are invariably anthropocentric: we select certain aspects of ecological phenomena to guide action, such as beauty or stability, over say chaos and upheaval. If it is inevitable that humans anthropocentrically select what to deanthropocentrise (equally, if it is inevitable that the more-than-human world has constructed us to be anthropocentric), then the problem for environmental educators can be stated with more precision. Rather than setting out to defend one pole of a crude dichotomy, the challenge lies in identifying concrete situations given thoughts, experiences, models, practices, and pedagogical experiences foreground and background the human, how these ways interact with one another, and what effects this has on people and the planet. We are left with a series of overlapping microdichotomies that shift and interact, forever banished from the a priori comfort of a totalising position by which to prescribe thought or pedagogy. While this does not make practice 'easy' for educators, it does have important consequences that will be explored in the discussion section of this paper.

Welcome to the (non)anthropocene.

I would like to propose a premise which I invite the reader to dispute. I suggest that in the very attempt to choose criteria by which to initiate deanthropocentrising, anthropocentric criteria invariably sneak in. Like a balloon, our attempts to grasp it merely squeeze the balloon out of our palm. The desire to rid our thoughts and activities completely of anthropocentrism fails from the outset for different reasons, but most basically because it denies a 'centredness' that preceded and heralded humanness. Pre-human apes were apecentric, a nonhuman condition for the possibility of the evolution of humans. The phenomenon of centredness is itself a creative and productive mode which matter takes on, and it is incoherent to dismiss it outright in favour of, say subjectless webs and processes of becoming. Tipping the scales in favour of the latter implies choosing the artefacts of a culturally located (be it scientific or otherwise) abstraction developed through human interests, pursuits and ways of knowing the world. Phylogenetics aside, on what basis are webs or processes of *becoming* permitted to extinguish an allegedly 'folk psychology' conception of the self that may sometimes tie us into more empathetic and responsive relationships with beings in the world? But on the other hand, when does that folk psychology get in the way? How would we know?

In the subsections below, I explore several (non)anthropocentric positions that can be considered in EE practice: multicentrism and noncentrism, process oriented materialisms, anthropocentrism vs anthropomorphism, and experiential (non)anthropocentrism achieved through meditation. This list is neither thorough nor exhaustive. One purpose is to show that anthropocentrism is not one thing (there are many different kinds with different implications), nor is it stable (one can shade into another or its opposite in different circumstances), and it is not necessarily to be avoided by posthumanist environmental educators. While this dynamism may remind the reader of Derrida Différance (1998), I do not attribute the instability of these concepts to human textuality. Instead, I assert it is the result of humans emerging and yet remaining embedded within complex webs of more-than-human causalities. Human meaning is unstable as a consequence of this originary and ongoing emplacement. Acquainting ourselves with the paradox of (non)anthropocentrism is part of accepting the way we are interconnected within the world, a nonreductive practice that deepens ecological understanding. However, if all available conceptions are in some way both anthropocentric and nonanthropocentric, we are called to ask what a given framing foregrounds and to what effect, and to draw out nascent alternative dimensions when such framings are problematic. Through accepting the fluid and productive interconnection between these modalities, educators can develop a lithe repertoire of modes of thinking that can be employed with students to dislodge sedimented thinking. I hope this interrogation can lead to more practice-oriented nuance for my own work as much as for others. May this binary be more fruitful with its fluidity released, our senses piqued to its prospects and challenges, and all our relations bettered thereby.

Multicentred approaches

There are a variety of multicentred ways of attempting to fly out of the orbit of anthropocentrism. It is helpful to consider them, in part because they foreground what is at stake when we talk about going beyond the human frame. For instance, Weston (2004) proposed a multicentric ethic to replace a human-centric one. His concern was that prominent approaches to animal and environmental ethics worked by expanding the circle of ethical consideration outwards to include other species (e.g. Regan 1983; Singer 1975). Doing this, he argued, kept the human firmly at the centre, as other species' considerability was always judged in reference to people.

Weston proposed instead that other beings can and should be considered their own centres of ethical consideration. Rather than expanding our circle we should recognise other ethical circles. Is Weston's multicentrism nonanthropocentric? It depends. Weston's attempt to think multiple centres certainly seems to privilege the idea of a 'centre' for ethical consideration. Is this then its own version of expanding a circle? To be centre-centred is still a form of anthropocentrism, the argument would go. It still ascribes value on the basis of a thing's capacity to even be a centre, which was grasped as valuable through appealing to our own self-centredness (so to speak). Perhaps. If so, should it be dismissed purely on that ground?

A different form of multicentrism comes from scholars working in autopoietic and related 'enactivist' traditions (Maturana and Varela 1988; Thompson 2007). Autopoiesis is the notion that an organism constitutes itself as a unity that relates to an environment-for-it, and this self/environment interface is both the process and the product of its ongoing activity. All living beings are centres, indeed this is what distinguishes life from nonliving beings. Autopoiesis, like its Uexkullian predecessors (e.g. von Uexküll 2011), challenges the perceptual and cognitive anthropocentrism articulated in Kantian epistemology, which continues to manifest in various subjectivisms from phenomenology through to postmodernism. Such approaches focus on how subjectivity is humanly constituted against the unknowability of the world 'itself.' This allegedly leads to a uniquely human epistemological split between the world of appearance and the real, but forever unknowable world (or between what Kant calls phenomena and noumena). Rather than rejecting this epistemology outright, autopoietic theory argues that this split is the process and product of all living beings. All organisms constitute their own species-specific (or even individual-specific) lived experience. The world is projected outwards from each biotic vantage point, with each organism coating the world through its unique sensorimotor engagements with various modes of space, time, and ways of categorising, all conditions for the very possibility of persisting in life. A panbiotically enacted Kantianism seems to immediately suggest multicentric (non)anthropocentrisms of its own.

One difference between the autopoietic approach and the Westonian approach is that the latter is largely concerned with ethical considerations, while the former has been more concerned with ontological and epistemological issues.[1] For example, autopoiesis might interrogate how a particular being makes sense of its world, what shows up as significant for it, and how it responds to such signification. In short, what is the lived experience of this being, be it a bacterium, a begonia, a doormouse? Weston's multicentrism does not require him to engage with such a bio-phenomenology (it does not forbid him either). On the other hand, many autopoietic theorists do not seem overly concerned with examining the ethical considerations of their models (the first major experiment that launched this theory is based on removing and replacing frog's eyes (Lettvin et al. 1959)). This seeming divorce between perspective and practice, with epistemological questions not necessarily implying ethical considerations -or vice versa- should give us pause. Both de-anthropocentrise through proposing multicentrisms. One is concerned with how things are, the other with how to best relate to those things, but neither seems to need to speak to the other.

Both likely also vary in their conception of a 'centre.' Weston may not hold so strongly to biological centres and might allow that rocks are centres in some sense.[2] (I have heard him suggest on occasion that *some* rocks were alive.) Autopoietic theory seems much more dogmatic on this point. For the latter, metabolism is a fundamental organisational difference between life and nonlife, and this difference brings with it a fundamentally different ontology. Living beings are those for whom ways of being and ways of knowing exist, and are those for whom ways of engaging matter. They are centres of concern, and what they are concerned with is directly related to their capacity to exist.

In any case, these two examples show how perspectives that seem anthropocentric on one level (in holding onto the concept of a centre) might themselves be in some fundamental way nonanthropocentric. In the case of autopoiesis, the argument might go: if it is the case that every human, like all living beings, constitutes their own being as a centre and this capacity is itself not a human production but something 'given' to humans by nonhuman processes (such as evolution and metabolism), then it is clear that to be anthropocentric is nothing more than to be autocentric, an achievement all living beings realise through autopoiesis. On the other hand, Weston's nonanthropocentrism takes seriously other being's claim for a place in the ethical landscape of the earth (and for him, no doubt, the galaxy), a power enabled by the very fact he retains his humane (or mammalian, or biotic) instinct to care for centre-like things.

The broader point, however, and one which we shall keep coming back to, is that neither category here -anthropocentric or nonanthropocentric- is stable through time. Each flip quietly into its opposite either immediately or under certain conditions. If the *temporal dimension* of living thought is absent from our educational considerations, the type of (non)anthropocentrisms we reify will not be nimble or responsive to the ongoing becoming of the world.

In my view, object oriented philosophy (e.g. Harman 2005; Bryant 2011; Bogost 2012), is a more recent form of multicentrism because it seeks to preserve objects as real entities. It is 'centric' because its proponents resist dissolving apparent objects into relations as is prevalent in most process-based ontologies (see below). For example, Harman (2005) opposes the undermining of things directly experienced by reducing them either to their component interactions, or to their role in larger ontologically constitutional entities such as ecologies or assemblages. This alone might seem anthropomorphic and quite likely anthropocentric to a process philosopher. However, he then forwards a worldview that collapses the subject/object dualism by asserting that Kant's phenomena/noumena distinction is present in all relations, even those between objects. Let's unpack this. For Harman (and Bryant), humans experience the things around them in certain ways, and part of this experience includes our awareness that these things are not completely disclosed to us. Harman argues all interactions between things have an analogous structure. Two celestial bodies interacting through gravitational interaction, to take a 'physical' example, respond to how each other's mass contorts spacetime[3] but not to one another's colour or smell. No object can present itself to another completely and only ever interacts with a subset of an object's attributes, according to its own nature. And so, every thing is an object not reducible to its relations. This is a much more radical pan-Kantianism than the autopoietic extension of transcendentalism to the living. All objects interact with each other's appearances (phenomena) and no interactions access another object's thing-in-itself (noumena). This position is radically nonanthropocentric in the sense that it undercuts the human as the alleged constitutor of the subject/object separation and the corresponding epistemological phenomena/noumena distinction. This is important because it posits continuity between 'knowing' and 'interacting.' But for this very reason it is also anthropocentric by ontologising and universalising an epistemological dualism grounded in human experience.

Harman escapes the anthropocentrism implicit in relational becoming such as we might find in modern science, Whitehead (1929), Latour (1993), and process oriented new materialists. But he anthropocentrises the cosmos in the opposite way. He asserts that things are more real than relations, and universalises a dualism gleaned through the structure of his own (human) experience -the appearance/reality distinction. Instead of denying the subject/object dualism he seeks to make it a special case of an object/object dualism, but the model of how the object/object dualism works was derived from his having subject/object dualism as a mental scheme, and using it as an analogue source to develop his alternative ontology. Again, the point is not that this is 'bad' simply because the ontology was born of and resides within a larger human frame. It is instead to indicate how paradoxical and unstable are the various forms of trying to go posthuman.

What about things that are *kind-of*-centres? Biology gives us many examples of various degrees of tightness, loosening but not eliminating the notion of a centre. From plants, somewhere between the interconnection of a central-nervous-system-endowed animal (such as Darwin's root-brain hypothesis (Baluška et al. 2009)) and a confederacy of parts (the strong modular claim made by Firn 2004), to slime moulds that spend parts of their lives as single-celled organisms and other parts in multicellularity, we are shown that being a centre is not always 'one' thing. While it may seem a paradox to a tightly centred being, some centres are less centralised than others. At the same time, the fact that our body is made up of germline cells and other microbial cells, each a centre of its own, but congregate into another -me or you- suggests centres within centres, each presumably different kinds of centres in turn. Is a virus a centre? On organ? What about a biome? What does it mean to be multicentric in a world where it is not always even necessarily clear whether or not something really is a centre? Do interactions in the process of becoming more centralised deserve more consideration than those where relations are dissociating? If there is a birth and death of centredness, how do we respond?

As one ponders such questions, it is difficult to ignore the power of empirical observation in engaging the human mind in more complex de-centring than can be accomplished solely through thought alone. The concept of a centre emerged from, is challenged by, and grows through events in experience. Close observation of the world (of any phenomenon within it) can show us and affect us with its different ways, fertilise our imaginations, and is therefore always (in part) a decentring. Through witnessing kind-of-centres, we decentre our very notion of centredness. Scholars produce different kinds of (non)anthropocentrisms depending on if and how they engage the empirical world. If our sense of a centre is continually informed by our embodied experience as tightly organised autopoietic entities, the encounter of those with different levels or modes of centring offers us the opportunity to stretch this conception. Without a tightly centralised mammalian body we might never have encountered a less tightly centralised slime mould as such. The point is to examine the various ways in which centres do and do not exist through our own centredness, how they come into and out of existence, how they interact together and with us, and to what effect.

But surely there are many 'things' not centred in *any* way. What of their ethical or ontological standing?

Science and process-centred approaches

With some important exceptions, the broad trend in Western science, from the Copernican revolution, to the theory of evolution, the discovery of DNA, the revelation of the massive expanses of the cosmos -from superclusters down to subatomic particles-, has been another kind of vast deanthropocentrising. The broad trajectory has been to diminish the importance of humans in the cosmos, by stripping the soul, denying the power of consciousness, ridiculing the notion that anything (even ourselves) was made for a purpose, and casting us into some random corner of an enormous universe chugging towards entropic heat death.[4] When looked at from an atomic or energetic level all things appear to be in the process of change, all mesophysical objects (including life and love and death) but tricks of consciousness reminding us of our bodily disconnect from the actual becomingness of the 'real' world. However, whether or not such de-anthropocentrisms re-anthropocentrise is a complex and shifty question. Again, the answers depend in part on what these various 'truths' do when entertained, believed, or transformed into action.

On the one hand, much of the classic scientific project, whether conceiving the world in terms of process and becoming, of mechanisms or atomic interactions, as material flows and webs, or in any other 'reduced' way (i.e. the set of activities Harman (2013) calls 'undermining'), obviously de-centres what we think of as human (and along with it, other organisms). Our struggles and dramas, thoughts and dreams, the sounds and smells that fill our sensory lives -each is relegated to epiphenomenon. Scientific discoveries have therefore been critiqued by some environmentalists for their implicit undercutting of the primacy and truth of 'lived experience.' Abram (1996) argues that scientific truths pull us away from our senses, which uniquely afford experience of things 'more-than-human.' It seems Abram wants to erect a dichotomy between conceptual and sensory decentring while favouring the latter.[5] However, because a concept-based decentring is made possible through the embodied emplacement of a person in their immediate lived experience (a phenomenological centredness), these seemingly opposed terms layer one upon the other in interesting but complex ways. Hence scientific decentring is sometimes accused of decentring a more significant decentring! (Sorry, I tried to erase that line but couldn't bring myself to it).

However a world without centres is still somewhat human centred. First of all, such an ontology devalues the centring projects of countless other beings (i.e.approaches such as those of Weston and the enactivists above) in deference to an abstract metaphysical position. On the other hand, scientifically informed decentring is possible in large part because of our conceptual and perceptual organisation in the first place (a point emphasised repeatedly since Husserl (1954)), and the various tools and techniques we have developed to encounter the world for specific human purposes. However detached it might seem, it is grounded in and always comes back to the living, feeling, sensing body. That is, it comes back to our corporeal centredness.

Scientific development was crucial for the rise of contemporary process philosophy even if many process philosophers are dissatisfied with scientific ontology. Atomic theory, evolution, relativity, quantum mechanics, and chaos and complexity theory have all featured as prominent muses in the development of process based ontologies (this is as true for early process philosophers such as Whitehead (1929), Peirce (Houser and Kloesel 1992), Dewey (1925), or Bergson (1946), as it is for contemporary ones like DeLanda (2002), Latour (1993), Prigogine and Isabelle (1984), Barad (2007), Dupré (2012), and Deleuze (see Villani 1999)). In various ways, each underemphasises the reality of beings (nouns) and lifts that becomings (verbs). This tendency continues in Deleuze-inspired 'new materialisms' in environmental education today (e.g. Clarke and Mcphie 2014).

While some might argue that becoming-centredness is 'truer' and that there persists somewhere some indigenous group that wisely does not 'noun' the world, an examination of lived experience questions this idea. While the piece of paper on my desk is in a slow process of transformation, –verbing an ongoing represencing and perishing event– it does not appear as such in unreflected upon experience. It is only through considering the paper through various lenses serviced by my prior knowledge (say about the nature of decay, or my understanding that the atoms inside it are continuously in motion reconstituting it as paper at each passing moment), that I can purge my naive phenomenological take on it and see it anew. Yes, the paper is continuously shifting in hue as the clouds float over the sun intermittently. But it appears as paper continuously changing in colour. No wonder that such nounless cultures have not been found to exist. It is arguable whether or not a deanthropocentrising project can be founded on the basis of the capacity to undermine direct experience through the careful injection of processy conceptualisations. It seems fairer to say that all sorts of things appear as nouns if their rate of change is slow enough that they behave consistently for the duration of the interaction. Different people, cultures, and contexts may elicit varied sensitivity to such rates of change without shaking off the dialectical interpenetration of constancy and change. This would seem to apply to other organisms too, who rely on consistency in their own *Umwelten* (as many animal studies on Piagetian object constancy suggest (see Reznikova 2007, 132–139)). Of course, it could be argued that this conceptualisation is itself anthropocentric too, and so it is. But it is a more-than-anthropocentric anthropocentrism, as it acknowledges a type of distortion that occurs pervasively throughout the biosphere. And it is an anthropocentrism supported by more-than-human evolutionary processes that simply do not allow plants, animals, people or languages to dice up the world any way they please. Living in the world restricts how things can be interpreted (while collaborating in those interpretations as well). Some of these may be false on a broader spatiotemporal scale, but true, and necessarily true for lived (or let's say, liveable) experience.

Some process-oriented new materialisms (e.g. DeLanda 2002; Barad 2007) assert that human thinking is born of constellations of interacting processes, foregrounding something more-than-human in their birth, persistence, or decay. But of course, achieving this realisation slips the human with its epistemologically supreme vantage point, back on stage. This is clear when we take note of the particular manner of some (often DeLandian) new materialist frontal attacks upon postmodernism (a project I heartily endorse). By firmly rejecting the notion that language is divorced from the world and limitless in its power to diversely represent 'reality', such new materialisms assert that the world is in fact in some way knowable -it really does bump up against us, interact and intra-act with us.[6] But one arrogance led to another. A seemingly humble agnosticism about our capacity to know the noumenal world morphs into a hubristic vision; the epistemological subject erecting at each moment the entire universe (as phenomenon) is now replaced by the equally incredible idea that we can actually know the world itself. We can glean the workings of the world that gave rise to the very distinction between subjects and objects. The workings are sufficiently human that they can be described in words and presented in arguments.

Barad's concept of 'agential cut' also illustrates the fluid nature of process-oriented (non)anthropocentric thinking. According to Barad, subject-object distinctions are enacted in what she calls 'cuts.' However, unlike various humanists, for her they are not enacted by a subject, but by the broader material arrangement of which the 'subject' and 'object' are consequences before they are causes (175–180). Barad introduces a posthuman perspective, but she can never escape the fact that this very conception of materiality was itself dependent on her human perspective (subjectivity). It is unlikely that a cat would conceive the rise of the subject-object cut as Barad does. Like Bennett and DeLanda, Barad's ontology therefore is and isn't anthropocentric. It is anthropocentric in the sense that it is produced by the very same assemblages as those ontologies she critiques: a human body, human reasoning, imagination, and language.[7] But it is not anthropocentric in that the condition she conceives for the possibility of such human situatedness lies in broader material dynamics.

Some readers will detect that Barad is vulnerable to a criticism levelled at Kant centuries ago. Kant (2007), for his part, was also interested in the conditions for the possibility for things like subjectivity (which he called experience). This is where he developed his infamous transcendental arguments. He showed that space, time, causality, being, becoming, and a handful of other categories are not necessarily in the world itself. They are merely the necessary ways in which we must perceive and conceive things in experience. However, Kant depends on these categories to describe this very position. In so doing, he imports apparently human ways of constituting appearances into the logical structure that underlies the emergence of these aspects of human subjectivity in the first place. How the world itself (which may or may not exist as it is not necessarily 'being' or 'becoming') gives rise to such a subject (given it may or may not operate causally) is quite a mystery. When Barad employs human conceptual structures to posit the causal origin of these very tools, she participates in an analogous Janus-faced transcendental argument.

The paradox is clearest when she goes normative. She calls for an ethics of responsibility that 'we' are supposed to adopt in response to this ontology, even though 'we' are co-constituted and entangled, not isolated moral agents: "[w]e have to meet the universe halfway, to move toward what may come to be in ways that are accountable for our part in the world's differential becoming" (353). There is something strange about this sentence. According to Barad's argument, of course we 'have to' meet the universe halfway because that is how we emerged in the universe in the first place. Why make an ethical commandment about something that one cannot not do? Why amend a posthumanist position with an ethical addendum that repositions the human as agent? The answer, I think, is that it is inevitable that an ethical person will humanise an ontological position that otherwise leads to destructive attitudes. Her origin story was insufficiently humane for her sensibilities.

Though very different thinkers, processy Barad and beingish Harman both seek to represent the nonrepresentational. When they do this, they invariably claim a congruence between their ways of conceptualising and the world itself, even if their conceptualisations deny any such allegiance. At root in any realism is the anthropocentric conviction that we are sufficiently similar to the world that our ways of thinking, observing, learning, and understanding are not wholly divorced from the universe that birthed us. Anthropocentric indeed, but an anthropocentrism that (not unlike deep ecology) brings with it a sort of identity relation however imperfect, and the humility that humans are created by the universe and are in complex ways in its image. My conviction is that this is to be embraced rather than avoided in environmental education (Affifi 2019).

Meditation and experiential decentring

Casting an eye upon the dynamics of the inner world, meditation is another way of decentring the human. In this case, it focuses and troubles the assumptions that come with being a self, -a seemingly unified, separate, and self-directing agent. For instance, meditation heightens awareness that thoughts and feelings are happening of their own accord. The self is challenged as a certain kind of causal centre (Affifi 2016a). It is increasingly experienced not as an ongoing source of causal power independent of the material universe but as instead wrapped into a multitude of interacting processes. The meditator experiences causality as the property of networks of changing interactions. This web of co-dependence leads to a pluralistic conception of the arising of any phenomena, but most especially that of the 'self,' a process known in Buddhism Pratītyasamutpāda[8] and commonly translating as 'dependent origination.'

Nevertheless, during meditation one *does* experience this multitude at some sort of centre. We conceptualise that 'thinking is happening' through an entangled web of sparks and connectivities as the (now *so-called*) inner and outer world reflect and ripple upon one another. This is a deanthropocentrising experience as it challenges the notion 'I think' -that a subject can be the one actively *doing* thinking- and replaces it with the experience that thinking is an ongoing relational process between various things in experience that 'I' do not really have control over. While there may be some discussion over whether all cultures and languages conceive of the relation between subjects and verbs in this way, it is clear that such a causal presumption is core to the type of experience asserted in English speaking people (and other similarly structured languages) and their associated conceptions of what humans therefore are. Such insights deeply trouble the Cartesian separation between the res cogito and the res extensa. In that sense, this might appear a very useful tool in a deanthropocentriser's toolbox. But this is where the now familiar paradox comes in: in order to 'have' the experience that thinking is happening in spite of us, this diverse flux of events needs to be held in some sort of unity. The dissolution of the ego requires an ego to experience it. We *think* that 'thinking is happening,' jump above it for a moment, and centrally organise this diverse manifold of events through the vantage point of a metalevel. While this metalevel immediately slips from our grasp, and dissolves into its own ongoing flux of thinking, a new totalising vantage point can and is routinely achieved. This relationship becomes more clearly experienced when we try to meditate because thinking itself becomes the thing we pay attention to rather than, say traffic, excel sheets, or Netflix. We turn and look at the looking instead of taking it for granted.

Meditation thereby reveals with particular empirical vigour the same unstable flux discussed elsewhere in this article, and proves an important practice for more deeply understanding the relationship between anthropocentrism and nonanthropocentrism (as well as many other dualisms). One important advantage it offers is that it allows people to experience directly how seeming oppositions depend upon, and transform into one another. It is not merely a conceptual scheme derived from another conceptual scheme, in a regress of playful, yet disembodied thought games. Alongside its many benefits, it provides us with ever more clear awareness of what is going on, and is therefore an important way to become a better phenomenologist (where 'essential structures' of experience is taken very loosely).

'Experience' and 'representation' are things that matter does when organised into certain types of bodies and social relationships. The antiphenomenological strain of new materialism (imbued in many interpretations of Deleuze and Guattari, but exemplified most specifically in the work of Brassier (2007)) is itself anthropocentric when it fails to accept that "experience" in its diverse forms is a product and process of materiality. Experience, and with it knowledge and representation, is only the synthesis of a Kantian subject if considered in isolation. But, as Barad suggests this very subject is the product of much broader complex spatiotemporal processes that give rise to it (and to which it eventually contributes). This means that while fallible, knowledge is not random or to be arbitrarily constructed at the whim of whoever holds power, and is rather of the world. (Of course, this does not mean we should not be critical of such knowledge. The world is up to some terrible things, and implicating us deeply in it). Collapsing materiality into that which can be third-person observed overemphasises one aspect of how matter creatively reconfigures itself through time. By rejecting phenomenology, some new materialists are implicitly betraying their anthropocentric Kantian framing of the origin and nature of subjectivity.[9] As educators, we need to be careful of blanket condemnations. Whether or not phenomenology is anthropocentric depends on many moving factors.

Anthropocentrism and anthropomorphism

Many observe (Plumwood 2009; Bennett 2009; Affifi 2011) that anthropomorphism and anthropocentrism can sometimes be a converse relationship. The methodological dogma common to much of science -thou shall not anthropomorphise- often leads to anthropocentric perspectives, ethics and behaviour. The idea is that the refusal to use one's own experience to try and imagine the experience of another cuts off possibilities for empathy, understanding, and relationship. In Affifi (2014b), I argued one should distinguish naive anthropomorphism from a more nuanced sort of anthropomorphism. Naive anthropomorphism assumes other creatures are identical to humans and thereby shuts off our capacity to acknowledge and respect their differences. A more engaged form of anthropomorphism participates in *relationship*: by assuming the other being is in some ways similar to us but also in some ways different, we can engage in interacting and understanding it, and be open to the fact that our ongoing conceptions of what the organism 'is' can serve as entry points into relationship but are necessarily temporary (Affifi 2014b). In this sense, there is a similarity between understanding other organisms and understanding other people, because we also continuously think we 'know' the other person but if the relationship is truly growthful this knowing is capable of allowing the other to jut through conceptions we had of them. In essence, knowing and coming to not know are in an ongoing dialectical relationship with one another through time.

That said, anthropomorphism *does* have a dimension of anthropocentrism to it. Each time we think we know what another being is, we assume that our model (consonant with our demands for particular forms of empirical corroboration, logic, coherence, etc.) is capable of representing another being. We centre our model and decentre how the other being violates it. Any anthropomorphism that does not open up its own representations to possible rupture by otherness is locked in such a conceptual anthropocentrism. However, we see that anthropomorphism and anthropocentrism can be dialectically entwined (and therefore so are anthropomorphism and nonanthropocentrism) through the way they tack back and forth through time. The rupture of the model is the thrusting of otherness into our schemes, made possible by the schemes themselves but transforming them in turn (see Affifi 2019, 2020). We could call this anthropomorphism's Gadamerian (1975) dimension, in acknowledgement of his working through the constraints and possibilities of 'prejudice'). For Gadamer, prejudices are inevitable for making meaning, given we are historical beings with a stock of prior knowledge front-loading the structure of possible experience. A prejudice, however is a virtue in disguise. Without such fore-structures we would be unable to enter into any relationships as we would not have any context that the otherness we encounter can modulate or transform. Anthropomorphism, in the same vein leads to anthropocentrism when it closes people to otherness but to nonathropocentrism when it invites schematic rupture. However, in these dialectical dynamics the disorder wrought by otherness is subsequently re-organised into a new scheme indicating otherness is unstable. We seal the fray. This suggests a general caution about the scope and possibility of deanthropocentrising. 'Who' is making the agential cut (Barad 2007), and who is cutting the cut is not clearcut.

A new empiricism should recognise that the ongoing interaction between a person (or any organism for that matter) and its environment leads inevitably to an unfolding co-constitution. Through ongoing relationship with a magpie I become magpiemorphised as I begin to consider and interact with myself through ideas, analogies, patterns of action and interaction, and feelings garnered from or generated through time spent with the magpie. In a more concrete sense, magpiemorphisation influences not merely my mental self but a more integrated self (that my mind arises from and contributes to): my thoughts and feelings magpieify as they emerge within embodied action that is itself inflected by this curious, hopping bird. When considering this embodied and enacted context, we see that I interact with a magpie that is already interacting in a way that was affected by my prior interaction with it. This infinite regress is not a fatuous philosophical abstraction but a very real dimension of the feedback nature of any interaction (see Affifi (2014a) for exploration of some of these consequences). It is *not* true that I 'become' magpie, nor that the magpie 'becomes' me; we each become ourselves but in a way that now incorporates the ongoing co-evolution of self and other, the openness to alterity and the transformations that occur. A new empiricism asserts and relies on the notion that neither our anthropocentrisms nor our anthropomorphisms are static. Rather, they are both continuously reshaped by our experiences with others in our world. When treated as generic concepts devoid of a contextual and temporally unfolding life, these concepts can be played around with, and subjected to various rebellions or conformations to logic. But this necessarily divorces them from the developmental nature of the world and in so doing reifies the rationalistic approach that remains (notwithstanding itself) the dominant undercurrent in philosophy. As Connelly (2010) puts it 'participation in a world of becoming teaches modesty about the powers of argument even while appreciating its pertinence' (10).

With a predominant focus on deanthropocentrising the human, there is the risk that we persist in anthropocentrising *our focus* despite (or rather because of) our efforts. In other words, it is not sufficient merely to focus on the human as a subject to be deanthropocentrised. We must devote ourselves to developing the sensitivities and capacities to understand, theorise and learn deeply from the rest of the universe beyond the deanthroposised anthropos. Our work is 'environmental education,' which (should) by definition involve a world of myriad creatures, beings, and processes. The logocentric obsessions inherited from postmodernism have taken environmental education theory and practice away from its traditionally sustained engagement with the nonhuman world and have explicitly or implicitly associated any such engagement with some sort of naive or corrupted scientism. New materialism could restore such an engagement with the diverse world, sharing and honouring the delight of both kinship and otherness, but without the conceits of earlier epistemologies that assert uncontroversial access to the world for the deified western subject (see Affifi 2019, 2020). But to do this, it needs to get beyond articulating what humans are and are not in generic ontological terms, and pay attention to the processes and transformations of the actual world. We should not be coy about taking science back. Environmental educators are perfectly situated to explore and experiment in this reconstruction.

Discussion

Why highlight anthropocentrism's fluid binary? Certainly not to enable business as usual through some new relativism! Without question, I too am sickened by activities my colleagues diagnose as "anthropocentric." I agree that many humans need to spend more time paying attention to other beings and have invested time reconstructing education theory and practice accordingly (e.g. Affifi 2015, 2016b). I am not aiming to shallow ecology, but to deepen it ever further by drawing attention to the extent of nature's interconnectedness. Any conception is anthropocentric insofar as it is framed by the human mind and heart, but equally nonanthropocentric because all human activity emerges from and persist within a vast field of beings and processes. The dialectic at the heart of this paper, that all anthropocentrisms are nonanthropocentric and vice versa, ultimately owes itself to an existential paradox: we are aware that we are, as Plumwood (1993) noted, *both separate from and interconnected with the world*. This is not just a language game, it is a fundamental binary manifest in all beings. The essence of "being" is the emergence of an entity sufficiently individuated from the flux to persist, but whose persistence depends wholly on that flux. There is a centralisation inherent in being, as much in flames and stones and ideas as paramecia and apes. Ultimately, the most profound nonanthropocentrisms are not those that deny centrism, but that recognise its unavoidability, work with its strengths, and are vigilant against it leading us astray. We cannot allow anthropocentrism to colonise non-centrism or vice versa, because both aspects exist. Wise decisions come from struggling with this paradox rather than concocting a metaphysics that collapses it, one side or the other.

Some questions we must continuously try to ask: what does a given deanthropocentrism do? Does it actually lead us to attend to the field of more-than-human presences around us? Does it invite us to relate to the birch trees and magpies around us, and to recreate our environments to enable such relationships? Does it actually compel us to stand in front of the bulldozers? Does it create beauty? Or, despite its claim to radical nonhumanness, does it instead merely invite us into more arguments, more time in front of computers, and more wandering around stuffy conference rooms? In the pragmatist tradition, Weston (1991) for example recognises that thoughts and environmental contexts are entertwined in an ecology. This means it does not make sense to focus on fixing the worldview without fixing the world. An abstracted deanthropocentrism easily parcels itself off and operate merely in a realm of discourse games without changing any actual (de)anthropocentrised relations. On the other hand, some physical intervention (say building a bat house) might be done for human reasons (e.g.we want bats around so we have less mosquitos), but inadvertently deanthropocentrise our thinking and feelings by acquainting us with the styles and personalities of bats, perhaps inviting us to consider what it might be like to be a bat (without Nagel (1974) coming first) in order to help it flourish. In other words, the direction a given (de)anthropocentrising manoeuvre is heading is often more relevant than its intention.

Ways of framing transcend themselves, not merely reproducing particular modalities of thinking, feeling and acting but also opening up new fields of possibility. We can deanthropocentrise towards biocentrism, towards ecocentrism (Callicott 1989), towards multiple centres (Weston 2004), or towards a complete lack of centre (which thereby creates its own process or material oriented centrism (Deleuze and Guattari 1972)). But each unfolds unexpectedly as well, giving rise to new mental and material events with their own unexpected trajectories in turn. This dynamic, which Kauffman describes as the movement into the "adjacent possible" (Kauffman 2000; Hoffmeyer 2004), pervades the universe, within and without. This is what we need to commit to painstakingly observing and engaging, through discussions and experiments, actively with students and in the context of real, lived encounters. Asking such questions is performative, even and especially when no definitive answers can be found. We are thrown into complex ecologies that change amidst our very attempts to understand them. Asking what a given belief does is more about opening a stance towards the world that fosters virtues necessary for reorienting ourselves within our ecologies, such as "ecological humility" (Kopnina, Sitka-Sage, et al. 2018). Ecological humility is not just a cultivation of epistemological uncertainty and openness before the ecosystems "outside" [sic] us. It is work holistic work that recognises our beliefs and ontologies emerge from, separate from and reunite with, and feedback into the very ecologies that call us to pause. To paraphrase Bateson (1979), the biosphere is a mental ecology that contains the human mind.

The question of where a given conception leads relates directly to another point already suggested in the introduction. Many attempts at nonanthropocentrism are not actually possible to live. For example, the idea that we treat all living organisms as equal centres of value (for example, Taylor's 'biospheric egalitarianism' (Taylor 1986) or Lupinacci and Harpel-Parkin's "nonanthropocentrism" (Lupinacci and Happel-Parkins 2016) can really only be briefly entertained conceptually and affectively. Within hours, the would-be egalitarian *will* value themselves over some other to acquire nutrients and calories. Indeed, they will need these calories to fuel their arguments, implicitly valuing the ends they hope their arguments will achieve over the particular life forms sacrificed to produce them. But this does not necessarily mean that 'impossible nonanthropocentrisms' are without value. They may still serve to guide action, to push people to be ever more compassionate or sensitive to their local and systemic impacts on the world. Let's call them *strategic nonanthropocentrisms* (in the vein of Spivak's (1990) 'strategic essentialism' and Bennett's (2009) 'strategic anthropomorphism'). In other words, nonanthropocentrisms may still function as ideals regulating action (the pragmatic notion of a regulative ideal comes from Kant's third critique (Kant 1790)). Of course, in keeping with the observations of this paper, holding such an ideal -or 'end in view' to use Dewey's (1922) term- as a way of directing action is itself a very human thing to do and valuing this particular way of teleologically investing behaviour with a future is therefore another form of anthropocentrism.

Similarly, some attempts at nonanthropocentrism are at best experienced or practiced only *sometimes*. As such, it is crucial that educators consider critically whether they actually have the potential to carry over into people's otherwise anthropocentric lives. For example, the experience of 'self realisation' characterised by deep ecology can be an aesthetic oneness or a deep universal empathy, but these states are not sustained. Educating for sustainability does not suggest we make some fleeting ontological experience universal but that we move beyond the tendency to search for a particular 'ism' that, if adopted, would carry us to salvation (for this reason, the important critiques of Kopnina (2012) are incomplete). It is an empirical question (to be settled through self-reflection and through observing others), whether deep ecology or any other ecstatic and temporary experience has the capacity to modify actions when people are back to their nondeep daily encounters, and if so whether it does so any better than some alternate conception or experience. It is my suspicion that we need to move towards a more flexible and pluralistic notion of what frameworks, experiences, and emotions are 'needed,' and to develop the general capacity to become more aware of when and how our various ways of looking at the world are motivating or demotivating to us and to others. Sustainability education might be a contemplative practice (see Eaton, Hughes, and MacGregor 2017) before it is a set of (metaphysical, ethical, or scientific) propositions to be accepted and transformed into action.

And still, I suspect some apparent nonanthropocentrisms are best taken in the smallest of dosages, if at all.

That 'taking action' is itself obviously anthropocentric (Taylor 2017), grounded as it is on the idea that we transform the world according to our own values, reveals another variation of the theme suffusing this article. Recognising anthropocentrism's fluid binary does not imply we put anthropocentrism and attempts to transcend it on equal footing (Wals and Jickling 2002). Anthropocentrism is an urgent problem. But nonanthropocentrism is an insufficiently ecological solution. We can take action to seek (non)anthropocentric values committed to biotic flourishing. This, to me, is the crucial issue. How do we help learners become more versatile at abandoning and restructuring the stock of approaches they deploy when their repertoire does not actually participate in the flourishing of people and the earth? Rather than seek to overcome anthropocentrism itself, we can inspire students in the arts of contemplation, guiding them in better observing, listening, reflecting upon, and experimenting with their various (non)anthropocentrisms in theory, in feeling, and in practice.

This foray into anthropocentrism's fluid binary is intended to illustrate the importance of philosophy for environmental education. I hope the reader will see this potency even if (or especially when) my particular arguments fail. It is easy to lock into beliefs that we think are 'part of the solution', and it is a normal to be less critical of those beliefs than those which contradict them. In a complex and ever-evolving world, sustainability would seem to require the capacity to *unsustain* habitual patterns of thought that are no longer helpful for understanding or engaging (sustain/unsustain: another fluid binary?). It is not sufficient to merely develop 'critical thinking' skills, as these often merely sharpen one's saber blade while keeping it pointed at illusory enemies. The exciting sense of wonder that comes through breaking down one's own cherished assumptions is an experience few people know how to embrace. Uncertainty is insecurity. These are fears the typical classroom culture is committed to continue quietly stoking. A flexible and responsive attention to any complex theoretical/material sustainability issue will require a completely reimagined pedagogy, one that delights in baffling journeys and the humility such journeys may bring. Revealing the instability and paradoxes inherent in (non)human reasoning is a part of this project.

Notes

1. Something like multispecies ethnography (Kirksey and Helmreich 2010) might be considered as another form of multicentric (non)anthropomorphism primarily aimed at empirical understanding
2. Physicists do treat them as acting as a single object with a centre of gravitation.
3. The range of ways that two such objects interact gravitationally is described in terms of relations between variables.
4. It is of course relevant to environmental educators that these discoveries remain abstract and in tension with a simultaneous counteractive trajectory. Demographic shifts mean humans are increasingly living in technologised spaces and cities that present themselves phenomenologically as generally human centric.
5. I question the sharpness of this dichotomy elsewhere. See Affifi 2016a.
6. I use the word intra-act with caution. Barad (2007) created this term but appears to use it in different ways. In one use (that I am sympathetic of), she contrasts it to interaction, in a way that is similar to Dewey and Bentley (1949) discussion of the difference between interaction and transaction. In transactions, the knower and the known are seen as aspects of a total situation rather than separate entities. A major difference seems to be that Barad wants to insist that all relationships are intra-active, whereas Dewey and Bentley assert that different phenomena can be seen to relate in different ways, and not all are transactive. I am more suspicious of a second use of the word intra-act by Barad, where she wants to extend a nonlinear temporal interpretation of quantum events at the electron level to the world writ large. I do not yet see how entanglement between the past and future, which she describes in quantum events, is needed for the sort of co-constitutive relationality most commonly associated with her term.
7. Moreover, the genesis of those ontologies she finds unsatisfying owe their existence to the same more-than-human entanglements that she asserts have given rise to her own views.
8. This word has a complex and evolving history. For some insight into the discussion of its meanings and uses in Buddhist practice, see Bucknell 1999; Shulman 2008.
9. Part of the problem is terminology. Many new materialists are committed to the term 'material' despite what I see as its problematic associations, while critical of associations that other suitable (or overlapping) terms might have. New 'materialist,' as the name suggests, emphasises matter and not mind. Even if some new materialists thereafter insist that 'everything' is matter is a broad term that includes thoughts and feelings, they are swimming against a current they themselves have created. Calling this domain of thought 'new naturalism' instead would avoid this problem (in the sense that any 'adequate naturalism must be able to account for all domains of being, including the transcendental' Grant (2011, p. 4), a point that the American pragmatists were also well aware of (as was Deleuze (2004) in what I understand to be the relationship between actualisation and counter-actualisation)). The cost, of course, is putting a foot on one side of the scales of another binary (natural/artefactual, etc.). Following scholars like Butler (2011), talk of nature and naturalism is itself deemed anthropocentric and is now (unfortunately, in my mind) avoided by many new materialists (Morton 2009; though see McPhie and Clarke (2018) for some important taxonomical work into our modes of thinking/rethinking 'nature').

Disclosure statement

No potential conflict of interest was reported by the authors.

References

Affifi, Ramsey. 2011. "What Weston's Spider and My Shorebirds Might Mean for Bateson's Mind: Some Educational Wanderings in Interspecies Curricula." *Canadian Journal of Environmental Education* 16: 46–58.

Affifi, Ramsey. 2014a. "Drawing Analogies in Environmental Education." *Canadian Journal of Environmental Education* 19: 80–93.

Affifi, Ramsey. 2014b. "Biological Pedagogy as Concern for Semiotic Growth." *Biosemiotics* 7 (1): 73–88. doi:10.1007/s12304-013-9178-4.

Affifi, Ramsey. 2015. "Educating in a Multispecies World." PhD diss.

Affifi, Ramsey. 2016a. "More-than-Humanizing the Anthropocene." *The Trumpeter* 32 (2): 155–175. doi:10.7202/1042991ar.

Affifi, Ramsey. 2016b. "The Metabolic Core of Environmental Education." *Studies in Philosophy and Education* 36 (3): 315–332. doi:10.1007/s11217-016-9555-y.

Affifi, Ramsey. 2019. "Restoring Realism: Themes and Variations." *Environmental Education Research, 1-15.* doi:10.1080/13504622.2019.1699026.

Affifi, Ramsey. 2020. "Between Will and Wildness in STEAM Education." In *Why Science and Art Creativities Matter: STEAM (Re)Configurings for Future-Making Education*, edited by Pamela Burnard & Laura Colucci-Gray. Leiden. Netherlands: Brill/Sense.

Attfield, Robin. 2011. "Beyond Anthropocentrism." *Royal Institute of Philosophy Supplement* 69: 29–46. doi:10.1017/S1358246111000191.

Abram, David. 1996. *The Spell of the Sensuous*. New York: Vintage Books.

Baluška, Frantisek, Stephano Mancuso, D. Volkmann, and P. W. Barlow. 2009. "The 'Root-Brain' Hypothesis of Charles and Francis Darwin: Revival after More than 125 Years." *Plant Signaling & Behavior* 4 (12): 1121–1127. doi:10.4161/psb.4.12.10574.

Barad, Karen. 2007. *Meeting the Universe Halfway*. Durham, NC: Duke University Press.

Bateson, Gregory. 1979. *Mind and Nature: A Necessary Unity*. New York: E.P. Dutton.

Bell, Anne., and Constance L. Russell. 2000. "Beyond Human, beyond Words: Anthropocentrism, Critical Pedagogy, and the Poststructuralist Turn." *Canadian Journal of Education* / Revue Canadienne de L'éducation 25 (3): 188–203. doi:10.2307/1585953.

Bennett, Jane. 2009. *Vibrant Matter*. Durham, NC: Duke University Press.

Bergson, Henri. 1946. *The Creative Mind*. Translated by Mabelle L. Andison. New York: The Citadel Press.

Bogost, Ian. 2012. *Alien Phenomenology*. Minneapolis: University of Minnesota Press.

Bonnett, Michael. 2004. *Retrieving Nature: Education for a Post-Humanist Age*. Cornwall, UK: Blackwell.

Brassier, Ray. 2007. *Nihil Unbound*. Basingstoke, UK: Palgrave MacMillan.

Bryant, Levi. 2011. *The Democracy of Objects*. London, UK: Open Humanities Press.

Bucknell, RoderickS. 1999. "Conditioned Arising Evolves: Variation and Change in Textual Accounts of the Paticca-Samupadda Doctrine." *Journal of the International Association of Buddhist Studies* 22 (2): 311–342.

Butler, Judith. 2011. *Bodies That Matter*. London, UK: Routledge.

Callicott, JohnB. 1989. *In Defense of the Land Ethic: Essays in Environmental Philosophy*. Albany, NY: SUNY Press.

Clarke, DavidA, and Jamie Mcphie. 2014. "Becoming Animate in Education: Immanent Materiality and Outdoor Learning for Sustainability." *Journal of Adventure Education and Outdoor Learning* 14 (3): 198–216. doi:10.1080/14729679.2014.919866.

Connelly, WilliamE. 2010. *A World of Becoming*. Durham, NC: Duke University Press.

DeLanda, Manuel. 2002. *Intensive Science and Virtual Philosophy*. London, UK: Continuum.

Deleuze, Gilles. 2004. *The Logic of Sense*. London, UK: Continuum.

Deleuze, Gilles, and Felix Guattari. 1972. *Anti-Oedipus: Capitalism and Schizophrenia*. London: Continuum.

Derrida, Jacques. 1998. *Of Grammatology*. Baltimore: Johns Hopkins University Press.

Dewey, John. 1925. *Experience and Nature*. New York, NY: Dover.

Dewey, John. 1922. *Human Nature and Conduct*. New York, NY: Henry Holt.

Dewey, John, and Arthur Bentley. 1949. *Knowing and the Known*. Boston: Beacon Press.

Drenthen, Martin. 2011. "Ecocentrism as Anthropocentrism." *Ethics, Policy and Environment* 14 (2): 151–154. doi:10.1080/21550085.2011.578365.

Dupré, John. 2012. *Processes of Life*. Oxford, UK: Oxford University Press.

Eaton, Marie, Hughes, Holly J., and Jean MacGregor (Eds.). 2017. *Contemplative Approaches to Sustainability in Higher Education*. New York: Routledge.

Firn, Richard. 2004. "Plant Intelligence: An Alternative Point of View." *Annals of Botany* 93 (4): 345–335. doi:10.1093/aob/mch058.

Gadamer, Hans-Georg. 1975. *Truth and Method*. London, UK: Continuum.

Grant, IainH. 2011. "Movements of the World: The Sources of Transcendental Philosophy." *Analecta Hermeneutica* 3: 1–17.

Grey, William. 1993. "Anthropocentrism and Deep Ecology." *Australasian Journal of Philosophy* 71 (4): 463–447. doi:10.1080/00048409312345442.

Hargrove, E., 1992. "The Weak Anthropocentric Instrinsic Value." *Monist* 75 (2): 183–207. doi:10.5840/monist19927529.

Harman, Graham. 2013. *Bells and Whistles*. Winchester, UK: Zero Books.

Harman, Graham. 2005. *Guerilla Metaphysics: Phenomenology and the Carpentry of Things*. Chicago, IL: Open Court Publishing.

Hoffmeyer, Jesper. 2004. "Uexküllian Planmässigkeit." *Sign Systems Studies* 32 (1): 73–97.

Houser, Nathan and Christian J. W. Kloesel Eds. 1992. *Essential Peirce Volume 1*. Bloomington, IN: Indiana University Press.

Husserl, Edmund. 1954/1970. *The Crisis of European Sciences and Transcendental Phenomenology: An Introduction to Phenomenological Philosophy*. Evanston, IL: Northeastern University Press.

Kahn, Richard. 2010. *Critical Pedagogy, Ecoliteracy, and the Planetary Crisis: The Ecopedagogy Movement*. New York: Peter Lang.

Kahn, Richard. 2007. "Anthropocentrism." In *Encyclopedia of Environment and Society*, edited by Paul Robbins. Thousand Oaks, CA: Sage Publications.

Kant, Immanuel. 2007. *Critique of Pure Reason*. Translated by Norman Kemp. New York: Palgrave MacMillan.

Kant, Immanuel. 1790/2000. *Critique of the Power of Judgment*. Cambridge, UK: Cambridge University Press.

Kauffman, Stuart A. 2000. *Investigations*. Oxford: Oxford University Press.

Kirksey, Eben, and Stefan Helmreich. 2010. "The Emergence of Multispecies Ethnography." from the Editorial Office." *Cultural Anthropology* 25 (4): 545–576. doi:10.1111/j.1548-1360.2010.01069.x.

Kopnina, Helen. 2012. "Education for Sustainable Development (ESD): the Turn Away from 'Environment' in Environmental Education?" *Environmental Education Research* 18 (5): 699–717. doi:10.1080/13504622.2012.658028.

Kopnina, Helen, Michael D. D. Sitka-Sage, Sean Blenkinsop, and Laura Piersol. 2018. "Moving beyond Innocence: Educating Children in a Post-Nature World." In *Research Handbook on Childhoodnature*, edited by Cutter-Mackenzie. Dorchtrect: Springer.

Kopnina, Helen, Haydn Washington, Bron Taylor, and John J Piccolo. 2018. "Anthropocentrism: More than Just a Misunderstood Problem." *Journal of Agricultural and Environmental Ethics* 31 (1): 109–127. doi:10.1007/s10806-018-9711-1.

Latour, Bruno. 1993. *We Have Never Been Modern*. Translated by Catherine Porter. Cambridge, Massachusetts: Harvard University Press

Lettvin, Jerome Y., Humberto R. Maturana, Warren S. McCulloch, and Walter H. Pitts. 1959. "What the Frog's Eye Tells the Frog's Brain." *Proceedings of the IRE* 47 (11): 1940–1951. doi:10.1109/JRPROC.1959.287207.

Lupinacci, John, and Alison Happel-Parkins. 2016. ("Un)Learning Anthropocentrism: An EcoJustice Framework for Teaching to Resist Human-Supremacy in Schools." In *The Educational Significance of Human and Non-Human Animal Interactions*, edited by S. Rice and A. G. Rudd. London, UK: Pagrave MacMillan.

Margulis, Lynn, and Dorion Sagan. 2000. *What is Life?* Berkely, CA: University of California Press.

Massumi, Brian, Gilles Deleuze, and Félix Guattari. (1980). [1987]. *A Thousand Plateaus: Capitalism and Schizophrenia. Capitalism and Schizophrenia. 2.* Translated by and foreword by Brian Massumi, xvi. Minneapolis and London: University of Minnesota Press.

Maturana, Humberto, and Fransisco Varela. 1988. *The Tree of Knowledge*. Boston, MA: Shambhala.

McPhie, Jamie, and David A. G. Clarke. (2018). "Nature Matters: Diffracting a Keystone Concept of Environmental Education research - Just for Kicks." *Environmental Education Research*. doi:10.1080/13504622.2018.1531387.

Morton, Timothy. 2009. *Ecology without Nature*. Cambridge, MA: Harvard University Press.

Mylius, Ben. 2018. "Three Types of Anthropocentrism." *Environmental Philosophy* 15 (2): 159–194. doi:10.5840/envirophil20184564.

Naess, Arne. 1973. "The Shallow and the Deep, Long-Range Ecology Movement." *Inquiry* 16 (1-4): 95–100. doi:10.1080/00201747308601682.

Nagel, Thomas. 1974. "What is It like to Be a Bat?" *The Philosophical Review* 83 (4): 435–450. doi:10.2307/2183914.

Nietzsche, Friedrich. 1966. *Beyond Good and Evil*. Translated by W. Kauffman. New York: Random House.

Passmore, J. 1974. *Man's Responsibility for Nature*. London: Duckworth.

Payne, Phillip G. 2010. "The Globally Great Moral Challenge: Ecocentric Democracy, Values, Morals and Meaning." *Environmental Education Research* 16 (1): 153–171. doi:10.1080/13504620903504115.

Plumwood, Val. 1993. *Feminism and the Mastery of Nature*. London, UK: Routledge

Plumwood, Val 2009. *Nature in the active voice. Australian Humanities Review*, 46, 113–129.

Prigogine, Ilya, and Stengers Isabelle. 1984. *Order out of Chaos: Man's New Dialogue with Nature*. New York: Bantam.

Rautio, Pauliina. 2013. "Being Nature: Interspecies Articulation as a Species-Specific Practice of Relating to Environment." *Environmental Education Research* 19 (4): 445–457. doi:10.1080/13504622.2012.700698.

Regan, Tom. 1983. *The Case for Animal Rights*. Berkeley, CA: University of California Press.

Reznikova, Zhanna. 2007. *Animal Intelligence: From Individual to Social Cognition*. Cambridge, UK: Cambridge University Press.

Spannring, Reingard. 2017. "Animals in Environmental Education." *Environmental Education Research* 23 (1): 63–74. doi:10.1080/13504622.2016.1188058.

Shulman, Eviatar. 2008. "Early Meanings of Dependent-Origination." *Journal of Indian Philosophy* 36 (2): 297–317. doi:10.1007/s10781-007-9030-8.

Singer, Peter. 1975. *Animal Liberation*. New York, NY: HarperCollins.

Spivak, Gayatri C. (Sarah Harasym, ed.). 1990. *The Post-Colonial Critic*. New York, NY: Routledge.

Taylor, Affrica. 2017. "Beyond Stewardship: Common World Pedagogies for the Anthropocene." *Environmental Education Research* 23 (10): 1448–1461. doi:10.1080/13504622.2017.1325452.

Taylor, Paul W. 1986. *Respect for Nature. A Theory of Environmental Ethics*. Princeton, NJ: Princeton University Press.

Thompson, Evan. 2007. *Mind in Life*. Cambridge MA: University of Harvard Press.
Von Uexkull, Jakob. 2011. *A Foray into the Worlds of Animals and Humans with a Theory of Meaning*. Minneapolis: University of Minnesota Press.
Villani, Arnaud. 1999. *La guêpe et l'orchidée: Essai sur Gilles Deleuze*. Paris: Éditions de Belin.
Wals, A. E. J., and Bob Jickling. 2002. "Sustainability' in Higher Education: From Doublethink and Newspeak to Critical Thinking and Meaningful Learning." *International Journal of Sustainability in Higher Education* 3 (3): 221–232. doi:10.1108/14676370210434688.
Weston, Anthony. 2004. "Multi-Centrism: A Manifesto." *Environmental Ethics* 26 (1): 25–40. doi:10.5840/enviroethics200426139.
Weston, Anthony. 1991. "Non-Anthropocentrism in a Thoroughly Anthropocentrized World." *The Trumper* 8 (3): 1–9.
Whitehead, Alfred N. 1929. *Process and Reality*. London, UK: Free Press.

Dark pedagogy: speculative realism and environmental and sustainability education

Jonas Andreasen Lysgaard and Stefan Bengtsson

ABSTRACT

This article draws on the emerging speculative realist philosophical movement in order to develop new understandings of the issues and content of education that needs framing and reframing within environmental and sustainability education (ESE) research. We argue for the potential of using speculative realist concepts such as correlationism, hyperobjects, the strange stranger, undermining, overmining and duomining in order to develop a dark pedagogy that could draw on insights from speculative realism and object-oriented ontology in particular in order to develop further key concepts and discussions within ESE. Based on inspiration from speculative realism, our conception of a dark pedagogy could strive to: develop an understanding of the hyperobjects and the strangeness of objects of education at the center of ESE. Our conception of dark pedagogy insists on the inherent withdrawnness of the objects at play in ESE, and envisions to develop a sensibility towards eco-, geo-, infra- and chrono-factors in ESE and make explicit the potential educational implications of speculative realisms.

Introduction

New "materialism" and "realisms" of various sorts seem to be all the rage these days (Bryant 2011; Cates, Bruner, and Moss 2018; Kraftl 2018; Morton 2011) The field of environmental and sustainability education (ESE) research is arguably a good place to invite new theoretical and practical concepts inside in order to further develop new perspectives on the interplay between materiality and education (Jickling and Sterling 2017; Poeck and Lysgaard 2016). Initial contributions to translate these new materialisms and realisms are emerging (Clarke and McPhie 2016; Mannion 2007; Mannion, Fenwick, and Lynch 2013; Payne 2016), including a number of articles published in this special Issue. The divergence of approaches and ontological outlooks is also reflected in work done in related educational fields that translates philosophical elaborations into an education context (Lewis and Kahn 2010; Oliveira and Lopes 2016; Snaza et al. 2014; Wallin 2014b). Pitfalls abound, of course. Stephen Sterling is quoted in arguing that the core concepts of ESE are "debilitated by a lack of philosophical clarity (Jickling and Sterling 2017). In this article we focus on the philosophical issues that arise when drawing on the emerging speculative realist philosophical movement, and object-oriented ontology (OOO) in particular, in order to develop our understanding of environmental and sustainability issues that need framing and reframing within ESE research. We argue that these perspectives can help us understand how we can deal with the *complexity* or *wickedness* of these issues at hand (Harman 2009; Morton 2013). This is done in order to develop the philosophical fundament of the field of ESE research, but also to push the link between research and practice through insights gained through contemporary and emerging philosophical possibilities. We will try to contribute to this by developing a conception of dark pedagogy that aims to constructively and reflexively engage with the complexity and wickedness of sustainability issues in an educational context.

This paper is a theoretical effort to draw on current developments within continental philosophy to philosophically interrogate and develop our understanding of the complexity and wickedness of the issues within ESE that, as we will show, will always remain hard to grasp and, as we will argue, in the dark. The ambition is not to dismiss any current or past work within the field of ESE, but to interrogate, substantiate and, thereby, to open up for discussions central concepts to the field. Given its relative content of education and issues concerned, we understand the ESE field as an obvious potential bridgehead of new philosophically informed perspectives on materialism and realism, as ESE is one of the many educational fields that deals with these issues most explicitly.

As mentioned above, we will frame the issues that ESE deals with as dark and based on this consideration of the dark aspects of the content of ESE to develop a dark pedagogy. Our use of the term 'dark pedagogy' is inspired by current philosophical perspectives on nature, realism and materiality, but also by an interest in the classic horror author H.P. Lovecraft (Harman 2012; Morton 2016; Oral 2015; Thacker 2011). His universe consists of a fragile human world constantly threatened and rearranged by uncanny powers that lie beyond the direct control of human powers, but always interfere, change and distort the world they inhabit. H.P. Lovecraft and his 'tales of the horrible' [1] emphasize how the great challenges to humanity always eludes petty human attempts to give meaning to them and to control them. Hence, our conception of dark pedagogy attempts to unsettle and de-center the autonomous and mature subject of education, as it has been framed by Enlightenment thought and still influences contemporary conceptions of education. This perspective and the writing of H.P. Lovecraft is being linked by several philosophers within the before mentioned speculative realist movement to analyses of issues such as environmental degradation, global warming, pollution, etc. (Harman 2012; Thacker 2011). While the inspiration from Lovecraft's stories are present in the development of Dark pedagogy, it is important to note that Lovecraft was both a misogynist and xenophobe and further that these regrettable personal tendencies and dispositions that blemish his biography inform his plots and narrative designs, especially his later work (Joshi 2013). Our decision for employing his horror stories as a backdrop of dark pedagogy was made on the basis that the regrettable and despicable biographical factors that informed some of his weird tales might be able to explain parts of his literature, but they do not overdetermine the potential meanings of his stories (Joshi 2013; Lysgaard, Bengtsson, and Laugesen 2019).

While this inspiration from Lovecraft and his stories are proving inspirational and influential in speculative realism, such uptake of these Lovecraftian philosophical approaches has only seen initial efforts to translate them into education and environmental education (Oral 2015; Wallin 2014a). These efforts have primarily been framed and limited to the scope of the fields or points of departure that the scholars have been trying to address, for example by a scope that focuses on the philosophical implications (Oral 2015) post-humanist perspectives (Wallin 2014b), or place-based education (Saari and Mullen 2018).

To specify and contextualize our contribution, our take on developing a dark pedagogy by drawing on speculative realism orients itself towards the German *Bildung*-tradition and a continental understanding of *Didaktik*. As a result we are primarily concerned with the implications of the content of education for the project of education or rather *Bildung*. While Taylor (2017) can here be seen to have tried in a similar effort to develop the conception of *Bildung* by introducing a post-humanist perspective, we will do so by introducing an object-oriented outlook on *Bildung* and *Didaktik* (Taylor 2017). Dark pedagogy relates in our understanding to a project of becoming a self and member of a collective as part of educational processes. The self in this sense, and as a consequence any social or human collective, bound up with or dependent upon its relation to an education content. Darkness enters this focus by relating to how we as human beings relate to the *horrible* things as contents of education. Such things are scaled beyond our comprehension, but still influence our lives and societies, and how education and learning could change in order to account for this. Our focus in conceiving of dark pedagogy focuses hence on what the darkness of a content of education means for processes of education.

The notion 'dark pedagogy' could lead to ideas of Darth Vader like types leading pedagogical activities, a romantic flirt with neo-Gothicism or perhaps even conjure images linked to the much berated German tradition of *Schwarze Pedagogik* (von Stechow 2010). This is, however, not the intent behind our use of the notion of dark pedagogy. We use the notion in order to signal the fact that educational processes also take place in darkened places, unlit domains and in not easily undisclosed ways. In that way we draw on the German adjective *dunkel* in our understanding of Dark (Merriam-Webster 2018). Dunkel as referring to deep, obscure, vague, gloomy, shady or opaque. For us, the subject of education resists in a sort of twilight zone, where learning is not an unfolding of a pre-given rational, human capacity but also an aesthetic process in which the subject is drawn out of itself and the light of reason by the object at the center of educational processes. By using dark and *dunkel* we want to argue that a range of the core issues in, and particularly the content of, ESE practice and research often resides exactly in such areas where we, on one hand, can develop ideas about their nature, but at the same time, constantly experience how we can never drag the totality into the probing light of either research or educational processes. Issues such as climate change, global warming, environmental crisis, sustainability etc. always harbor characteristics that escape our best efforts to fully understand them and engage them. This dark, or *dunkel* area remains an area that is hard to understand and traverse, but it nonetheless has huge influence on our societies and how we understand both education and pedagogical practices. Further, darkness is for us associated with an emotional disposition, where that what we encounter in darkness is reverberating a feeling of uncanniness. In the dark, even the most familiar things seem strange and somewhat challenging. Dark pedagogy is an effort to think philosophically and structurally about the inherent dunkelness of the content of ESE research and practice.

Our use of the term "dark" in dark pedagogy is also to explicitly signal our theoretical debts to speculative realism and OOO in particular. In 2016, Timothy Morton's book *Dark Ecology* came out (Morton 2016). In it, he elaborates the implications for ecological thinking that he takes to follow from his previous books. Thus, for instance, he had already employed the term in his *The Ecological Thought* (2010) and *Hyperobjects* (2013). In all three publications, 'dark ecology' appears to be Morton's umbrella term for several, semantically associated phenomena. Dark ecology is in one sense about becoming aware of how implicated you yourself are in the environmentally destructive processes of global warming. This involves, hence, us as individuals to delve into a melancholic *Stimmung* in commemoration of the somber aspects of late modernity and the hidden costs of its accelerating consumer societies. To become attuned to this *Stimmung,* involves, as Morton argues, a seeking out alien intimacies with strange co-inhabitants of your local environment, e.g.your own backyard, thus shedding light on the otherwise dim strangeness in your very vicinity.

Pro-actively bringing about this kind of finding, delving and seeking perspectives of dark ecology is the practical goal of our conception of dark pedagogy. Whereas Morton has given us a philosophically inspired, yet prophetic vision of the contemporary entanglement of mind and matter, culture and nature, linked with the concept of the Anthropocene, dark pedagogy attempts to put forward an account of the educational processes and didactical points of intervention that could be inspired by this. Thus, dark pedagogy tries to provide a dark and ecologically oriented notion of *Bildung* and *Didaktik.* The aim of this paper, in its effort to specify a tentative dark pedagogy, is to outline some of the philosophical underpinnings of a dark and ecological *Bildungs*-oriented outlook in ESE. Such an outlook is compatible with the enlivened and everyday implications of speculative realism, where the ethical implications of speculative realism "consists in the demand to embrace that which is 'weird', i.e.that which exists outside of the usual horizon of experience and the common frame of reference." (Laugesen et al. 2017, 20).

In order to develop the notion of dark pedagogy and the potential for ESE practice and research we start with accounting for some of the key philosophical underpinnings of speculative realism and OOO, before bringing into play the concepts of hyperobjects, under-, over-, duo-mining and the strange stranger in order to illustrate what a return to realism might offer for environmental and sustainability education.

Speculative realism and environmental and sustainability education

Modern societies and individuals have made a habit out of thinking about nature as an untapped resource. Nature has therefore been thought of as something that is for the sake of modern humans. However, perhaps that is not an everlasting truth. What if the world does not primarily exist to serve our needs and ensure our continued survival and growth? Such a cosmic pessimism has made inroads into contemporary continental philosophy (Harman 2009; Morton 2011; Morton 2016). That nature is to be categorized and handled in order to make sense to us, human beings, points to the simple fact that nature is not meant to serve humans, but remains whatever it is, whatever it will be, independent of particular human interests (Harman 2009; Morton 2011).

As a current form of philosophical response to the world's objective autonomy and indifferent attitude toward human purposes, speculative realism has, since 2007, managed to send shock waves through not just western, but even the global, intellectual and academic life (Shaviro, 2014). Speculative realism is a loose association of thinkers and one that harbors very different and at times mutually exclusive positions (Harman 2009). The *realism* of speculative realism consists of the basic ambition of thinking reality in itself; to explore the world independently of human techno-scientific interventions and interpretations mediated through language. The particular *speculative* part of this realism is that the efforts to recognize the world itself are managed in the form of a revitalization of metaphysics. 'Metaphysics' as the name of the philosophical discipline draws on the reflexive elements of human thinking in order to understand the very way in which things exists (Harman 2009). With regards to the understanding of how things exist, the French philosopher Quentin Meillassoux (2009), who occupies a central position in speculative realism, argues that, in spite of its critique of earlier notions of idealism, continental philosophy itself has ended in an *anti-realistic position* in terms of what Meillassoux names 'correlationism'. By 'correlation' he refers to the ontological position that we only ever have access to the correlation between thinking and being, and never to either term considered apart from the other (Meillassoux 2009). The critique of this correlationism is at the very center of not only Meillassoux project, but also the speculative realist project. To sum it up, the return to realism claims that things do not only exist in or for a human thinking mind.

Consequently, the basic premise of speculative realism is "not a doctrine, but an umbrella term for a variety of research programmes committed to upholding the autonomy of reality [...] against the depredations of anthropocentrism" (Brassier et al. 2007). Common to those thinkers meeting under the umbrella of speculative realism is, thus, a resistance to anti-realism, which the speculative realists argue can be understood as the dominant position within continental philosophy (Harman 2009). Any anti-realistic position consists in rejecting the notion of an objective reality, as it is, independent of and indifferent to how humans recognizes it and, incidentally, deal with it. Such an objective reality has been for the most part barred from continental philosophical discussions and consequently also from educational thought (Harman 2009).

Instead, the focus or universal point of departure has been human beings understood as subjects, that is as the conditional entity in relation to the object. Human relations with the environment, and in particular how these relationships are conveyed through the language, have been the central pivotal points for a significant part of the continental philosophy that the speculative orientation is critical of. Thus, phenomenology (Husserl, Heidegger and Merleau-Ponty), hermeneutics (Gadamer), structuralism (Levi-Strauss, Lacan), post-structuralism (Foucault, Deleuze) and deconstruction (Derrida) can all be understood as examples of an anti-realistic tradition, which have been central to continental philosophy during the 20th century (Laugesen et al. 2017). This anti-realistic tradition per se has also been fundamentally shaping educational thought with its focus on the human subject at the center of education and educational processes.

It is against the background of imminent climate crises, global issues regarding resource shortages, the impact of technology in the everyday lives of ordinary people and similar pressing conditions, anti-realism seems insufficient to both theoretically recognize and practically handle the world we are in, and the challenges it is witnessing (Levi Bryant and Harman 2011; Meillassoux 2009). Moved by this thinkers associated with speculative realism tie these philosophical discussions to contemporary global challenges and sustainability in order to reconsider the most basic human concepts of our relation to the world (Brassier 2007; Bryant 2011; Morton 2016). In relation to this speculative realism can be said to have the ambition to nurture and develop the metaphysics of the anthropocene era, as all the movement's representatives - in continuation of, among others, Bruno Latour – argues that the basic distinction between 'nature' and 'culture' is impossible to maintain (Harman 2011; Latour 1993). Human beings are on a very basic level interwoven with what we, from inside 'culture' call 'nature' and what the speculative realism wants us to systematically think of is the end of this divide and to explore the 'great outdoors' (Harman 2009). Thus, what we interpret speculative realism to offer ESE research and practice is a rethinking of what it means to be a human subject engaging with environmental and sustainability in educational processes.

Linking concepts from speculative realism and ESE raises the question of the significance of content for the project of education as a form of developing humanity: What consequences does the object at the center of educational processes have for how subjects and collectives are understood? The link questions the ontological significance of the object for educational processes. Consequently, dark pedagogy examines, according to a *Bildung*-tradition, what kind of self engages in and is produced in educational processes? What is the relation of the self to other objects, objects that are not only there for a human to apprehend but that remain *dunkel*.

We argue, given the emergence of speculative realism, that it is possible to re-approach major current challenges in education, such as climate challenges of geopolitical, agricultural and legal (and educational) nature by moving beyond a correlationist approach to education that focuses solely on how these challenges are reduced to phenomena to be known in and controlled through educational processes. Dark pedagogy asks the question: what does it mean to be a self engaging in and a teacher structuring educational processes in the anthropocene? The anthropocene is here exactly understood as a process when the human subject becomes aware of its limits of ability to know and control (Morton 2016). We might have thought we were merely driving our car to work and do this without causing harm to others, but find out that we were creating something called global warming. Thus, dark pedagogy reflects on how the *dunkelness* and somewhat opacity of the content of education undermines the notion of a self at the center of education that is wholesome, both as a unity and given its intellectual capacity. This potentially changes our notion of self and education once we consider how the production of self and educational processes is caught up with the *dunkelness* of the content of education to be addressed.

While the project of dark pedagogy is in itself explorative (Lysgaard, Bengtsson, and Laugesen 2019) in the following section we will briefly explore what OOO, as a strand of speculative realism, has to offer for education. We will introduce the concept of hyper objects and strange stranger provided in the work of Morton (2011, 2016) in order to try and draw some of the insights of speculative realism, and OOO in particular, into the field of ESE research and see how we can build upon a conception of education that, as (Jickling and Sterling 2017) point out, needs to take place against the background of a rapidly changing global context and in the face of a need to move beyond what speculative realism has labeled correlationist conceptions and critiques. We make this argument for a need to engage with speculative realism against the background of a field that deals explicitly with the great outdoors, and argue that analytical strategies that focus only on how issues and problems of the great outdoors appear to us humans need to be supplemented with new perspectives derived from materialism and realism.

The issues at the core of ESE research – hyperobjects as objects of education

The foci of ESE research has been widely discussed under the umbrella of environmental education (EE), education for sustainable development (ESD) and now ESE (Gough and Scott 2003; Jickling 1992; Jickling and Wals 2008; Poeck and Lysgaard 2016; Sterling 2017; Sund and Lysgaard 2013). Efforts to underline specific topics or issues of special concern to the field have often been sidelined by different conceptualizations of how to address the field, what the field encompasses and discussions of the responsibilities of any given research field within education. Besides the obvious headers of 'environment', 'sustainability' and 'education', there seems to be a very broad range of understandings of what those different notions cover, and rarely does that include and a positive definition of the specific privileged concepts that defines a field such as ESE research and what they entail and potentially exclude. Returning to Scott and Gough (2003) they settle on the borrowed term (from political science) of understanding concepts such as environment, sustainability and in effect the liaison educational fields as contested, but also very inclusive in their character (Scott and Gough 2003). Reviews of the research within the field also underline how ESE research covers much ground and shares general, but seldom particularly specific understandings of the core issues and challenges that the field addresses (Aikens, McKenzie, and Vaughter 2016; Hart and Nolan 1999; Poeck and Lysgaard 2016; Rickinson 2001). Other efforts have been focusing on framing the overall focus of the field within established theoretical movements relative to educational research such as critical pedagogy and Klafki's epochal core problems (Mogensen and Schnack 2010) or Freire's focus on oppression and how that feeds into education (Jickling and Wals 2008). Thus, the relationship between issues and education has often been approached from either a focus on the issues or educational processes. Dark pedagogy aims to take a middle road, by focusing on the objects of education. In particular we argue that hyperobjects are at the core of ESE and hence see them as constitutive to the field and of special significance to our understanding of educational processes.

ESE and hyperobjects

While we enjoy the freedom to develop our research focus within a field such as ESE research, it is also fruitful to frame some of the discussions of how we share or do not share understanding of the core of ESE research. Drawing on Timothy Morton's concept of *hyperobjects* it is possible to inform some of the discussions of the general focus of ESE research with a critical perspective on how we understand the reality of the issues that govern the agendas of ESE research. Morton draws on Graham Harman's understanding of objects, as expressed in Harman's concept of OOO, when he analyzes how hyperobjects are challenging the concept of the human centered word view (Harman 2011, Morton 2013). Morton's hyperobjects has some interesting features that in many ways can be understood as being at the very core of what the field of ESE tries to address (Morton 2013). In his 2013 book on hyperobjects, Morton introduces the concept in order to make sense of the challenges that we face today and how they are bound in scales that surpasses our human senses, but at the same time influence and change how we think, live and act:

> Hyperobjects have numerous properties in common. They are *viscous*, which means that they 'stick' to beings that are involved with them. They are nonlocal; in other words, any 'local manifestation' of a hyperobject is not directly the hyperobject, they involve profoundly different temporalities than the human-scale ones we are used to. [...] And they exhibit their effects *interobjectively*; that is, they can be detected in a space that consists of interrelationships between aesthetic properties of objects. (Morton 2013, 1–2).

An example of a hyperobject could be oil. As the result of biological and geographical processes with its immense impact on modern day society, the environment and leaving remains for many future generations to deal with, oil is everywhere, but not possible to limit or capture in its totality as an object. Another example could be global warming. It relates to an object that appears and changes on a global scale, that individual humans have no access to in its totality, yet that produces specific consequences at a local level. Global warming cannot be directly seen, but it can be thought and computed. The consequences however are extremely tangible and challenging to understand in both their scope but also their range of impact. Morton argues in the context of hyperobjects that we live in an age of asymmetry, where the challenges we face are rooted beyond the scope of us as human beings, but at the same time these issues influence our daily lives to an accelerating degree (Morton 2013). The issues and challenges we face seem to lack a proportionality that humans can comprehend and work around. They can be described and even partially apprehended on an individual level, but staying true to Harman's OOO they always escape total representation and operate on scale, that are below, above and beyond human comprehension.

The concept of hyperobjects underlines that there are objects to be addressed as content of education that are of concern for education even though they defy the pedagogical principle that they must become apparent to learners in educational processes. Further, it is rather the fact that they remain in the dark that they become of significance to humans as well as the educational project. They challenge the very core idea of Enlightenment that is inherent to the education project; that of fostering a maturing rational subject (Deligiorg 2005). This project of maturing, as the development of onés capacity to reason about a content of education, is clearly linked to the empiricist tradition of rendering nature transparent to the human mind and to uncover its inner workings. The hyperobject, however, withholds this possibility given that it is nonlocal and its effects only emerge in inter-objective relations. Yet, given the possibility of massive data generation and analysis humans have become aware of the existence of such objects that are not reducible to their local manifestations. They are real, yet, remain in the dark for the apprehending human subject.

Rendering climate change a legitimate issue and content of education can, hence, be seen as to challenge a key assumption underlying the didactical construction of educational processes. It is the assumption that what counts as real and significant content to be addressed in education needs to be apprehensible in educational processes by the human subject. Given the trend that inaccessible climate is increasingly rendered a valid content in education (UNFCC 2015), the violation of the principle of "apprehensibility" can be seen to open up the possibility of an influx of other contents normally deemed as non-real, non-significant and beyond reason, for example ghosts.

Further, despite their non-locality, hyperobjects are able to produce a certain *Stimmung*, as characteristic of our understanding of darkness. They are able to produce a certain emotional reverberation in the self at the center of educational processes; it might be the experience of beauty or the feeling of alienation. In the encounter of such emotions, education is able to tune into the viscosity of hyperobjects. Thus, dark pedagogy highlights the aesthetic dimension of educational processes as crucial for addressing environmental and sustainability issues in education.

A dark pedagogy requires in this sense an educational outlook that moves beyond a metaphysics of presence (Derrida 1982; Heidegger 1927). It renders significant contents of education that are in a traditional didactical outlook denied the status of being real and hence deemed not worth considering in education. Further, giving that hyperobjects as content of education have the qualities of being non-local and viscous, dark pedagogy undermines the notion of place that provides a setting for educational process (Gruenewald 2003; Stevenson 2008). The nonlocality of the hyperobject that, yet, remains viscous undermines the notion of education as process of subjectification (Biesta 2014) and identification (Stevenson 2008; Gruenewald 2003) as taking place *within* a Newtonian absolute space. As the hyperobject is both not-here (nonlocal) and very near (viscous), it undermines the very identity of the self engaging with it. The displacement of the self takes place in the emotional resonance of the hyperobject, for example the feeling of alienation from notions of nationhood in the face of globalization or migration. Yet, OOO and dark pedagogy do not highlight a form of distance that the viscosity of a hyperobject entails but the very nearness (Heidegger 1927; Irigaray 1977) that challenges and ruins the process of producing an autonomous and unified self in educational processes (Bengtsson 2018). The self at the core of education finds itself entangled with and in hyperobjects unable to distance itself from them emotionally.

The concept of hyperobjects could inspire and infuse ESE research and practices once we rethink the specific issues as contents of education that it focuses upon. It offers both a way of rethinking the scope of the issues and their characteristics, but it also insist on the inherent collapse of that thinking, in as much as it needs to insist on the withdrawnness of the hyperobject, and of all objects. To recall, dark pedagogy is a speculative pedagogy as it consists of an attempt to break out of an idealist paradigm and to return to the content of education as relating the reality of objects. We argue this can be a highly generative way of approaching issues such as climate change, migration, environmental issues and social issues as it resists the temptation to clean cut reliance on current scientific facts, and potential instrumentalization of these through correlationist reasoning. At the same time, it does however also open up a literal abyss under key assumptions about education and the potential outcomes of education.

The content of education is a strange stranger

OOO has not only something to offer to the conception of hyperobjects, so massive and distributed in time and space that they exceed human apprehension, but also about the most familiar objects that education deals with. What OOO offers is an outlook on what it means to be a self engaging with a content of education where the processes of education are not explained by either an upward nor downward reductionism.

Harman offers through his conception of undermining, overmining and duomining a stimulating critique of reductionism that is also applicable in education research (Harman 2011). The concept of Undermining emphasizes the tendency to reduce the explanation of a phenomenon to a universalist account of the most basic components of that phenomenon. We might here think of behaviorisms explanations of learning processes in terms of elementary forms of human behavior, neurosciencés account for learning in terms of the working of neurons, or cognitivism's account of learning in terms of schemata in cognition. Overmining on the other hand moves the other way, relating to a macro-perspective and idealism in order to account for what is really taking place in education. Sociological accounts of reproduction a la Bourdieu or Foucauldian archeology and genealogy applied to account for specific encounters in education share the overmining step of relating something particular to the social or society as a whole. The concept of duomining is an example of when both practices of overming and underming are combined in order to account for what education is really all about. The problem with these approaches is twofold. First, they cannot account for the excessive nature of any given object taking part in the educational process. Harman argues that all these strategies lose sight of the basic withdrawnness of all objects (Harman 2011), that is that one certain characteristics of the things taking part become apparent while other characteristics and other objects involved remain withdrawn. Second, given that they cannot account for this excess, these strategies withhold the condition of possibilities for change

As a result, these accounts are providing static notions of education as they withhold the possibility of things becoming different as they claim to account for things in their entirety given their under-, over- or duomining. When Carl Anders Säfström argues that education has been, first and foremost, the effort of any society to reproduce itself we can be inspired by speculative realism to argue that all such efforts ultimately fail but also *should* fail in order for change to be a possibility (Säfström 2017). As the transfer never succeeds fully, something different always can and will happen. This something can be seen as an unavoidable kernel of excess, embedded within education. The notion of excess - that something always eludes total illumination and remains withdrawn and in the dark - is central to speculative realism, OOO and of course the different but linked concepts of excess found in psychoanalysis (Evans 2010; Harman 2009, 2011). It is not possible to understand education as processes of causality or direct transference of skills, instead education can be understood as imbued with the central points of the speculative realist call to arms: Education is at its core not causal, but *open*.

Education is in this way, as a process on both structural and individual level, not strictly correlationistic. Education is more than us as subjects and *our* relationship to the world. Quentin Meillassoux's critique of *correlationism* and thus of the Kantian dictum that we do not have access to the thing in itself, but only to how our senses perceive it, can arguably also apply to education and ESE (Meillassoux 2009). A dark pedagogy sees education not only to limit itself to a correlationistic logic but sees education as relative to things, objects in themselves and the hyper complex multiverse that permeates everything.

This means for a dark pedagogical outlook that the self, engaging in learning processes, is bound up with a content of education that is subverted by objects that remain withdrawn. Thus, when engaging with a content, the objects that are relative to that content remain somewhat withdrawn. Education in the understanding that Säfström is putting forward above can be seen as an attempt at normalizing the way we perceive objects, reducing the *Unterrichtsgegenstand* (object of education) into a content of education as it is apprehended from a particular human subject-position to be attained through the didactical encounter (Bengtsson, forthcoming). However, the withdrawnness of the object of education can subvert any normalization, rendering strange again that which was taken as normal. Thus, accordingly the *dunkelness* of the object of education subverts the didactical attempt of reducing it to a *content* of education. This process of subversion takes place in both how an object of education appears to a self in educational processes as well as in an emotional resonance such as alienation or a feeling of uncanniness that is counter to any form of normalization. In the prior, the self becomes aware of an excess of qualities of an object that, so to speak, break out of content, where content is referring to the notion of things being held or included in something. In this encounter we are becoming aware that the object of education is what Morton (2010) calls a *strange stranger*. The object of education in its strangeness is striking us in its unfamiliarity, its uniqueness and our inability to reduce it to the category of a particular kind of object.

In this sense the excess as a form of strangeness of the object reminds us that there is always something more, something that we should and do relate to, consciously or subconsciously. It is important to stress this excess and strangeness of the object of education in order to highlight what is at stake not only for the content of education but also the production of self and social collectives. Given that the object of education is withdrawn, the self that is to produce itself fails to do so as well. It is itself caught up in the strangeness of the object (Bengtsson, forthcoming). The self emerges in this sense only in inter-object relations and as a result it is also strange to itself. The encounter of this strangeness and the emotional reverberation of other objects in their strangeness is dark, yet, this darkness is also an entry point for breaking out or rather through an educational position where the self is a subject distinct and different from objects. What dark pedagogy aims to offer is a form of paranoid self in education. Am I really a self? Am I a person?

Hence, dark pedagogy holds that we first have to embrace the insight that we cannot grasp objects or hyperobjects in their totality. Yet, staring into the abyss will only get us that far. We have to move through it. Accepting that the world is dark, but that different hues of dark exist and that some of this darkness might hide insights that we can partially illuminate, while others will be moored outside of our reach, is a powerful first step towards critiquing both societal use of education as a functionalistic means of behavior modification and as individual coping strategies in order to make sense of this changing and contingent world we live in. However, beyond critique, and once we start to move through this darkness, we need to embrace a certain degree of paranoia; we need to engage in fostering a somewhat paranoid self in educational processes. Paranoia, as an engagement with or partial acknowledgement of hidden forces that might be influencing the self, is a dark strategy for breaking through correlation's reasoning and to open up both aesthetics and reasoning in education about the great outdoors. Moving towards paranoia is a way to open up towards thinking and feeling that does not want to reduce the object of education to a content, as an object appearing to humans and their history of engaging with it. Thus, moving towards paranoia, dark pedagogy wants to return to objects and unbind "... the spells that preserved them intact likewise prevented them from making an initial move, and they could only lie awake in the dark and think whilst uncounted millions of years rolled by" (Lovecraft, 1928). By including such a position into the thinking that shapes our understanding of the core issues at play in ESE research and practice, it might be possible to better understand how the reality of withdrawn objects is influencing us humans and why our individual, societal and educational strategies again and again fall flat in efforts to try and tame what cannot be tamed.

Conclusion - towards a dark pedagogy

The problem with speculative realism in its current form is that it still remains, amid countless ongoing efforts to apply it in a whole range of various fields, a philosophical and ontological project. If we live in a democracy of objects, as Levi Bryant puts it, where does it put the assumed exclusively human project of education? How do we incorporate speculative realism into education and pedagogy in order to develop a constructive and critical dark pedagogy? In education, human beings are not only objects among objects, but also given human disposition of the self towards other objects privileged objects among less privileged objects. Developing an understanding of, and sensibility towards the darkness of objects and hyperobjects is a fine educational intention, but it has also significant ethical and political implications. Do we really want to open up children and other people to feelings of uncanniness, depression and alienation as part of educational processes?

Instead of answering the question head on, we would like to ask a counter question: Are children and people in general not already exposed to feelings of uncanniness, depression and alienation? With Heidegger (1927) we argue that these feelings resonate an underlying existential condition of being in the world. Instead of treating anxiety, depression and boredom as outcomes of education gone wrong, a dark pedagogy engages with these feelings as possibilities of breaking out of thinking that wants to normalize and engage in the fantasy of rendering the world a cozy place sugar coated in valium and Prozac. We argue that the endeavor of dark pedagogy to tune into a more-than-human *Stimmung* resonating in the self at the core of education is an ethical project worth considering, given that tuning out such a *Stimmung* is seen as amplifying the suffering and destruction of the more-than-human and as a consequence also the suffering and potential destruction of humanity. The continued normalization of how humans engage with objects through education is in this sense not eradicating darkness but exacerbating suffering through its denial.

Accordingly, a dark pedagogy plays into the project of Levi Bryant (2013) to translate speculative realism into the realm of politics and outlining an expanded form of democracy, that is a democracy of objects. Bryant does so by emphasizing not only human or social factors in politics and societies, but also the importance of developing an acute awareness of other factors that reside in the world without us (Bryant 2013). Such factors could be ecological factors and how they influence and change how we see and approach the world, geological factors and how they limit and develop our scope, factors of infrastructure and how they decide our ways of moving through the world and especially chrono factors emphasizing both the proportionality of some aspects of time, but also the deeply asymmetrical aspects of e.g. deep time that has a huge influence on how we as human objects live and experience in the world. Our take on a dark pedagogy aims to open up education to the aesthetics of these factors, factors that will partially remain in the dark.

As we tried to show in this paper, drawing on inspirations from speculative realism it is possible to both reframe central discussions within the field of ESE research related to how we understand the very core of the field and how we theoretically and philosophically develop our conception of education and its relation to environmental and sustainability issues as content of education. The project continues, as the emerging materialist and realist perspectives of the 00's and 10's solidify and are mainstreamed. The ambition here is to outline some key characteristics of a dark pedagogy and to open up for an alternate understanding for engaging with environmental and sustainability issues in education by infusing some insights and concepts from speculative realism. A potential ambition for further development of a speculative realist inspired dark pedagogy could be to:

- Develop an understanding how hyperobjects as environmental and sustainability issues could become objects of education (Morton 2013, Morton 2016).
- Address the inherent contingency and risk of correlationism at play in ESE research and practice (Meillassoux 2009).
- Exploring phenomenologically the withdrawnness of objects of education (Harman 2011) and the attunement of the human self to other objects in ESE practice (Morton 2016).
- Develop a pedagogy that sensibilitizes towards eco-, geo-, infra- and chrono-factors in ESE practice (Bryant 2013).
- Develop anti-correlationist outlooks on education that move beyond a nature/culture divide (Bengtsson 2018; Wallin 2014a; Oral 2015).
- Discuss the educational implications and how to address the issues at hand, without overburdening the individual (Lysgaard 2018).

Given that the conception of a dark pedagogy is a pedagogy in the making, the ambition of this article is not to close down discussions or dismiss past or current work within ESE research, but to open up discussions related to core and fringe concepts within ESE and how we ensure that we do not reach comfortable plateaus, but continue to strive for insights of both theoretical and practical value. Speculative realisms represent emerging philosophical thoughts and perspectives. As tradition dictates, the immigration of concepts and insights from one academic field to another is a long process, but we would like to argue that the emphasis on materialism and realism is especially obvious for a field like ESE. To be inspired by current developments within philosophy and develop them further in our field not only offers new pathways for the field of ESE, but for education in general. A philosophically explicit darker pedagogy, both depressing, sweet, uncanny and alienating, would be able to develop current regimes of thought and practice to include new perspectives on the central issues, such as global warming or environmental crisis, that only become more and more pressing, both, inside and outside of ESE.

Acknowledgements

This paper is part of a greater project focusing on linking speculative realism and pedagogical and educational research under the header of Dark Pedagogy. The project has many mothers and fathers. One of them, Martin Hauberg-Lund Laugesen, is not credited as an author of this paper, but is substantial to the development of the thoughts behind Dark Pedagogy and also directly involved in other publications within the project (Lysgaard, Bengtsson, and Laugesen 2019).

ORCID

Jonas Andreasen Lysgaard http://orcid.org/0000-0001-8687-0805
Stefan Bengtsson http://orcid.org/0000-0001-7052-1088

References

Aikens, K., M. McKenzie, and P. Vaughter. 2016. "Environmental and Sustainability Education Policy Research: A Systematic Review of Methodological and Thematic Trends." *Environmental Education Research* 22 (3): 333–359. doi:10.1080/13504622.2015.1135418.

Bengtsson, S. (forthcoming). *The Uncanny Content of Education*.

Bengtsson, S. L. 2018. "Outlining an Education without Nature and Object-Oriented Learning." In *Research Handbook on Childhoodnature*, edited by K. Malone, Amy Cutter-Mackenzie, and E. Hacking Barratt. Springer.

Biesta, G. 2014. *Beautiful Risk of Education*. Boulder, Colorado: Paradigm Publishers.

Brassier, R. 2007. *Nihil Unbound: Enlightenment and Extinction*. London: Palgrave McMillan.

Brassier, R., I. H. Grant, G. Harman, and Q. Meillassoux. 2007. *Speculative Realism*. London: Goldsmith College.

Bryant, L. R. 2011. *The Democracy of Objects*. London: *Open Humanities Press*.

Bryant, L. R. 2013. "Politics and Speculative Realism." In *Speculations IV: Speculative Realism*, edited by M. Austin, P. J. Ennis, F. Gironi, T. Gokey, and R. Jackson. Brooklyn, NY: punctum books.

Cates, C.,. M. L. Bruner, and J. T. Moss. 2018. "Recuperating the Real: New Materialism, Object-Oriented Ontology, and Neo-Lacanian Ontical Cartography." *Philosophy & Rhetoric* 51 (2): 151–175. doi:10.5325/philrhet.51.2.0151.

Clarke, D. A. G., and J. McPhie. 2016. "From Places to Paths: Learning for Sustainability, Teacher Education and a philosophy of becoming." *Environmental Education Research* 22 (7): 1002–1024. doi:10.1080/13504622.2015.1057554.

Deligiorg, K. 2005. *Kant and the Culture of Enlightenment*. New York: State University of New York Press.

Derrida, J. 1982. *Margins of Philosophy*. Brighton: The Harvester Press Limited.

Evans, D. 2010. *An Introductory Dictionary of Lacanian Psychoanalysis*. London: Routledge.

Gough, S., and W. Scott. 2003. *Sustainable Development and Learning: Framing the Issues*. London: Psychology Press.

Gruenewald, G. A. 2003. "The Best of Both Worlds: A Critical Pedagogy of Place." *Educational Researcher* 32 (4): 3–12. doi:10.3102/0013189X032004003.

Harman, G. 2009. *Towards Speculative Realism*. London: Zero Books.

Harman, G. 2011. *The Quadruple Object*. London: Zero Books.

Harman, G. 2012. *Weird Realism: Lovecraft and Philosophy*. London: ZERO books.

Hart, P., and K. Nolan. 1999. "A Critical Analysis of Research in Environmental Education." *Studies in Science Education* 34 (1): 1–69. doi:10.1080/03057269908560148.

Heidegger, 1927. *Sein Und Zeit*. Frankfurt Am Main: Vittorio Klosterman.

Irigaray, L. 1977. *This Sex Which is Not One*. New York: Cornell University Press.

Jickling, B. 1992. "Wy I Don't Want my Children to Be Educated for Sustainable Development." *Journal of Environmental Education* 23 (4): 5-8.

Jickling, B., and S. Sterling. 2017. "Post-Sustainability and Environmental Education." In *Post-Sustainability and Environmnetal Education*, edited by B. J. S. Sterling. London: Palgrave.

Jickling, B., and A. E. J. Wals. 2008. "Globalization and Environmental Education: Looking beyond Sustainable Development." *Journal of Curriculum Studies* 40 (1): 1–21. doi:10.1080/00220270701684667.

Joshi, S. T. 2013. *I Am Providence: The Life and Times of H.P. Lovecraft*. New York: Hippocampus Press.

Kraftl, P. 2018. "A Double-Bind? Taking New Materialisms Elsewhere in Studies of Education and Childhood." *Research in Education* 101 (1): 30–38. doi:10.1177/0034523718791880.

Latour, B. 1993. *We Have Never Been Modern*. Cambridge: Harvard University Press.

Levi Bryant, N. S., and Graham Harman. 2011. "Towards a Speculative Philosophy." In *The Speculative Turn*, edited by N. S. Levi Byant, Graham Harman. Melbourne: re.press.

Lewis, T. E., and R. Kahn. 2010. *Education out of Bounds: Reimagining Cultural Studies for a Posthuman Age*. New York: Palgrave Macmillan.

Lysgaard, J. A. 2018. *Learning from Bad Practice in Environmental and Sustainability Education*. Pieterlen: Peter Lang.
Lysgaard, J. A., S. Bengtsson, and M. H.-L. Laugesen. 2019. *Dark Pedagogy*. New York: Palgrave.
Mannion, G. 2007. "Going Spatial, Going Relational: Why 'Listening to Children' and Children's Participation Needs Reframing." *Discourse: Studies in the Cultural Politics of Education* 28 (3): 405–420. doi:10.1080/01596300701458970.
Mannion, G., A. Fenwick, and J. Lynch. 2013. "Place-Responsive Pedagogy: Learning from Teachers' Experiences of Excursions in Nature." *Environmental Education Research* 19 (6): 792–809. doi:10.1080/13504622.2012.749980.
Meillassoux, Q. 2009. *After Finitude*. New York: Continuum.
Mogensen, F., and K. Schnack. 2010. "The Action Competence Approach and the 'New' Discourses of Education for Sustainable Development, Competence and Quality Criteria." *Environmental Education Research* 16 (1): 59–74. doi:10.1080/13504620903504032.
Morton, T. 2011. "Here Comes Everything: The Promise of Object-Oriented Ontology." *Qui Parle: Critical Humanities and Social Sciences* 19 (2): 163–190.
Morton, T. 2013. *Hyperobjects*. Minneapolis: University of Minnesota Press.
Morton, T. 2016. *Dark Ecology*. New York: Colombia University Press.
Oliveira, T. R. M. s and D. B. Lopes. 2016. "On the Limits of the Human in the Curriculum Field." *Curriculum Inquiry* 46 (1): 110–125. doi:10.1080/03626784.2015.1113512.
Oral, S. B. 2015. "Weird Reality, Aesthetics, and Vitality in Education." *Studies in Philosophy and Education* 34 (5): 459–474. doi:10.1007/s11217-014-9447-y.
Payne, P. 2016. "What Next? Post-Critical Materialisms in Environmental Education." *The Journal of Environmental Education* 47 (2): 169–178. doi:10.1080/00958964.2015.1127201.
Poeck, K. V., and J. A. Lysgaard. 2016. "The Roots and Routes of Environmental and Sustainability Education Policy Research." *Environmental Education Research* 22 (3): 305 - 318. doi:10.1080/13504622.2015.1108392.
Rickinson, M. 2001. "Learners and Learning in Environmental Education: A Critical Review of the Evidence." *Environmental Education Research* 7 (3): 207–320. doi:10.1080/13504620120065230.
Snaza, N., P. Appelbaum, S. Bayne, D. Carlson, M. Morris, N. Rotas, … J. Weaver. 2014. "Toward a Posthumanist Education." *Journal of Curriculum Theorizing* 30 (2): 39–55.
Sterling, S. 2017. "Assuming the Future: Repurposing Education in a Volatile Age." In *Post-Sustainability and Environmental Education*, edited by B. Jickling and S. Sterling. London: Palgrave.
Stevenson, R. 2008. "A Critical Pedagogy of Place and the Critical Place(s) of Pedagogy." *Environmental Education Research* 14 (3): 353–360. doi:10.1080/13504620802190727.
Sund, P., and J. A. Lysgaard. 2013. "Reclaim "Education" in Environmental and Sustainability Education Research." *Sustainability* 5 (4): 1598–1616. doi:10.3390/su5041598.
Säfström, C. A. 2017. "Livable Life, Educational Theory and the Imperative of Constant Change." Paper Presented at the ECER 2017, Copenhagen.
Saari, A., and J. Mullen. 2018. "Dark Places: Environmental Education Research in a World of Hyperobjects." *Environmental Education Research* 1–13. doi:10.1080/13504622.2018.1522618.
Taylor, C. A. 2017. "Is a Posthumanist Bildung Possible? Reclaiming the Promise of Bildung for Contemporary Higher Education." *Higher Education* 74 (3): 419–435. doi:10.1007/s10734-016-9994-y.
Thacker, E. 2011. *In the Dust of This Panet*. London: Zero books.
UNFCC. (Producer) 2015. UNFCC Article 6. Retrieved from https://unfccc.int/news/article-6-climate-education-and-training
von Stechow, E. 2010. "Rückkehr Zur Schwarzen Pädagogik? Von Super Nannys Und Anderen Erziehungsnotständen." In *Unkulturen "in Bildung Und Erziehung*, edited by M. Dörr and B. Herz, 135–149. Wiesbaden: VS Verlag für Sozialwissenschaften.
Wallin, J. 2014a. "Dark Pedagogy." In *Animal Catalyst*, edited by P. MacCormack. New York: Routledge.
Wallin, J. 2014b. "Dark Posthumanism, Unthinking Education, and Ecology at the End of the Anthropocene." In *Posthumanism and Educational Research*, edited by N. Snaza and J. Weaver. New York: Routledge.

Dark places: environmental education research in a world of hyperobjects

Antti Saari and John Mullen

ABSTRACT
Escalating ecological events like global warming haunt our lived experience of place and demand thinking the concept anew in environmental education research. Using Timothy Morton's notions of hyperobjects and dark ecology as a springboard, we articulate the uncanny—the withdrawn, unknowable, excessive and anxiety-provoking—aspects of place that events like global warming disclose for place-based pedagogies. Given the advent of hyperobjects as a theoretical model coupled with the ontological and ethical attunement of dark ecology, we suggest conceptualizing place through the strategies of juxtaposition and working-through in order to begin thinking through the implications of how to live in dark places that are haunted by the reality of hyperobjects.

Introduction

Recent discussions converging around post-structuralism and new materialism in environmental education research problematize place as a foundation of meaning-making experienced as present, familiar and knowable. Escalating ecological events have made it increasingly difficult to think about place in terms of belonging and dwelling 'in' an 'environment', 'nature' or 'world' without acknowledging the gaps, disparities and lines of flight therein. Instead, concepts and metaphors of becomings, paths and networks have been mobilized to open novel vistas for thinking about places as unbounded, non-present and inherently unknowable (see e.g. Clarke 2017; Nakagawa and Payne 2014). In an effort to partner with these problematizations of place in environmental education research, we attempt to highlight the uncanny abjection that haunts places both within and without.

Consider for example the advent of global warming: we can experience its effects and compute its existence, yet we are unable to point to it in any meaningful sense as if it exists hauntologically; that is, producing spectral effect that are both present and yet absent as if inhabiting our climatic systems disrupting and distabilizing them and consequently producing an uncanny aspect to our experiences of place (Caputo 2016b, 116–120).

This uncanny aspect became visceral on the morning of October 17th, 2017 in southern Finland when the rising sun was exceptionally red, and this, we were told, was due to the sands of the Sahara and the fine particles from the Portuguese forest fires that hurricane Ophelia carried to the Irish coast and then over Finland. These particles filtered out blue light from the sun and left only red hues. Was this singular moment haunted by the specter of global warming? Our guesses can only be probabilities as we cannot see the relation between the red sunrise and global warming in any causal sense, nor can we capture it in any distinct geographic location: it is, in short, non-local and phased, aspects among others, which we will explore in greater detail below. Moreover, in an uncanny way, global warming is also 'in' us: it is inside our thoughts, our ways of life and the objects we engage with daily; and therefore, it is impossible to externalize in our traditional spatial categories like 'nature' or the 'environment'. The event of global warming, in short, does something to places that our concepts of it are not prepared for: *it haunts*.

Foregrounding global warming as a case example, we use Timothy Morton's concepts of *dark ecology* and *hyperobjects* to tease out the uncanniness involved in thinking about place-based pedagogies without the symbolic support of 'place', 'nature' or 'world' that have aided in compartmentalizing ecological phenomena and awareness into cultural and natural, local and global and human and non-human categories. We argue that thinking about dark ecology and hyperobjects aids in articulating the dimension of the *uncanny* in place-based pedagogy. We suggest *juxtaposition* and *working-through* as pedagogical strategies in order to begin thinking about *how* to live in dark places.

Place-based pedagogies

Since the early 2000s many environmental and social justice education scholars have converged around the notion of 'place'. Place-based pedagogies[1] are an index for a heterogeneous group of educational theories and practices that seek to 'go local' by foregrounding our immediate experience of interconnectedness with the human and the more-than-human in local communities (Gruenewald 2003a, 2003b; Smith 2007; Hill 2013).

Although precise definitions vary, the concept of place in place-based pedagogies refers to a bounded symbolic, social and material domain, to which humans are emotionally attached, and the meanings of which are constitutive of their identities (Kudryavtsev, Stedman, and Krasny 2012, 230–233; Garrard 2010; Bonnett 2013; Wattchow and Brown 2011, xxi). Place evokes notions of an immediate, deeply meaningful experience and imbuement with the more-than-human world (Hill and Brown 2014, 228; see also Nicol 2014; Thorburn and Marshall 2014). Subsequently, pedagogies of place highlight the importance of teaching 'by-means-of-an-environment' in order to gain an understanding of the inextricable connection of the human with the more-than-human and to cultivate solidarity towards them (Mannion, Fenwick, and Lynch 2012, 803).

Recognizing these various modalities, David Gruenewald (2003a) provides a typology of place in five dimensions: The first is the *perceptual* dimension wherein place is understood phenomenologically as 'the ground of direct human experience' (623). Gruenewald argues that in this dimension 'people are capable of perceiving places and learning from that direct experience' and that 'our ability to perceive places can be either thwarted or fostered by educational experience' (625). The second dimension of place is *sociological*: 'place is where the world manifests itself to human beings' (625). Place in this modality is understood as a social construct that is produced and reproduced through culture. Third, places are *ideological*; that is, they instantiate 'ideologies and relationships of power' (628). Fourth, and closely associated to the third dimension, places are *political* as in forming social identities. And finally, places are *ecological*. This dimension acknowledges the dependency of the human on the more-than-human lifeworld, the limits thereof and the spectrums of diversity that not only should be preserved, but that make life itself possible (633–636).

The categories of 'place' and 'local' also need an outside, a contrastive space in order for them to make sense. In other words, there must also be spaces that are not 'places'. This division gives place-based pedagogies their critical edge as the latter spaces pose a host of problems that these pedagogies seek to address. This is especially so in discourses employing critical pedagogy and/or social justice theories, where these other spaces are often characterized with prefixes 'colonial', 'global' and 'capitalist'. Accordingly, global capitalist logic produces ways of ordering and mobilizing human and more-than-human communities into standardized, monocultural spaces of production and consumption, thereby dissolving traditional cosmologies, values and ways of life that may have upheld a balance with the surrounding more-than-human community (Gruenewald 2003a, 2003b; Hill and Brown 2014; McInerney, Smyth, and Down 2011; Wattchow and Brown 2011).

Following this logic of uprooting and mobilization, modern schools have inculcated modes of splitting human experience into different schooling subjects and organizing them into isolated grids of time and space, which further dissolve the relationship with the local (McInerney, Smyth, and Down 2011). As a result, ecological problems like global warming may become isolated and intellectualized by framing them in abstract biological and geographical discourses and by dealing with decontextualized macro-scale concerns that have little to do with how a student might experience their immediate human and more-than-human community (Graham 2007; Smith 2007).

In sum, the concept of place in place-based pedagogies is often connected to notions of transformative experiences of connectedness, social justice and sustainability, while simultaneously constructing opposing notions of alienation, instrumentalism, and a compartmentalizing orientation to education (see e.g. Hill and Brown 2014).

Yet many approaches in place-based pedagogies have also sought to problematize the very notion of 'place' and the meaning of 'local' (McInerney, Smyth, and Down 2011; Stevenson 2007; Ruitenberg 2005) as these may sometimes inadvertently smuggle in universalist, decontextualized and de-materialized meanings. Moreover, some critics have pointed out that the relationships between what are considered 'local' and 'global', 'culture' and 'nature' need to be more carefully traced in order to understand current economic and ecological problems (Gruenewald 2003a; Bowers 2008; Nespor 2008; McInerney, Smyth, and Down 2011, 9–11; Nakagawa and Payne 2014, 149–150). This is especially the case for settler colonialism wherein an anthropocentric, racialized, gendered and capitalistic conception of place has violently marginalized, inferiorized and supplanted other understandings of place (Seawright, 2014; see also Tuck, McKenzie, and McCoy 2014; Korteweg and Russell, 2012).

In recent new materialist discourses of environmental education in particular, place has come to indicate a heterogeneous set of material and symbolic trajectories and becomings that constitute reciprocal relations (Mannion, Fenwick, and Lynch 2012, 793--794; Clarke and Mcphie 2016; Gough 2016; Clarke 2017).

What is persuasive in these conceptualizations is their problematization of 'place' as a circumscribed site of identifying ecological concerns, pedagogical initiatives to be taken, and the influences that education has upon the human and the more-than-human environment. Moreover, they also resist holistic and systemic arguments (Weakland 2012) about place and make undecided and unstable the difference between local and global, nature and culture (Clarke 2017, 312).

While we heartily acknowledge the vast contributions that have been made in these problematizations of place, we sense that there is there is still space for elaboration in the different stipulations of place in the extant discourses of place-based pedagogy; that is, something that would account for the aforementioned hauntologically unsettling nature of experiencing global warming. For this, we turn to Timothy Morton's conceptions of dark ecology and hyperobjects. Rather than engaging with a detailed commentary on Morton's works, we use these two concepts as incitements to unpack the concept of the *uncanny* in order to ferret out its significance in conceptualizing place in environmental education research. While the uncanny aspects of these two concepts resonate somewhat with Gruenewald's (2003a) aforementioned dimensions of place, we focus here on how it unsettles the first and the last—the perceptual as connecting to a place with our immediate experience, and the ecological as signifying the more-than-human lifeworld. We argue that Morton's thoughts further the post-stucturalist and new materialist critique of bounded, stable places through introducing an ontology of finitude, withdrawal, and excess.

The uncanny

Morton argues that global warming discloses that what was once understood as a morally neutral and cyclical background (read: nature) has gained a capricious agency, confounding our ethical aptitudes (2013; see also Jenkins 2013). Our particularly 'modern' way of differentiating human and non-human and culture and nature have been thrown into disarray, haunting our lived experiences of place:

> *Space* in this sense [i.e. the Euclidean concept of space as abstract and stable] has collapsed, and *place* has emerged in its truly monstrous uncanny dimension, which is to say its nonhuman dimension. How? Now that the globalization dust has settled and the global warming data is in, we human find ourselves on a very specific planet with a specific biosphere… Our sense of planet is not a cosmopolitan rush but rather the uncanny feeling that there are all kinds of places at all kinds of scales… And these places, as the concept of spacetime, are inextricably bound up with different kinds of timescale… So many intersecting places, so many scales, so many nonhumans. Place now has nothing to do with good old reliable constancy. What has dissolved is the idea of *constant presence*: the myth that something is real insofar as it is consistently, constantly 'there.' (Morton 2016, 10)

With the dissolution of space as present, knowable and controllable, ecological experience today is inevitably *uncanny*. Here, Morton follows Heidegger's (2000) use of German *das Unheimlich*: the not-at-home, alienated, or wandering (Morton 2012, 11). Yet it is Freud's (1919) definition of the term that provides the most import: the uncanny as that what once seemed safe and familiar, but which has assumed an air of strangeness and foreboding (Morton 2012, 11–12)—like seeing an environmental catastrophe in the beautiful color of a Finnish sunset. In other words, the experience of the uncanny results from a strange juxtaposition between things that, in the dominant cosmology, should have remained separate, like culture and nature, human and non-human, safe and dangerous (Saari 2018).

The concept of the uncanny also resonates with *abjection*, that is, the repressed objects of our experience. When isolated and repressed contents surface in consciousness, they may shatter existing coordinates of reality, thereby inducing the experience of the uncanny. This is not only related to personal experience, but resonates also with collective ontological imageries that delimit our experience. The experience of the uncanny is indelibly etched in the philosophical Enlightenment tradition, which accentuates knowing, reason, lucidity, control and planning and, inversely, represses unreason, darkness and unknowing (Dolar 1991).

Consider global warming as an event most of us would rather not think about: our lives and thoughts are often organized around isolating and insulating that ominous, uncontrollable, non-local reality into the 'ecological unconscious' (Morton 2007). Engaging with global warming, therefore, requires sensitivity to forms of recoil, isolation, and 'passionate ignorance' that often emerge in our encounters with ecological catastrophes and degradation (Nicholsen 2002; Norgaard 2011; Dodds 2011). These aspects, we argue, are the aspects often absent from place-based pedagogies. And this is where Morton's work on hyperobjects and dark ecology come to bear for environmental education researchers.

Dark ecology

Morton (2010a, 2010b, 2016) proposes 'dark ecology' as an ontological and ethical attunement to the uncanny event of global warming and our ecological reality more generally. Dark ecology, in the shortest of definitions, is a model of ecological awareness (Morton 2016, 5). And ecological awareness, in short, is coexistence (Morton 2010a, 2). And coexistence, in short, is dark (Morton 2010a, 15–16). Dark, as we understand it, alludes to at least three interconnected threads that run throughout Morton's work. In the first place it presses up against the 'positive, sunny "Zippity Doo Da"' sentimentality that has characterized 'green' environmentalist thinking: one that says, 'we're all interconnected; and isn't it wonderful!' Morton posits that this sunny, holistic, hearty and healthy sentimentality occludes other dimensions like negativity, ambiguity, darkness, irony and fragmentation (Morton 2010a, 16). Dark in this first understanding reveals the 'other' side, the shaded side, of ecological awareness and coexistence.

Dark refers also to a veil of unknowing as opposed to the anthropocentric models of knowing and mastery that characterizes environmental phenomena, aspects of which are present even in—actually, especially in—scientific knowledge about our environment; that is, the more we acquire data about global warming, for example, the more ambiguous and uncanny it becomes. It defies, in other words, our inherited and common sense notions of space, time and matter (Morton 2010a, 17), which we will elaborate on in further detail in our discussion on hyperobjects below.

In the third place, dark is explicitly connected to object-oriented ontology (OOO), which for our purposes is a philosophy that highlights *finitude* and *withdrawal*. Finitude refers here to our inability to access the thing in itself: 'when I try to find the thing in itself, what I find are thing data, not the thing in itself. And I grasp that data in such a way that a thing does not (meaningfully) exist (for me) outside the way I ... correlate the data' (16). This irreducible rift, according to OOO, between thing and appearance reveals the 'profoundly limited and incomplete' access of human knowledge, which is characteristic of nonhuman knowing as well (17). This epistemic finitude follows a withdrawn and excessive ontology wherein 'things exist in a profoundly "withdrawn" way: they cannot be splayed open and totally grasped by anything whatsoever, including themselves' (16). Reality, according to OOO, exists outside the mind, yet, and this is crucial, we cannot know it except for 'indirect, allusive, or vicarious means' (Harman 2016, 17). In short, we know only through a glass darkly as things withdraw and remain in excess of any and every entity's access of anything.

The advent of global warming prominently draws our attention to the withdrawn character of objects as a ghostly 'pres/absence' that we must attempt to come to grips with as educational researchers (Caputo 2016b, 116). How these three aspects of Morton's *dark* ecology matter for place-based pedagogy in the event of hyperobjects is the contribution we would like to provide for environmental education research. But before we discuss this, we will need to elaborate on how Morton describes hyperobjects.

Hyperobjects

As we mentioned above, Morton sees global warming as an event that haunts our stabilized notions of place. As such, global warming is understood as a *hyperobject*—an entity massively dispersed in space and in time. Global warming is by no means the first or only existing hyperobject—Morton mentions styrofoam cups, radioactive plutonium, and the Florida Everglades among a variety of examples—but it condenses for us the multiple temporal and spatial scales and ethical concerns imbued in them in relation to our experiences of places. Hyperobjects can be identified by the following five qualities: (Morton 2010a, 2013a, 2013b): First, hyperobjects are *viscous*: they stick, bleed and stretch despite our attempts to categorize and insulate them. The more you try to study them, the more strange they become. The more you try to externalize and compartmentalize them, the more you find your own self being tangled in them. Second, hyperobjects are *non-local*: they are not reducible to a separate place and time but are indeterminately dispersed, as the above example of the red sunrise serves to testify. Third, hyperobjects are *temporally undulated*: they are time-stretched to a vast extent, assuming multiple rhythms.

A hyperobject like global warming not only covers planetary distances but industrial revolutions, the nature of the evolutionary process itself, and the migratory and mass extinction patterns of both flora and fauna to name but a few: 'Recall the three timescales of global warming: the horrifying, the terrifying, and the petrifying—500, 30,000, and 100 years respectively. These *very large finitudes* collapse my cliched ideas of time from within' (Morton 2013a, 136). Fourth, hyperobjects are *phased*: they are transdimensional such that we can only ever see a part of them while other parts withdraw; that is, they 'occupy a high dimensional *phase space* that makes them impossible to see as a whole on a regular three-dimensional human-scale basis' (Morton, 2013a, 70). The problem with using weather and climate synonymously is a helpful illustration of this: climate is the transdimensional system of which weather is its sensual impression (74). The one we see and feel—and even its raindrop exceeds our access—while the other withdraws when we try. Lastly, hyperobjects are *interobjective*: they form a heterogeneous human and non-human assemblage and are not reducible to but exceed the human-world correlation. This means they are never experienced directly but are always mediated or translated through some entity's access and yet they have a history of their own apart from any human, non-human mediation (Morton 2013a, 27–95).

In sum, Morton (2010a, 14, 2011, 216–217) characterizes hyperobjects as 'strange strangers'; that is, groups of objects that are strangers even to each other, which also conveys the idea that they are irreducible in our attempt to make them intelligible as in our totalizing and homogenizing concepts of 'nature' or 'place' (see also Wark 2015, 200). This radical alterity inherent to hyperobjects runs contrary to the theories of 'relationality' and 'environmental' views that characterize ecological spaces as (tacitly) homogenous, knowable planes or 'ecosystems' (Harman 2005; Morton 2011, 2012). The concept of hyperobjects brings to the surface the short-circuits, objects and sources of anxiety inherent in ecological awareness, thereby challenging our notions of 'place' as bounded, identifiable and (intellectually) safe.

A dark pedagogy of place

Accordingly, we argue that Morton's theory of dark ecology and hyperobjects opens up an understanding of the uncanny aspects of place in environmental education, especially with regard to its perceptual and ecological dimensions. We condense this in the form of the following theses:

1. The impossibility of drawing geographic and temporal contours around the notion of 'place', leading to conceptualizing and researching 'place' in terms of excess and non-locality rather than an organic and romantic whole as is usually the case;
2. Re-imagining the ethical values and aims of pedagogies of place in terms of an irreducible alienation and experiences of the uncanny rather than a return to blissful harmony and belonging.

Using the dark qualities of global warming, we argue that these registers can be activated with the conceptual strategies of *juxtaposition and working-through.*

Juxtaposition

The Western imaginaries of nature and the more-than-human environment have for long provided a grid of intelligibility for environmental education: organizing identities and differences therein and drawing external borders with the human and the cultural. This, in turn, has often framed place-based approaches and outdoor education in tracing untrammeled spaces for 'nature experiences' that express holistic and immersive contact with nature (Bonnett 2004, 2007; Clarke 2017, 308–309).

This imagery is founded on a metaphysics positing the existence of nature-as-a-whole: as a clear-cut organic or systemic unity (Clarke 2017, 309). In the event of hyperobjects, the metaphysics of the 'whole' is contrasted with an ontology of withdrawal, excess and out-of-jointedness—being as always more than what we are able to capture in conceptual classification or perceptual experience. Being, in other words, does not constitute a container holding homogeneous matter. Instead, it is a plethora of objects that are inaccessible to each other and therefore escape classification and reduction into an organic whole (Morton 2007; Harman 2005, 251). This is why Morton (2007) demands that we do away with the concept of nature altogether, as it only buttresses the unacknowledged, taken-for-granted spacing of mainstream ecology.

Moreover, Morton claims that our notions of space and nature and solidarity towards 'nature' are founded on a metaphysics of presence (Morton 2016) or an ontological figure-ground distinction (Morton 2007)—equating what is near, immediately perceived and controllable with what is 'true' and 'real', and as such, worthy of our solidarity. As Massey (2004) notes, often our notions of solidarity and empathy follow certain 'geographies of responsibility' that are structured like concentric circles going from the parent–child relationship as the nucleus of caring to the local and to the global. Yet, when encountering the uncanniness of hyperobjects, the coordinates that organize things into near-far, local-global become ambiguous (Morton 2007, 2016).

We argue for conceptual strategies of disrupting the fantasies of organic wholeness in the uses of nature, environment or place in education. *Juxtaposition* (Morton 2007, 143–150) means bringing into uncanny proximity, play or comparison objects and events that are conventionally seen as distant, incompatible or altogether abject in environmental discourse (see also Saari 2018). This aligns closely with Mitchell Thomashow's (2002) call to cultivate an open-ended imagination of places outside of those regarded as 'nature' or those planned and meticulously designed 'learning environments' (see also Rautio 2014). This might include broadening the focus from pristine, 'natural' places in place-based pedagogy to including their crossings and layerings with urban, industrial and polluted spaces (Garrard 2010).

This is especially the case when dealing with hyperobjects like global warming where juxtaposition articulates place-based pedagogy without place being bounded and subject to intellectual mastery. Juxtaposition may start with an immediate perceptual experience of seeing a beautiful red sunrise, for example, and moving on from there to studying climate science, the advent of the human species, the industrial revolution, and then coming to realize the geological force our species has become and our own individual implication in it (Morton 2016, 7). There are no privileged space-times in nature: here, too, the uncanny is ever-present.

Juxtaposition as a pedagogical strategy resonates with Jason Wallin's (2015) 'pedagogical dark ecosophy' wherein we recognize that the 'forces of planetary decay, mutation, and violent upheaval [are] obfuscated by education's subjugation of the unconscious in an image of romantic humanism' (139, 142-–143; see also Wallin 2014). Wallin argues that 'education must become capable of *unthinking* anthropocentrism by taking seriously both the collapse of human and non-human 'things' as well as that which is already inhuman within the human' (142). The merit of tethering the strategy of juxtaposition with Wallin's dark ecosophy is their mutual interest of bringing encounters with abjects into the fore of pedagogical experiences, and certain forms of art carry unparalleled potential in this regard.

Consider as an example the recent 'Ice Watch' installation in Paris by Icelandic artist Olafur Eliasson and geologist Minik Rosing where 'Ice Watch' is meant as a double entendre: the clock-like shape of icebergs on a Parisian street refers to the haunting timescales of global warming and also the haunting reality of watching the ice melt right in front of the viewer. And yet as Eliasson explains, 'It is a mistake to think that the work of art is the circle of ice—it is the space it invents' (Zarin 2017). Inventing spaces like 'Ice Watch' are powerful pedagogical tools for juxtaposition and dark ecosophy that have the potential to draw to the surface aspects of our haunting ecological unconsciousness, or those aspects we intentionally suppress and would rather not dwell on.

In sum, the particular aim of juxtaposition in the mode of dark ecology is not to make these events fit into a larger scheme of things, but to highlight the inexorable uncanniness of place. It should also be noted that dark ecology differs from some Latourian applications of tracing a 'hybrid' (Latour 1993) or an actor-network (Latour 2005). As mentioned above, objects that are inherently strange to themselves cannot compose an empirically characterizable network. Instead, they express non-localizable gaps and disparities.

Working-through

Global warming is an event which reveals a deep ambiguity not only in the places outside us, but also questions our own possibility of dwelling within them: there is an uncanny undecidability between the subjective and the 'objective', the human and the non-human, the far away and the deeply personal. Our own existence is, in a strange way, imbued with that which always has remained at a comfortable distance as impersonal and non-human (Morton 2013b).

Morton refers to the literary and cinematic trope of *noir* to characterize this dramatic turn. In a typical film noir, the protagonist detective all the while looking for the murderer finds out that he himself is the culprit (Morton 2016, 9). In a similar fashion, researching and teaching about global warming—investigating melting ice caps, CO_2 and methane emissions—inevitably closes in on the sensitive area of our own implication in it. However, this does not mean singling out ourselves as being in control of the environment, but instead the uncomfortable discovery of finding ourselves inside a massively dispersed hyperobject, which we cannot control.

Ecopsychoanalytic theorists have recently pointed out the 'climate trauma' inflicted by the uncomfortable reality of global warming and our own implication in it. With global warming they argue that we are not only having difficulty relating to a 'place', but we are also, to some extent, in danger of losing some of the basic coordinates of experiencing reality like planning ahead, remembering and the emotional attachments that we bear on our environments (Orange 2017, 14–18). The uncanny, disorienting experience of global warming may lead to both personal and collective defensive strategies of denial, splitting, compartmentalization, and emotional freezing (Dodds 2011; Lertzman 2014; Orange 2017; Weintrobe 2013). What is often seen as lacking in our relation to ecological concerns is the ability to express loss and the crippling empathy of suffering, and the guilt of our own implication in ecological degradation (Orange 2017, 55–56; Lertzman 2015).

Paying close attention to the way spatial disorientation is imbued with manifold psychic distresses, Glenn Albrecht coined the term solastalgia in order to name the longing for places where a person had once felt at home, but have since been experienced as disorientation, senses of homelessness, depression, memory loss, and various modes of estrangement from self and others. Albrecht identifies this among indigenous people displaced by colonial settlers as well as those Americans born and grown up in industrial environments who have felt the lack of connection to our ever-changing geographical environment (Albrecht 2005, 43–44).

Albrecht's solastalgia is palpable in Gaston Gordillo's (2014) ethnography of the Andes where the livelihoods of the rural poor are obliterated by companies creating soy fields in the area. Fields are filled with rubble from bulldozed infrastructures from different eras and cultures, and this rubble becomes a metaphor for the objects and debris constantly produced by global capital. Gordillo notes how the operation of agribusiness in the area together with global capital and the layers upon layers of violent colonial rule and the deep anxiety and mental distress among the peasant population destroys the gaucho way of life, obliterates markers that provide geographic orientation, and suppresses their personal and shared memories and identities.

Bodnar (2008) and Lertzman (2014) attest that strong emotions, affects, and identity are mutually embedded not only with humans but also the more-than-human world. Lertzman claims that 'certain objects are like 'psychic keys' through which we articulate ourselves' (Lertzman 2014, 88). This is documented in her study of Green Bay inhabitants: their life histories amid industrialization and ecological degradation in the area provide insights into how global ecological and economic trajectories are imbued in psychic trauma that is 'shot through with powerful and inchoate affective investments, memories, desires and losses' (101). In her interviews, Lertzman found a recurring narrative of a loss of innocence, purity, and sanctuary in increasing ecological degradation. This innocence was projected into elements of the environment—clear blue skies, sunlight reflected in a lake surface (78–81). When these elements are taken away, the experience of anxiety, loss, and disorientation becomes pressing, yet difficult to articulate.

Studies like this show that uncanny experiences of places may engender denial, projection, and isolation, which are not only psychological but also political phenomena, bearing on the questions of social justice and environmental concerns. Acknowledging and expressing difficult experiences of dark places rather than suppressing them in education is crucial for ecological engagement. Dodds (2011, 69, 100–101) uses the analogy with the psychoanalytic practices of 'working-through' as necessary in thinking about the task of overcoming anxiety and guilt in relation to ecological disasters. This term refers to the psychoanalytic interpretive 'work', the articulation of one's own thoughts, memories, dreams, and emotions, by the analysand. The 'through' here also may be read as a direct confrontation, naming and being with, rather than avoiding, suppressing or even trying to 'solve' anxiety caused by the uncanniness of place (see also Lertzman, 2014, 101, 129–130).

This finds resonance in the transformative aspects of dark ecology, which traces out the shifts from dark-depressing to dark-sweet that transpire in the articulation of the uncanny aspects of ecological experience (Morton 2016, 160). Without denying the embodied, psychological or political benefits that 'nature experiences' may have, a dark ecological ethics of working-through might contribute in highlighting a certain uncanniness as an inherent quality of existence, not only with humans, but objects in general (see Morton 2010b, 255–256). This approach will not do away with our predicament of 'alienation' from, or uncanny relation to, our environment, as in seeking a holistic experience of nature. Instead, it changes our attunement to the predicament (Morton 2016, 264) by creating pedagogical environments where working-through can take place; where loss, unknowing, and anxiety can be witnessed and articulated instead of denial or repression (see also Lertzman 2014, 150). This finds resonance in Rebecca Martusewicz's (2015) call to 'let our hearts break:' to articulate the grief and anxiety over the injustice and pain inflicted on the more-than-human world and to assume responsibility for it (see also Pihkala 2017; Morton 2016, 161, Saari 2018).

This is point is uniquely and poignantly drawn out in Wallin's (2015) analysis of zombie narratives that seem to be only gaining in popularity: 'Zombie-life', he writes, 'deterritorializes the anthropocentric facialization of the planet by diagramming an unthought dark ecology'; Wallin argues that 'the subterranean ecology fabulated in zombie fiction questions not only how a life might go, but also what thinks and of what things might think *where we are not* [and one] in which humans are divested of their presumed status as dominant planetary actants transcendent to material life' (p. 139). Speculative fictions like zombie narratives and dystopian futures can be a helpful mode of beginning to work-through the social and psychic imaginaries in which we humans are ontologically and ethically privileged and prioritized.

This also points to the need of dealing with the ambivalence and ambiguity of experiencing place—the undecidability of the locus of responsibility, personal and global, human and more-than-human nature of environmental crises and the mixed affects these crises engender. One can be a beneficiary of an economic system that produces jobs, income and a decent standard of living, while sustaining grave social injustices and inflicting ecological damage (Lertzman

2014). If our affects, desires, hopes and fears are projected into material and more-than-human environments, working-through can also take place in the form of recognizing sites of mourning, ambiguous affects and the corresponding senses of injustices (150). And all the while recognizing these emotions as uncanny and ambivalent rather than splitting them into good and bad or local and global.

Our point here is not simply to dwell on what is repugnant or horrendous, but to enable engagement by not sustaining a culture of denial that occludes these essential aspects. Ideas from place-based pedagogies are already used with arts education—visual and performing arts and drama—to articulate a sense of hope, fear, or responsibility in relation to global warming (see e.g. Popov 2018). These can give expression to aesthetic and emotional aspects of the uncanny and sustain rather than efface ambiguity like the 'Ice Watch' installation in the streets of Paris.

Conclusion

Throughout this article we have been referring to hyperobjects like global warming as events, and as is the nature of events, they tend to shock and/or undermine our prevailing conceptual systems or frameworks of which our expectations are found lacking (Caputo and Dickinson, 2012; Caputo, 2016a, 2016b). Our commonplace understandings of place are destabilized by the haunting events like global warming thereby revealing the uncanny and dark dimensions of place. Understanding the nature of hyperobjects as an event has consequences for thinking about and living into 'place' in place-based pedagogies.

We should not go without mentioning that the darkness we are proposing in this study is not intended to inculcate guilt, depression or apathy in the face of an overbearing event like global warming. Instead, we propose that it opens up a heretofore undisclosed dimension for thinking about places as deeply uncanny. Accordingly, this thinking problematizes two extant ways of addressing ecological threats in environmental education: First is the tendency to see ecological degradation as a result of our alienation from 'nature', which is conveyed in the practices of cultivating holistic 'nature experiences' in place-based pedagogy as a cure. As discussed above, the notion of place as deeply uncanny is founded on the realization that more we investigate a place, the weirder it becomes as it reveals the infinite multiplicity of temporal rhythms and spatial trajectories. Our strategy of juxtaposition aims to disrupt this sense of alienation that is produced in a 'nature-as-whole' ontology. Instead, because of the advent of hyperobjects, reality is revealed as excessive and disjointed, which calls for an open-ended imaginary that can live in the conceptual cracks and fissures that the haunting event of global warming creates for our experience of living into dark places.

The other tendency is to understand educational responses to ecological threats as an issue of the correspondence of knowledge to attitudes to action (see also Lertzman 2015, 146–149). If students are witnesses to environmental degradation in their local community and begin to understand the effect their actions have then this should motivate them to change their habits into more sustainable ones. While this might sound like a straightforward way of engaging with environmental education, this has the tendency to obscure the anxiety and denial that occurs when faced with the reality of global warming and our complicity in it. This model of immediacy 'wants to be relieved of the burden of anxiety and uncertainty' that ecological threats pose, but it is precisely these uncanny aspects of place that we need to learn how to live with if we are to take coexistence seriously (Morton, 2018, 18). It is for this reason that we suggest the pedagogical strategy of working-through the senses of loss and uncertainty, denial and anxiety that hyperobjects produce. The realization that we are 'part of an entity that is now *a geophysical force on a planetary scale*', like the detective that is the criminal in a noir film, highlights our entanglement and culpability that we cannot gloss over in our attempt to do something about it (Morton 2016, 9, 29). Instead, working-through our complicity and bearing its burden is a phase we cannot go without confronting in our efforts at coexistence in the uncanny multiplicity of places (9–12). Both of these strategies are pedagogical possibilities for hopefully opening up a conversation on how to live in dark places that are haunted by the reality of hyperobjects (Morton, 2018, 15).

Note

1. While acknowledging the plethora of these approaches—place-based education, place-responsive education, and a critical pedagogy of place, to name but a few—we refer to them here as 'place-based pedagogies' for the sake of simplicity.

ORCID

Antti Saari http://orcid.org/0000-0003-3777-6948

References

Albrecht, G. 2005. "'Solastalgia'. A New Concept in Health and Identity." *PAN: Philosophy Activism Nature* 3: 41–55.
Bodnar, S. 2008."Wasted and Bombed: Clinical Enactments of a Changing Relationship to the Earth." *Psychoanalytic Dialogues* 18(4): 484–512.
Bonnett, M. 2007. "Environmental Education and the Issue of Nature." *Journal of Curriculum Studies* 39 (6): 707–21.
Bonnett, M. 2004. "Lost in Space? Education and the Concept of Nature." *Studies in Philosophy and Education* 23(2): 117–30.
Bonnett, M. 2013. "Sustainable Development, Environmental Education, and the Significance of Being in Place." *Curriculum Journal* 24(2): 250–71.
Bowers, C. A. 2008. "Why a Critical Pedagogy of Place is an Oxymoron." *Environmental Education Research* 14 (3): 325–35.
Caputo, J. D. and T. W. Dickinson. 2012. "Education as Event: A Conversation with John D. Caputo." *Journal for Cultural and Religious Theory* 12 (2): 25–46.
Caputo, J. D. 2016a. *Truth: The Search for Wisdom in the Postmodern Age*. London: Penguin Books.
Caputo, J. D. 2016b. "Teaching the Event: Deconstruction, Hauntology, and the Scene of Pedagogy." In *The Pedagogics of Unlearning*, edited by Aidan Seery and Eamonn Dunne, 111–30. Earth, Milky Way: Punctum Books.
Clarke, D. A. G. 2017. "Educating Beyond the Cultural and the Natural: (Re)Framing the Limits of the Possible in Environmental Education." In *Reimagining Sustainability in Precarious Times*, edited by K. Malone, S. Truong, and T. Gray, 305–18. London: Springer.
Clarke, D. A. and J. Mcphie. 2016. "From Places to Paths: Learning for Sustainability, Teacher Education and a Philosophy of Becoming." *Environmental Education Research* 22 (7): 1002–24.
Dodds, J. 2011. *Psychoanalysis and Ecology at the Edge of Chaos: Complexity Theory, Deleuze/Guattari and Psychoanalysis for a Climate in Crisis*. New York: Routledge.
Dolar, M. 1991. ""I Shall Be with You on Your Wedding-Night": Lacan and the Uncanny." *October* 58: 5–23.
Freud, S. 1919. "The 'Uncanny'." In *Standard Edition 17*, 217–56. London: Hogarth Press.
Garrard, G. 2010. "Problems and Prospects in Ecocritical Pedagogy." *Environmental Education Research* 16(2): 233–45.
Gordillo, G. R. 2014. *Rubble: The Afterlife of Destruction*. Durham: Duke University Press.
Gough, N. 2016. "Postparadigmatic Materialisms: A "New Movement of Thought" for Outdoor Environmental Education Research?" *Journal of Outdoor and Environmental Education* 19: 51–65.
Graham, M. A. 2007. "Art, Ecology and Art Education: Locating Art Education in a Critical Place-Based Pedagogy." *Studies in Art Education* 48: 375–91.
Gruenewald, D. A. 2003a. "Foundations of Place: A Multidisciplinary Framework for Place-Conscious Education." *American Education Research Journal* 40: 619–54.
Gruenewald, D. A. 2003b. "The Best of Both Worlds: A Critical Pedagogy of Place." *Educational Researcher* 32(4): 3–12.
Harman, G. 2005. *Guerrilla Metaphysics: Phenomenology and the Carpentry of Things*. Peru, IL: Open Court.
Harman, G. 2016. *Immaterialism: Objects and Social Theory*. Malden, MA: Polity Press.
Heidegger, M. 2000. *Introduction to Metaphysics*. New Haven: Yale University Press.

Hill, A. 2013. "The Place of Experience and the Experience of Place: Intersections Between Sustainability Education and Outdoor Learning." *Australian Journal of Environmental Education* 29 (1): 18–32.

Hill, A. and M. Brown. 2014. "Intersections Between Place, Sustainability and Transformative Outdoor Experiences." *Journal of Adventure Education & Outdoor Learning* 14 (3): 217–23.

Jenkins, W. 2013. *The Future of Ethics: Sustainability, Social Justice, and Religious Creativity*. Washington, DC: Georgetown University Press.

Korteweg, L. and C. Russell. 2012. "Editorial: Decolonizing + Indigenizing = Moving Environmental Education Towards Reconciliation." *Canadian Journal of Environmental Education* 17: 5–14.

Kudryavtsev, A., R. C. Stedman, and M. E. Krasny. 2012. "Sense of Place in Environmental Education." *Environmental Education Research* 18 (2): 229–50.

Latour, B. 1993. *We Have Never Been Modern*. Cambridge, MA: Harvard University Press.

Latour, B. 2005. *Reassembling the Social: An Introduction to Actor-Network-Theory*. Oxford: Oxford University Press

Lertzman, R. 2014. *Environmental Melancholia: Psychoanalytic Dimensions of Engagement*. New York: Routledge.

Massey, D. 2004. "Geographies of Responsibility." *Geografiska Annaler: Series B, Human Geography* 86 (1): 5–18.

Mannion, G., A. Fenwick, and J. Lynch. 2012. "Place-Responsive Pedagogy: Learning From Teachers' Experiences of Excursions in Nature." *Environmental Education Research* 19 (6): 1–18.

Martusewicz, R. A. 2015. "Letting Our Hearts Break: On Facing the "Hidden Wound" of Human Supremacy." *Canadian Journal of Environmental Education* 19: 31–46.

McInerney, P., J. Smyth, and B. Down. 2011. "'Coming to a Place Near You?'The Politics and Possibilities of a Critical Pedagogy of Place-Based Education." *Asia-Pacific Journal of Teacher Education* 39(1): 3–16.

Morton, T. 2007. *Ecology Without Nature: Rethinking Environmental Aesthetics*. Cambridge, MA: Harvard University Press.

Morton, T. 2010a. *The Ecological Thought*. Cambridge, MA: Harvard University Press.

Morton, T. 2010b. "The Dark Ecology of Elegy." In *Oxford Handbook of Elegy*, edited by K. Weisman, 251–71. Oxford: Oxford University Press.

Morton, T. 2011. "Sublime Objects." *Speculations* 2: 207–27.

Morton, T. 2012. "The Oedipal Logic of Ecological Awareness." *Environmental Humanities* 1(1): 7–21.

Morton, T. 2013a. *Hyperobjects: Philosophy and Ecology After the End of the World*. Minneapolis, MN: University of Minnesota Press.

Morton, T. 2013b. "Poisoned Ground: Art and Philosophy in the Time of Hyperobjects." *Symploke* 21(1): 37–50.

Morton, T. 2016. *Dark Ecology: For a Logic of Future Coexistence*. New York, NY: Columbia University Press.

Morton T. 2018. *Being Ecological*. London, UK: Pelican Books.

Nakagawa, Y. and P. G. Payne. 2014. "Critical Place as a Fluid Margin in Post-Critical Environmental Education." *Environmental Education Research* 21 (2): 149–72.

Nespor, J. 2008. Education and Place: A Review Essay. *Educational Theory*, 58(4): 475–489.

Nicholsen, S. W. 2002. *The Love of Nature and the End of the World: The Unspoken Dimensions of Environmental Concern*. Cambridge: MIT Press.

Nicol, R. 2014. "Entering the Fray: The Role of Outdoor Education in Providing Nature-Based Experiences That Matter." *Educational Philosophy and Theory* 46 (5): 449–61.

Norgaard, K. 2011. *Living in Denial: Climate Change, Emotions, and Everyday Life*. Cambridge: MIT Press.

Orange, D. M. 2017. *Climate Crisis, Psychoanalysis, and Radical Ethics*. New York: Taylor & Francis.

Pihkala, P. 2017. "Environmental Education After Sustainability: Hope in the Midst of Tragedy." *Global Discourse* 7 (1): 109–27.

Popov, B. 2018. "The Uncle Vanya Project: Performance, Landscape, and Time." In *Art, EcoJustice, and Education: Intersecting Theories and Practises*, edited by R. Foster, J. Mäkelä and R. Martusewicz, 82–94. New York: Routledge.

Rautio, P. 2014. "Mingling and Imitating in Producing Spaces for Knowing and Being: Insights From a Finnish Study of Child–Matter Intra-Action." *Childhood* 21(4): 461–74.

Ruitenberg, C. 2005. "Deconstructing the Experience of the Local: Toward a Radical Pedagogy of Place." *Philosophy of Education*, edited by K. Howe, 212–20. Urbana: Philosophy of Education Society.

Saari, A. 2018. "The Experience of the Uncanny as a Challenge for Teaching Ecological Awareness." In *Art, EcoJustice, and Education: Intersecting Theories and Practises*, edited by R. Foster, J. Mäkelä and R. Martusewicz, 36–46. New York: Routledge.

Seawright, G. 2014. "Settler Traditions of Place: Making Explicit the Epistemological Legacy of White Settler Colonialism for Place-Based Education." *Educational Studies* 50: 554–72.

Smith, G. A. 2007. "Place-Based Education: Breaking Through the Constraining Regularities of Public School." *Environmental Education Research* 13(2): 189–207.

Stevenson, R. B. 2007. "Schooling and Environmental Education: Contradictions in Purpose and Practice." *Environmental Education Research* 13(2): 139–153.

Thomashow, M. 2002. *Bringing the Biosphere Home: Learning to Perceive Global Environmental Change*. Cambridge, MA: MIT Press.

Thorburn, M. and A. Marshall. 2014. "Cultivating Lived-Body Consciousness: Enhancing Cognition and Emotion Through Outdoor Learning." *Journal of Pedagogy* 5 (1): 115–132.

Tuck, E., M. McKenzie, and K. McCoy, 2014. "Editorial: Land Education: Indigenous, Post-colonial, and Decolonizing Perspectives on Place and Environmental Education Research." *Environmental Education Research* 20 (1): 1–23.

Wallin, J. 2014. "Dark Pedagogy." In *The Animal Catalyst: Towards Ahuman Theory*, edited by P. MacCormack, 145–62. London: Bloomsbury.

Wallin, J. J. 2015. "Dark Posthumanism, Unthinking Education, and Ecology at the End of the Anthropocene." In *Posthumanism and Educational Research*, edited by N. Snaza and J. A. Weaver, 134–47. New York: Routledge.

Wark, M. 2015. *Molecular Red: Theory for the Anthropocene*. London: Verso.

Wattchow, B., and M. Brown. 2011. *A Pedagogy of Place: Outdoor Education for a Changing World*. Monash: Monash University Publishing.

Weakland, J. P. 2012. "Queering Ecology: Interrogating 'Seductions to Organic Wholeness' in Popular Environmental Rhetoric." In *Learning for Sustainability in Times of Accelerating Change*, edited by A. Walsh & P. Corcoran, 121–32. Wageningen: Wageningen Publishers.

Weintrobe, S. 2013. "The Difficult Problem of Anxiety in Thinking About Climate Change." In *Engaging with Climate Change: Psychoanalytic and Interdisciplinary Perspectives*, edited by S. Weintrobe, 33–47. New York: Routledge.

Zarin, C. 2017. "The Artist Who is Bringing Icebergs to Paris." The New Yorker, December 5, 2015.

Environmental end game: *ontos*

Chris Beeman and Sean Blenkinsop

ABSTRACT
In this paper, we attempt to do a kind of theorizing that we think is compatible with new materialisms. To do this we explore the idea of what it might be to separate *ontos*- from *-logos,* and give suggestions to readers for ways of experiencing this idea. We posit that it is not only possible to make diffractive *readings* and re-readings of texts, but also that experiences themselves can provide a diffractive context for *ontos.* Thus, not only can different concepts held in the forefront of our minds allow us to encounter, interpret, and understand differently but so too can positioning and entangling ourselves differently. As such we will ask the reader to do things beyond simply reading the text as they engage this paper. We also tell stories, two from an Anishinaabe knowledge keeper of Chris's acquaintance, to provide another way for shifting conceptualizing towards a different ontological position compatible with an immanent materialist position. We explore the capacity of words to tell a story about experiences that are inherently non-verbal, in which human I-ness shares a being state with what are normally thought to be separable objects in the natural world. The paper concludes with a short discussion of pedagogy in light of this discussion/action.

Introduction

The goal of this paper is to bring attention to a potential risk, a missed opportunity even, we see arising in the understanding and employment of new materialisms in current theorizing about environmental education and, by extension, education writ large. The risk we see is the reduction of new materialisms to another academic theory, or just a new way of thinking about the world, or a novel set of intellectual tools and concepts. Our worry is that any of these moves runs the risk of ignoring the materiality of new materialisms and, for environmental educators, potentially ignores the actual physical entanglements[1] (Ingold 2008, 2010) and the role they play in thinking, knowing, theorizing, and being. If enactment is a necessary part of theorizing across new materialism, then we see an opportunity – a possibility to extend both what it means to theorize and what it is possible to theorize about. While we see evidence (Ross and Mannion 2012; Rautio 2014; Clarke and Mcphie 2014, 2016) of the combining of new materialisms with work in the field, we see less evidence of an attempt to bring an enactment of ontological change, rather than a report of the enactment, into academic theory itself, in papers like this one. In this paper, we propose to track and register these potentialities by trying to focus our discussion not solely on the *conceptualizations* of being but on the *practice and state of being* itself: experiences of entanglement (in this paper we will also use, *feltings* and *enmeshments*: see previous footnote) that, if enacted, might change understanding.

Thus, in the somewhat problematic form of an academic paper,[2] we propose to try not a *new materialisms-inspired diffractive reading* (see, for example Geerts and van der Turin 2016; Hill 2017) but a*n attempt to move the reader, themselves into a diffractive enmeshment*, in order to see what possibilities might be on offer. For if entanglements can flex and differ then, we postulate, so too might the thinking, reading, and understanding that arises therein. In this paper, we challenge the suggestion that new materialisms are a direct offspring of post-structuralism.[3] We worry that there are some understandings or readings of new materialisms that under-value and even foreclose the possibility that we, as humans, might have, be witness to – and perhaps be materially part of – encounters with the more-than-human world that are before, outside, even beyond text and theorizing as currently understood within the framework of the modern global West and its concomitant cosmologies. We argue that in addressing environmental and ecological issues in educational practice, when all of the moves are taken – when all of the foreseeable critiques and counter-critiques in discussions around ecological world views have been made – then the end game[4] will always require a reference to, even immersion in, the ontological underpinnings of a practice and a way of life, an *ontos*. And we suggest that a new materialist framework, if enacted in ways similar to those we are pointing to here, might be a way to encounter, even support, a different state of being-in-the-world.

We understand the diverse field of new materialisms to be intertwined as part of a broad response to the intense focus on subjectivities of 'post-' era studies. With regard to environmental education, during that era, the focus on the role of human subjectivity is viewed to have diminished consideration of and engagement with other aspects of the world (Hart 2005). Certain philosophers, such as Graham Harman, have developed entire philosophical systems that emphasize not merely the more-than-human, but the world-not-human. One of the points of Object Oriented Philosophy (1999) and later Object Oriented Ontologies (2009) was to understand a world with the potential for interactions that were not solely mediated nor completely known through human subjectivity. In the context of environmental education, the purpose of this paper is to expand the field of new materialisms, by – perhaps paradoxically – taking a look at stories that emerge from an emphatically 'pre-' perspective: in this case, stories from Alex Mathias, a Teme Augama Anishinabe Elder/knowledge keeper who retains a way of interacting with the world – and a corresponding worldview – that long pre-dates 'new' materialisms. This way of being de-centers human participation, expects complex communication between other beings, whether or not humans witness it, and is compatible with a state of being that might be viewed by the modern West as less linguistically focused. Alex's understanding of the world is not primarily dualistic but one much more akin to Clarke and Mcphie's version of animism, deriving from Deleuze and Guattari's immanent materiality (2004).[5]

As we write, we are aware that the dust has not yet settled on the 'post-post' discussions in environmental education raised in this journal more than a decade ago (Hart 2005; Payne 2005; Russell 2005). Nor has it settled upon the various takes on a reorientation toward the material world, be it through Karen Barad's (2007) work, Jane Bennett's (2009), Graham Harman's (2002, 2016), or Quentin Meillasoux's (2008). We draw from each of these to varying degrees, aware that, in at least some cases, proponents define themselves as distinct and in some cases incompatible with one another (see Harman's (2016) incisive exploration of Barad's agential realism and how it is distinct from his own speculative realism).

What remains central to the work here is that something out there in the world ought to be given greater import than it has hitherto been given. And the something out there might not be *out there* (in the sense of being separable from human arising), and might not be a 'thing' (in the sense of being primarily a separate object rather than an integrated locus of becoming) (Clarke and Mcphie 2014; Barad 2007). In some sense Alex and other Elders represent a worldview that was prior to the above, without the language of such movements (Beeman 2006). This enacted *ontos*, way of being in the world, however carries with it the ontological import the above works appear to be interested in and as such looking at it from a new materialist perspective seems apt. We note that it would be misleading to suppose that the worded explorations alluded to above were conceived of earlier in Anishinaabe cosmology. Our point is that the language of new materialism might find vital examples in some lives actually being lived.

A major challenge in writing this paper is this: if what we are proposing is another being state that may be incommensurable or at least inconsistent with the assumptions upon which academic language and much of western theorizing is built, then we need to find a way we can 'speak' from, to, and even out of, that state and not simply stay stuck in the territory of language *about* that state. To remain there would be, from the start, to deny what makes a different being state distinct, while also running counter to the argumentation resulting therefrom. While we do address some current literature in this paper, were we only to do this in the languages and theoretical lineages of western ontological traditions then some of the possible state-of-being differences of new materialisms would be *de facto* lost.

We therefore propose to employ two alternative strategies. The first is to give guidelines for reading the paper that involve a series of actions in the world that we think may help to frame its understanding. We encourage the reader, whatever their schedule or reasons for not doing so, to try this out. For if we are right that our immersions and positionings influence how we read then materially situated academic work demands attention to these. Barad writes, 'Language has been granted too much power. The linguistic turn, the semiotic turn, the interpretive turn, the cultural turn: it seems that at every turn lately every "thing" – even materiality – is turned into a matter of language or some other form of cultural representation.' (2003, 801) Barad's complaint is against a lack of rootedness in materiality. Academe is inherently embroiled in *wordedness*, but we think this way is limited: it by definition cannot adequately function in cases in which a state of being that is possibly incommensurable with a primarily English-language-based,[6] modern, global Western, linguistic stance may be at play. Thus, we suggest that to give this paper a proper 'read,' things in the world must be *done* in the hopes that the doing itself might shift our entanglements, if even slightly, such that our reading is diffracted. If we are felted, enmeshed beings then, we argue, understanding can change not only through re-positioning ourselves intellectually (i.e. attempting to put ourselves in a position in which our *logos* is temporarily displaced, positioning new concepts to do different work, etc.) and reading diffractively, but also through diffractively re-entangling (i.e. potentially shifting our *ontos*) and reading anew.

We think that fully encountering what new materialisms propose will necessarily involve not just a thought experiment such as this one, but a material change in the being of the interpreter. Hence the request that you *do* as well as *read* this paper: you act in the world in ways that help the understanding of ideas herein. We suggest the reading of the section on positing *ontos* to be preceded or accompanied by a challenging route-finding activity, such as back-country or cross-country skiing or hiking in difficult climatic conditions through unfamiliar terrain. The section comprised of stories and analysis, we see as accompanying or reading after a concerted activity in meeting life needs, such as preparing wood for burning by sawing or splitting it, or working in a food garden. The conclusion could be read in front of the fire (indoors or out) made from the wood prepared earlier or while eating the meal that is a result of the earlier preparations. We note that in at least two of the contexts suggested above, to read from a screen would be challenging.

Second, in this paper, we employ stories in a particular way to get at being and meaning differently. We offer three stories, two from Alex Mathias, an Anishinaabe knowledge keeper, as a way of describing or perhaps encountering what an enacted change in *ontos* might be. This will be our attempt to read the world differently using new materialism and will require, from the reader, an attendance – or attunement – that asks one to step into a vital, vibrant, and agential world (Bennett 2009). We believe that all of this will help to point in the direction of how a shift in position, in one's actual state of being, might lead to a different reading of the world, rather than the other way around.

Section 1: positing *ontos* (to be considered while route-finding in difficult conditions)

This section introduces the concept we call *ontos* (*being-)* as separated from its usual partner *logos* (*-wording)*. This separation allows us to position *ontos* within the larger ontological conversation (Heidegger 2008; Barad 2003; Harman 2016; Haraway 2016). We then attempt to make the case for why, in addressing environmental and ecological issues in educational practice from the standpoint of new materialisms, an *ontos* that is conceptually and interpretively always constrained by *logos* forecloses the possibility of something that is unfamiliar to the modern being state.

Current theorizing work addressing environmental and ecological issues in educational practice in the global, modern West has grown to include, we think astutely, questions of state of being itself (Lynch and Mannion 2016, Mcphie and Clarke 2015). To this end we propose to introduce the term *ontos*[7] instead of staying with the usual *ontology*, because of what we see as the English language's relative impoverishment around words relating to being and the conceptual limitations that arise therefrom. English has only one term, *ontology*, to describe both the study of being and the being state itself. Thus, given this linguistic relation of both being and the study of being, by necessity, any reference to the being state tends to invoke the *logos*-tistic description and theorizing about it. Take a summer rain for example. We can discuss its origins as warm moist clouds encountering mountain slopes, or determine the possible safety hazards as creek levels rise, or note the difference in cultural responses to its arrival, or statistically chart the time of its arrival according to existing data and date, but in none of these interpretations are we getting wet and in none of these do our material surroundings actively play a role. *Logos*, its implied separation from being-with, and its predilection for commentary-on, necessarily accompanies the way of being and its performative 'immersion,' by definition. We suggest that in separating this overlay we can diffract the standard ontological discussion that we believe obscures the lived interaction with the world. Thus, we propose *ontos*, temporarily separated from *logos*, to emphasize the being state itself. In this diffractive act, we mean to focus on the being state, *ontos*, with relatively less emphasis on the linguistic interpretation of such.

While it might be argued that the two are inseparable, we think that an attempt to – preemptively, as it were – shift focus to the material is necessitated by the current context in which we live, where so much is *languaged* and so little is *enacted* in our relationships with the material world. In *Meeting the Universe Halfway (2007)*, this is one of Karen Barad's main ideas. We are referring to the enacted life – the actual playing out of a relationship with, the entanglement of getting wet and responding – or better still, *a related-being-with* the more-than-human. The challenge here then, which parallels our concern with a perhaps unintentional default to focus predominantly on the *logos,* even when intending to approach the *ontos,* is that although many theorists acknowledge the necessity of including the ontological in this discussion, they end up not actually reconstructing a related-being-with the more-than-human through taking too small a step and remaining, anthropocentrically,[8] in the *logos*.

This work emerges from earlier research suggesting that there may be differences between the *ontos* of *homo mobilis* and the *ontos* of Autochthony/Indigeneity (Beeman 2006). In earlier work, Chris noted a possible distinction between the ethnicity of First Nations, Metis, Intuit and Indigenous people globally, and what may be posited as an ontological circumstance of Autochthony or Indigeneity (Beeman 2014). In the mode of being of Autochthony, also called *attentive receptivity*, a moment-by-moment awareness of the intricate interactions of place and its needs, an awareness that is both global and particular, through an attunement to its signals, replaces the will-directedness characteristic of the being state of *homo mobilis*. Such a distinction makes possible one of the virtues of conceiving of Autochthony as an ontological condition: though not itself directly addressing historical, political and economic circumstance it may facilitate a bridging of ethnic divides by making possible ways-of-being alliances that may accommodate and speak to (in the example of Canada) not only First Nations, Metis, Inuit but also people who are not from these groups. In doing so, the very existence of an Autochthonous mode of being challenges the dominance of the *ontos* of the modern global West. The argument recognizes the global project of colonization (in Canada, of predominantly First Nations, Metis and Inuit people) through political, cultural, linguistic and other means, but it also invokes an ontological aspect. That a colonization of the natural world has also occurred is understood (Blenkinsop et al. 2017)

Section 2: enacting *ontos*: three stories (to be considered while meeting life needs through concerted activity, for example in cutting wood or digging in a food garden)

In this section, we offer stories as another way to make meaning. We think they offer some things not possible in standard academic reckoning. First, considering these particular stories through a new materialist onto-epistemology allows us to consider the possibility that there might be an aspect of human being, which, when engaged with the more-than-human, is pre-language or outside of particular human language. The first story of the cry of the coyotes is particularly resonant in this aspect. Second, these stories potentially point beyond concepts of language that are currently trapped in the anthropocentric. The second story of mother moose and calf figures here. This story suggests the possibility not just of communication between animals, but of what might be thought of as intentional miscommunication. Thus, the complexity of other-than-human being communication is noted. This leads to the third point, that these stories may offer a way of encountering moments of intra-active communication. They may present ways of coming to know from another being state. The third story, involving Alex Mathias's account of how knowing comes to occur in the context of a spirit place, echoes with this aspect. The stories we offer do not only do these things; they overlap and flow in many directions at once. Stories work in other ways than simply offering aspects of linear 'proofs'.

The cry of the coyotes that awakened me two nights ago

> I am in the little tent that is a kind of refuge for me, a long bicycle ride away from my university. Some of the ride goes through horse pasture, to a place I stay, in-between the world of academe and wilder places. Not in "wilderness," not in the city, but a meeting place, somewhere in between. I am awakened in the darkest of night in the deepest of sleep by a cry nearby. Yet, at the moment it occurs, it is not yet "a cry". For a moment, here is only the absolute being of the cry. "I" am with it, and in it, nameless and wordless. There is no differentiation between where the cry ends and the I begins. There is only the cry, which forms a bond with all being. The sheer exhilarating aliveness, a being-ness-in-me-that-is-also-in-the-world, is not communicating language. It is just is-ness and being-ness, being. I do not make a sound nor am I conscious of myself as a being that could make sound, but take it all in, as a kind of person-less absorber.... Awake, now, as I become alert, I hear the end of the cry disappearing in the night. By now, it is "a cry": the perceiver perceiving is once again: "I." (Chris).

For the purpose of this paper we are treating this experience as text because it is a written representation of a memory. And we are distinguishing between the I-ness, coyote-ness, and world-ness to make the writing less cumbersome. But that this is text now does not preclude the possibility that it was not then, *even if what occurred was intended to be communication by the coyotes,* and even if, in the moment after its perception, immediately following the instant of awake-ness, Chris was able to interpret it as a kind of text, as 'cry'. The very separation of knowing and being in this moment of awakening is troublesome for Barad 'who like Haraway, … wants to move away from the traditional representationalist ontology and epistemology where the knowing subject, the object that is being represented, and the produced representations or knowledge are seen as separately existing entities.' (Geerts and van der Turin 2016)

This incident is also in key with the work of Clarke and Mcphie (2014). This tiny moment of transition between being-ness and meaning-making seems to bridge the void between a world composed of intermingling energies, lacking the usual divisions between discernible and discrete entities. In such an understanding, the human perceiver is 'human-ing,' and the coyote is 'coyote-ing,' while the world is 'world-ing.' (vis Deleuze and Guattari 2004, 265, quoted in Clarke and Mcphie 2014, 203). From Deleuze and Guattari, the materiality of the plane of immanence precedes 'structure and genesis.' Materiality, and the form it takes, is immanent ('indwelling, inherent, permanently pervading the universe' in Oxford.) Within this view, being is perhaps better described as 'becoming' because the latter suggests a continual re-emergence at each moment. Clarke and Mcphie propose the term animism as a way of seeing founded on an ontology of immanent materiality … we use the term animism … to describe a mode of being (becoming) that embodies both 'seeing' and 'acting' within a world *ontologically understood* by its inhabitants (animists) to be constituted by immanent materiality; to be *whole, alive* and *forever becoming.* Immanent materiality is opposed to the prevailing transcendental conception of a world of objects set in space that we experience as we go about our lives. Rather, it suggests a world of affect where the boundary between objects is dispelled, thus dispelling objects (205).

This story also resonates with the work of Ross and Manion (2012) who, with Ingold, propose a world of relationships, in which meaning only inheres in relationships, and not in beings that could theoretically be separable from these relationships.

The thought of the gutsiness of the coyotes to act in this (very proximate) way comes later. The thought of this communication as a kind of text emerges later. The thoughts of whether coyotes in this area breed with dogs, if there is a wild dog population, if there have been any reports of attacks on humans nearby, all of these thoughts emerge later. What occurs in the moment itself is an encounter with Being in which the edges of Chris merge with those of the world. 'I-ness' changes; being-ness changes. In this instant, there is no *I* to interpret. Repeatedly, in working with Indigenous knowledge keepers, Chris has heard about the limits of the (human) self not stopping at the border of the skin (Beeman 2006; Evernden 1985). And about the agency of the more-than-human world, interpreted but not determined by human participants in the relationship of communication.

If, using standard linguistic interpretation, text is anything that can be 'read' then it is possible to think of the cry of the coyote as text. In that case we would suggest that there are also myriad readers of that text (e.g. humans, other coyotes, roughed grouse) that are all aspects of understanding the cry and knowing coyote. (Harman 2002) However, our sense is that both the cry of coyote and the immediate response of 'aliveness' referred to in the story seem to exceed everyday interpretations, and even our proposed expansions, of the term 'text'. Way yonder, preceding the considering of, interpreting of, and storying about, it may be possible that there is something that is *out there* in Richard Rorty's phrasing (noted in Gough 2016), which is perceivable and connectable with, but whose being is also not simply encapsulated through *descriptions.* And this entity that is out there involves both the being of the world and the being of the human perceiver. This being state can be considered as one in which the *logos* does not enter or is at least playing a secondary role to *ontos.* This being state, like the border of the tent to coyotes, is transparent, but keeps the *logos* at bay. The considering, interpreting, rendering of the cry of the coyote as text is not what we are focused upon here: it is the immediate and responsive state: it is being meeting being, with nothing to distinguish and no articulable mediation between: a kind of enmeshment.

This way of being-knowing – again, it is a question of degree – seems to point in the direction of a being-without-human-language. We are uncertain of this, but it is almost as if, without sliding into a simple dualism, a being state can emphasize *ontos* or *logos*. Neither is lost but one or the other is foregrounded (Plumwood 1993) The being state that emphasizes *ontos* brings us more directly into relation with the material world.

To some extent, what we are positing now pushes against Barad's concept of intra-agency, interpreted succinctly by Rautio (2014) as something which 'co-emerge[s]' and suggests simultaneous and mutually co-creating being. Barad writes, 'The relationship between the material and the discursive is one of mutual entailment. Neither discursive practices nor material phenomena are epistemologically prior.' (2007, 152) We agree with this possibility, but think that it may not be applicable in all cases. And if we expand the somewhat anthropocentric concept of the discursive then maybe even less than initially thought. One of the concerns we have about assuming an equitable co-emergence of material being-ness and discursive knowing-ness (*ontos* and *logos*) is that we might fail to notice those encounters that are led by, or are leaning more heavily on, one or the other. What we are attempting to refer to in these stories is something that is more on the *ontos* edge of the continuum – were continuum something commensurable with what we are attempting to suggest.

How to lose a moose calf & caring well for a spirit-place

We wish to present two more stories, both from Alex Mathias, with whom Chris has worked for more than 15 years. The first is a story of moose and calf. It describes how a mother moose intentionally loses its calf from the previous year, when it is about to give birth to another. The Elder describes this event:

> In the spring of the year after the calf is born – [9]
> if the mother knows that she will give birth again –
> well, she cannot take care of two so
> she finds a time when both she and the calf are browsing
> near a creek or something
> and when there is some distraction
> she walks off up the creek
> hiding her scent and sound with the water
> so that her calf cannot find her.
> That way, she is free to give birth
> and to take care of the new calf.[10]

At the logo-tistic level of interpretation, this story shows a strong degree of interpretive skill on Alex's part, despite the fact that he is interpreting the experience across the borders of language (i.e., English is not his first language) and culture (i.e. he was born and raised on the land in an Indigenous community). He has seen the event he describes happen. He has watched a moose intentionally lose its own calf when the moose was about to give birth to another calf. And perhaps he has heard about this kind of thing through his parents. Or perhaps, he has checked this story of interpretation with others. And in the examination of the text, there is much that can be done from interpretive perspectives. There is the action of the moose breaking the relationship with her kin. There is the simple ecological necessity, which humans might understand, of a mother making way for new offspring. There is the action of Alex telling this story and what it means in the context of a working relationship with Chris who was not raised in the same environment and worldview. And there is another experience of moose and calf in which Chris witnessed a battle of wills between another cow moose and calf. This was in some ways the antithesis of the story Alex told: In this case, the mother moose had to coerce the calf to come with her, because she knew the potential danger of humans being near. The calf's interest in food was expressed beautifully in a bawling against the mother's insistence that she leave.

So many aspects can be interpreted and so many different kinds of texts can be visited. But to be diffractive, it might be argued that, in quite an inconvenient manner for textual interpretation, there was a moment of Alex 'just knowing': of suddenly coming to understand, through standing-in-being-with, something that had not been previously understood. One might imagine this moment from Alex's perspective: not taking an image or story home and pondering its meaning, but suddenly realizing 'meaning' without attempting to understand. One might also imagine it through a lens held to the 'eye' of world: of human 'consciousness' being brought to the 'world.' From Deleuze & Guattari's perspective, it might simply be that, in Alex's way of being in the world in this moment, consciousness is brought to the plane of immanence.

This kind of event might also be viewed as evidence of a deep, ecological, related being bringing knowledge about a particular place, informed by immersion, a lifetime of observation and the knowledge of others. A post-structural stance, or a hermeneutic stance, or any other kind of interpretive act such as the one presented in this text can later occur – placed *upon* the experience, as it were – because the story then exists in language, as text. Alex's observation and involvement, and the stories from others that inform his life, all form part of his interpretation. And yet, we would suggest that at least some of it was not gathered as text, within the western definition, in the first place. And the knowing that occurs, occurs because of a particular kind of life lived: one in which a being-with state is more enacted, and one in which it is possible to 'speak moose.'

This resonates with Rautio's (2014) lament over how outdoor teaching styles often get in the way of, predetermine even, the kinds of interactions children might naturally have on their own, with 'things'. For example, she notes, '... psychologically oriented research on matter and children has tended to take up child-matter interaction insofar as it can be discussed as serving a purpose.' (463) But, from her observations, children also can learn with, rather than about, the world (462). That is, in very simple ways, the active and agential aspect of what the global modern West normally interprets as inert, can actively shape meaning, direction, and understanding.

In addition to this, the story above de-centers language from humans to other beings. It shows how one more sophisticated being – in this case the cow moose – may be able to literally mislead the calf. This suggests a kind of complexity in communication and intent. As important as it may be for the cow moose to clear the way for new offspring, the interpretation Alex gives (which is similar in complexity to the descriptions of the actions of animals of many First Nations people with whom we have worked) suggests a degree of sophistication not usually attributed to this animal. The way the story is told, the mother moose may be acting out of necessity, but her acting also involves what might be called 'deception.' And deception requires some meta-comprehension of communication. In working with First Nations communities, Chris has frequently heard detailed understandings of animals that can only come from an ontological position of non-dominance and non-assumed dominance.

If an understanding of this story is situated in an ontology in which intelligence of the animals one encounters is assumed, we are left wondering, is this need to abandon offspring felt as a physical need in the cow moose? Or is there some kind of thinking about it? For herbivores, avoidance of a predator is first nature. At some point, does the calf become an irritant and is the mother avoiding the calf as if it were a predator (in the sense of predating upon the resources needed for another calf)? Is there pre-consideration to what she is about to do? Perhaps she learned this from her mother, because she herself was deceived in this way, and the experience has stayed with her. Or perhaps it is an occurrence that is triggered not by her own body, but by the place? A certain plant comes into bloom, and this is part of a sequence of events making possible the birth of her new calf, but requiring her to abandon her offspring. All of these questions begin to emerge as soon as we begin to think about communication between animals and amongst beings and their places as potentially being significant and complex, as well as non-human/non-logo-centric. It becomes even more intricate if the fabric joining all being, of the kind suggested by Clarke and Mcphie, prohibits simple differentiation between beings, because all are joined as part of the fabric of being, the plane of immanence, a la Deleuze and Guattari (2004).

Beginning to think of even the possibility of complexity in moose communication is a little like the 'two women' test in films which feminist theorists have proposed to determine whether film roles have managed to escape conventional sexism. In this test, if two women are having a conversation that relates to the unfolding of plot and that is not about a man, conventional 'Hollywood' film norms have been transgressed. When we look at films in this way, we come to understand both the concerns of feminist theorists and the limited acting roles women have historically been given. But unless we apply the test, we could easily be duped by other aspects of plot and by our own enculturation. Similarly, when we approach communication and enmeshments between other-than-human beings expecting to find complexity and agency, we are more likely to be open to its possibility.

Spirit-place-knowing

But let us propose something else about this story, by way of another story. When Chris asked Alex about the spirit place for which he was responsible to care for, and what happens there, he said:

> people getting answers
> maybe not by hearing voices
> but in their mind
> the answers coming to them[11]

This too is a text first offered by Alex and then reported by us, and requires interpretation, which we are about to try. But let us be clear again that the interpretation we are about to do *upon* this text, and which, it could be argued, is demanded by any perceived text, is different than the moment that the text describes, the *ontos* of those 'getting answers'. The telling of the story exists for the purpose of communicating knowledge, in particular, in this case, to Chris, who is not of the same worldview. And the point Alex was making in this passage was spoken in the midst of other words about words. But the words-about-words passages are precisely about a state of being – and a way of knowing – that is not one of words, at least those kinds and ways of words that are normally shared between Knowledge Keeper and Chris.

According to the patterns of the global modern West, Alex does not have a lot of formal education. Instead, he has an immense articulateness that is rooted in acute perceptions and understandings of the natural world drawn, we think, from his ontological orientation therein. And yet, we think this passage contains more than Alex simply using a different interpretative framework: the interpreter himself is different. It is noteworthy that in this particular case, the words presented did not form conventional sentences. His words remind us of Wade Davis's description of when he asked a friend about a hunt. The friend, who had just come back with the hunt responded: 'The hunt was five days long; have you got five days?'[12] Davis's point was about the precision with which stories in some Indigenous traditions were relived in order to be retold. In order for stories to be accurately told the teller is focused not just on the particulars of the tale but on rendering authentically their being-in as the story unfolds. Alex's state when he said the words quoted above was a kind of reliving of the mode of being entered in the spirit place. They were a hint for us as listeners of a state outside language. The state, itself, is not reachable with words. When Alex says, 'maybe not by hearing voices,' he is suggesting that something like understanding may emerge through just being in that place. The metaphor being offered is communication but framed with a caveat that undercuts, diffracts even, what Alex knows will be obvious assumptions made by Chris, who, as an 'educated' listener, was not raised on the land. It is likely that he will think that words and speakers are necessarily involved.

This work of translation and interpretation was extremely difficult for Alex. He visibly struggled to present these things and we suggested above that part of the challenge was to 'translate' from, using Barad's terms, an intra-active being state into the *logos* of a predominantly inter-active way of being. Geerts and van der Turin, from a slightly different perspective than Rautio, describe the distinction between interactive and intra-active this way, 'an interactive model supposes that interaction takes place between ontologically separated objects, an intra-active model tackles the boundaries between these objects, and proposes that phenomena are always already entangled.' To which we would add, 'not theorizing while it does so.' (Beeman, 2006,133) Alex had the difficult task of translating experience and being into an inadequate language that did not have the capacity to do full justice to the entangled way of being he was describing.

Conclusion: (to be considered while resting/enjoying leisurely in front of a fire)

Ultimately, we believe no reasoned argument will ever fully convince *homo mobilis* out of one ontological space and into another. Thinking and ideas may prepare the way for an ontological shift, but this will only ever be effected by living out a different kind of life as a being that is in and of the world. Many environmental educators intuit this. The richest possibilities of new materialisms for environmental education might be in naming and allowing for the enactment of this continuing project of actually shifting *ontos*, rather than simply focusing on the logo-tistic description of such a shift. Thus, the end game is not only ontological, but *ontos*, itself.

In conclusion and making more direct references to educative practice, we wish to briefly again invoke Clarke and Mcphie (2014), who articulate at least one understanding of another kind of *related-being-with* the more-than-human, which includes an understanding that the more-than-human world operates with agency quite apart from the limits of human interpretation. This is the 'immanent materialism' of Deleuze & Guattari, noted earlier, and expressed in a kind of animism, which Clarke & Mcphie describe. As an enacted view of the world it would likely lead to an immense educational step away from the anthropocentrism and logo-ism that infuses so much schooling today.

But how might this discussion actually affect the practice of environmental education? Perhaps an example might help. Our pedagogical intuitions are that environmental educators might use this opening towards *ontos*, this diffractive reading resulting from diffractive entanglement, this distinction between intra- and inter-active that we have been discussing through experiences already familiar to their practice. These ideas might help explain the power and possibility of some well-known outdoor experiences. The solo or quiet spot may come to be framed and understood as an opportunity to enfelt or enmesh differently, to prioritize an immersive being state while quieting the logo-tistic. This could also be an opportunity to attune oneself to the 'languages', 'customs', and ways of being of the diverse and agential places in which we find ourselves. We may consciously start to position ourselves in the world as conduit and potential ally. Environmental educators may also re-think key metaphors being used, especially if many lean towards the more logo-tistic or anthropocentric. 'What did you learn?' might become 'What was happening?' 'How did you change?' might become; 'What enmeshments were you part of'? And did these open any other ways of understanding the place?' 'Where was understanding coming from?' 'Who and what was speaking? And what was being said?' are differently entangled questions from 'what did you notice?' Even these, in light of Clarke & Mcphie's cosmology, presuppose separation between perceiver and what is perceived. Perhaps a focus on what unfoldings played out, rather than the beings doing the unfolding, or learning, or teaching, might occur. Finally, we see this discussion as challenging environmental educators to remember and consider our entanglements and to seek ways to diffract not only the thinking and reading of students but also the places that gather them into the world.

While immanent materiality, diffractive entanglements, and shifting of *ontos* are concepts which are potentially useful to situate new understandings of (the necessarily named-separately) human/place interactions, the educational work of actually enacting and also supporting learners into a different kind of relationship with place still needs to be done. Such work entails not merely a set of actions, or intra-actions, and a deep attunement, but a way of 'human' being that is of 'world' such that the 'being-ness' between bridges the separation of *ontos* and *logos* while acknowledging understanding arising through onto-epistemological entanglement. This is a challenge to us as educators and humans. What might we do to notice our own enmeshments while supporting our students to notice states of being, to inch beyond the edges of language, thoughts, and culture? How might we lose our moose calves to another way of being in and with the world?

Notes

1. For defining *entanglements*, we refer back to Ingold and his earlier use of it (2008). While we begin with this term to permit readers to orient themselves, something about what has become the static-ness of such as central term that purportedly describes not only emphatically non-static instances, but moments in which subjectivity/objectivity distinctions are rendered meaningless, concerns us. So, we also use *enmeshments* and *feltings* and variations thereof to get at a broadly similar idea.
2. While we recognize the argument that academic papers cannot and ought not to attempt to do everything, to not attempt to make such an argument in this form might be considered tantamount to leaving the field to others. If the argument we make herein is justified, precisely the opposite course is needed.
3. In an earlier version of this paper, a substantial section was devoted to a critique of Noel Gough's excellent paper, 'Postparadigmatic materialisms: A "new movement of thought" for outdoor environmental education research?' We consider this to be one of clearest delineations of a post-structural position as it relates to new materialisms. However, it concerned us that the full difference, indeed the sometimes wordless aspect of new materialisms could be apparently so seamlessly incorporated into a primarily linguistic stance.
4. The term *end game* is used here as it is in chess, to refer to the final stages of the game in which few pieces remain, the risks to poor strategic moves are high and the range of possibilities is limited. We think this corresponds well to current planetary conditions, with one exception: in the game of chess, defined "moves" are well established. What we propose is a different kind of move, which takes the game from the board to the world. This case was initially made, in nascent form, in a presentation given at the WEEC conference in Gothenburg, Sweden, in July, 2015.
5. See Clarke and Mcphie (2014), Mcphie and Clarke (2015) and Clarke and Mcphie (2016).
6. English appears particularly to be at odds with the way meaning is made in many verb-based languages, especially Indigenous ones.
7. For earlier discussions see: Beeman (2012, 2014). We prefer to use *ontos* rather than the more obvious *being,* because the English carries with it so many other connotations, even when italicized.
8. We include an inclination toward anthropocentrism here because of the way theorizing about *logos* often includes assumptions based on humans as the model for what language, knowledge, and proof are. This tends to then be skeptical of things like trees speaking or ravens knowing.
9. In an attempt to honour the rhythms of speech of the Elders pauses in phrasing were marked with line-breaks. Thus appears a poetic form, relating to Davis' notion of how 'the poetry of the shaman gave way to the prose of the priest.' (Davis, personal communication, April, 2010)
10. Alex Mathias in Beeman, forthcoming.
11. Beeman (2006).
12. Wade Davis, personal communication.

Disclosure statement

No potential conflict of interest was reported by the authors.

References

Barad, Karen. 2003. "Posthumanist Performativity: Toward an Understanding of How Matter Comes to Matter." *Signs: Journal of Women in Culture and Society* 28(3): 801–831. doi:10.1086/345321.

Barad, Karen. 2007. *Meeting the Universe Halfway: Quantum Physics and the Entanglement of Matter and Meaning.* North Carolina: Duke University Press.

Beeman, Chris. 2006. "Another Way of Knowing and Being: Opening Spinoza and Heidegger in the Company of Elders." Doctoral diss., Queen's University at Kingston.

Beeman, Chris. 2012. "Autochthonous Education." *Encounters in Theory and History of Education* 13: 27–40. doi:10.24908/eoe-ese-rse.v13i0.4473.

Beeman, Chris. 2014. "Autochthonous Ed.: Deep, Indigenous, Environmental Learning." *Teacher Education in a Transnational World*, edited by R. Bruno-Jofre and J. S. Johnston, 191–207. Toronto: University of Toronto Press.

Bennett, Jane. 2009. *Vibrant Matter: A Political Ecology of Things*. North Carolina: Duke University Press,

Blenkinsop, S., R. Affifi, L. Piersol, and M. Derby. 2017. "Shut-up and Listen: Implications and Possibilities of Albert Memmi's Characteristics of Colonization upon the "Natural World." *Studies in Philosophy and Education*, 36(3): 348–365. doi:10.1007/s11217-017-9565-4.

Clarke, David, and Jamie Mcphie. 2014. "Becoming Animate in Education: Immanent Materiality and Outdoor Learning for Sustainability." *Journal of Adventure Education and Outdoor Learning* 14(3): 198–216. doi:10.1080/14729679.2014.919866.

Clarke, David, and Jamie Mcphie. 2016. "From Places to Paths: Learning for Sustainability, Teacher Education, and a Philosophy of Becoming." *Environmental Education Research* 22(7): 1002–1024. doi:10.1080/13504622.2015.1057554.

Deleuze, Gilles and Guattari, Felix. 2004. *EPZ Thousand Plateaus*. Bloomsbury Academic: New York.

Evernden, N. 1985. *Natural Alien*. Toronto: University of Toronto Press.

Geerts, Evelinen, and Iris van der Turin. 2016. "The Feminist Futures of Reading Differently: How Barad's Methodology Replaces Conflict-Based Reading of Beauvoir and Irigaray." *Rhizomes: Cultural Studies in Emerging Knowledge* (30). doi:10.20415/rhiz/030.e02.

Gough, Noel. 2016. "Postparadigmatic Materialisms: A "New Movement of Thought" for Outdoor Environmental Education Research?" *Journal of Outdoor and Environmental Education* 19(2): 51–65.

Haraway, Donna. 2016. *Staying with the Trouble: Making Kin in the Chthulucene*. Durham, NC: Duke University Press.

Harman, Graham. 2002. *Tool-Being: Heidegger and the Metaphysics of Objects*. Peru, IL: Open Court.

Harman, Graham. 2016. "Agential and Speculative Realism: Remarks on Barad's Ontology." *Rhizomes: Cultural Studies in Emerging Knowledge* (30). doi:10.20415/rhiz/030.e10.

Hart, Paul. 2005. "Transitions in Thought and Practice: Links, Divergences and Contradictions in Post-Critical Inquiry." *Environmental Education Research* 11(4): 391–400. doi:10.1080/13504620500169270.

Heidegger, Martin. 2008. *Being and time*. New York, NY: Harper Perennial.

Hill, Cher. 2017. "More-than-Reflective Practice: Becoming a Diffractive Practitioner." *Teacher Learning and Professional Development* 2(1): 1–17.

Ingold, Tim. 2008. "Bindings against Boundaries: Entanglements of Life in an Open World." *Environment and Planning A: Economy and Space* 40(8): 1796–1810. doi:10.1068/a40156.

Ingold, Tim. 2010. "Bringing Things Back to Life: Creative Entanglements in a World of Materials." NCRM Working Paper. Realities/Morgan Centre, University of Manchester (Unpublished).

Lynch, Jonathon, and Greg Mannion. 2016. "Enacting a Place-Responsive Research Methodology: Walking Interviews with Educators." *Journal of Adventure Education and Outdoor Learning* 16(4): 330–345. doi:10.1080/14729679.2016.1163271.

Mcphie, Jamie, and David Clarke. 2015. "A Walk in the Park: Considering Practice for Outdoor Environmental Education through an Immanent Take on the Material Turn." *The Journal of Environmental Education* 46(4): 230–250.

Meillasoux, Quentin. 2008. *After Finitude: An Essay on the Necessity of Contingency*. Translated Ray Brassier. New York, NY: Continuum.

Payne, Phillip. 2005. "Lifeworld and Textualism: Reassembling the Researcher/ed and 'Others." *Environmental Education Research* 11(4): 413–431. doi:10.1080/13504620500169411.

Plumwood, Val. 1993. *Feminism and the Mastery of Nature*. London: Routledge.

Rautio, Paula. 2014. "Mingling and Imitating in Producing Spaces for Knowing and Being: Insights from a Finnish Study of Child–Matter Intra-Action." *Childhood* 21(4): 461–474. doi:10.1177/0907568213496653.

Ross, Hamish, and Mannion, Greg. 2012. "Curriculum Making as the Enactment of Dwelling in Places." *Studies in Philosophy and Education* 31(3): 303–313. doi:10.1007/s11217-012-9295-6.

Russell, Constance. 2005. "'Whoever Does Not Write Is Written': The Role of 'Nature' in Post-Post Approaches to Environmental Education Research." *Environmental Education Research* 11(4): 433–443. doi:10.1080/13504620500169569.

Words (*are*) matter: generating material-semiotic lines of flight in environmental education research assemblages (with a little help from SF)

Noel Gough and Chessa Adsit-Morris

ABSTRACT
We address the aims of this Special Issue by exploring, critiquing, and responding constructively to the emergence and potential significance of new materialist thought in environmental education research and the broader theoretical landscape in which such research is situated. We offer some productive possibilities for advancing postparadigmatic materialist theorising in environmental education research by deploying concepts selected from Deleuze and Guattari's 'box of tools' (namely, 'machinic assemblage' and 'lines of flight'), supplemented by concepts associated with SF (science/speculative fiction)—namely 'cognitive estrangement' and 'object-orientated thought experiments'—to generate new lines of flight in material-semiotic environmental education research assemblages.

Introduction: where we begin

The textual performance we enact here explores productive possibilities for thinking with what one of us (Gough 2017, 53) characterises as *postparadigmatic* materialisms, following John Caputo's (1987, 262) suggestion that a term he coins, 'postparadigmatic diaspora', might help us to escape, exceed, and complexify Kuhnian structures/strictures and engage in less adversarial approaches—'this postmetaphysical lack of foundations, generates a new morality of civility and fair play'—and to avoid the dangers of Kuhnian rhetoric identified by Atkinson, Delamont, and Hammersley (1988, 233), which 'frequently leads to intolerance, fruitless polemic, and hypercriticism'. As Atkinson et al. (1988, 243) conclude: 'classifying research and researchers into neatly segregated "paradigms" or "traditions" does not reflect the untidy realities of real scholars… and may become an end in itself, displacing researchers from gathering data on important problems and/or building theories'.

We also seek to extend Adsit-Morris's (2017, 2) efforts to formulate a research practice that redraws 'the boundaries of subjectivity and objectivity… that moves from the individual to the collective, from static objects to material-semiotic generative nodes'. We proceed by deploying concepts from Gilles Deleuze and Félix Guattari's 'box of tools'[1] that (from our standpoint) generate lines of flight in the assemblages of researchers, empirical materials,[2] methods and milieux that constitute the intellectual and imaginative terrains of environmental education research. Deleuze and Guattari (1987, 4) conceive an assemblage as a collection of machinic concepts that can be plugged into other machines or concepts and made to work, which they explain with regard to their understanding of a book:

> As an assemblage, a book has only itself, in connection with other assemblages and in relation to other bodies without organs. We will never ask what a book means, as signified or signifier; we will not look for anything to understand in it. We will ask what it functions with, in connection with what other things it does or does not transmit intensities, in which other multiplicities its own are inserted and metamorphosed, and with what bodies without organs it makes its own converge. A book exists only through the outside and on the outside. A book itself is a little machine... We have been criticized for overquoting literary authors. But when one writes, the only question is which other machine the literary machine can be plugged into, must be plugged into in order to work.[3]

Elsewhere, Deleuze (1994, xx–xxi) asserts that a philosophical work 'should be... in part a kind of science fiction' in the sense of writing 'at the frontiers of our knowledge, at the border which separates our knowledge from our ignorance and transforms the one into the other'. In a somewhat similar vein, Burton (2015, 9) traces intersecting genealogies of science fiction and philosophy by referencing both historical and ahistorical interpretations of their enactments and materialisations. Historical interpretations posit philosophy as beginning at a certain time in ancient Greece, and science fiction as beginning during the modern (18th and 19th century) European industrial and scientific revolutions. Burton suggests that materialisations (in diverse media) of both philosophy and science fiction can be understood as manifesting 'an experience of wonder': 'Might it not be argued that this sense of wonder, in that it gives rise to speculation and narration about what *may* be, is the source of a kind of science fiction as much as it is of philosophy?' Burton describes how the (post-)enlightenment development of modern science via rationalism and empiricism separated out what was considered unscientific, unnatural and unknowable (i.e. the supernatural and metaphysical) from the knowable and measurable (i.e. natural laws and facts). As he explains: 'Anything which does not belong within the sphere of the knowable so defined... acquires the status of what is commonly termed fiction [and] gradually relegated to this inferior, denigrated realm on the wrong side of the real' (Burton 2015, 8).

Therefore, as this essay unfolds, we privilege texts and concepts associated with a genus of 'the literary machine' known as SF, drawing on Haraway's (2016, 2) hybrid conception of SF as 'science fiction, speculative fabulation, string figures, speculative feminism, science fact, so far'. We embrace SF as a boundary bending genre with many 'attachment sites' (Haraway 2018) for affective, cognitive and material thought practices. Thus, in parallel with Deleuze and Guattari, our question is which other machines the SF machine can be plugged into in order to work. We offer examples of SF stories, and concepts closely associated with their critique—namely *cognitive estrangement* and *object-orientated thought experiments*—as 'little machines' than can be plugged into an environmental education research assemblage to produce new material-semiotic conjunctions and configurations of researchers, empirical materials, methods and contexts.

In Deleuze and Guattari's (1987, 9-10) lexicon, a 'line of flight or deterritorialisation' is a means by which we move away from spaces regulated by dominant systems of signification. Our approach to deterritorialisation is similar to Roy's (2003, 31):

> ... we make small ruptures in our everyday habits of thought and start minor dissident flows and not grand 'signifying breaks,' for grand gestures start their own totalising movement, and are easily captured... small ruptures are often imperceptible, and allow flows that are not easily detected or captured by majoritarian discourses.

Such 'signifying breaks' are what Haraway (2007, 9), following Bruno Latour (1993, 97–99), describes as leaping over the 'Great Divides', or the cognitive 'switching' that Thomas Kuhn (1996, 114) describes in his reference to the duck-rabbit illusion. These ruptures and flows can also be imagined as what Karen Barad (pers. comm. see also Barad 2014) describes as *quantum tunnelling* in which particles pass through a classically forbidden energy state. Such particles defy classical mechanics as they relationally travel through barriers.

Here we make three 'small ruptures' in the systems of signification that we see as pervading the discourses of new materialist theorising in educational research. Our first rupture concerns the alleged (and somewhat taken-for-granted) 'newness' of new materialism, which diminishes the long history of materialist thought in place-conscious literatures. Our second and third ruptures offer two different but complementary ways of exploring the generative possibilities of speculative fiction (and other literatures encompassed by the initialisation SF) for informing material understandings of the objects of environmental education research, namely, SF as *cognitive estrangement* and SF as *object-orientated thought experiments*. We do not intend the sequence of ruptures 2 and 3 to suggest their relative value or significance, but neither is the sequence arbitrary. Rather, their sequence reflects our understanding of the logic of deploying these concepts in environmental education research.

Rupture 1: how 'new' are new materialisms?

Our first 'small rupture' in the systems of signification that pervade the discourses of new materialist educational research concerns the alleged (and somewhat taken-for-granted) 'newness' of new materialism, which might serve to diminish the long history of materialist thought in place-conscious literature (including much of what is commonly called 'nature writing', such as field guides, homages to solitude and wilderness, accounts of rambles in remote areas, and other reflections on experience in nature) as well as indigenous cosmologies (see, e.g. Todd 2016).

We write from positions that refuse to take for granted neither conventional wisdoms, dominant paradigms, nor the givenness of the material world, among which we include the stances taken by several prominent scholars associated with new materialist thought. For example, the first three words of our title, 'words (*are*) matter'—inspired, in part, by the title of Le Guin's (2016c) book, *Words Are My Matter*—deliberately and emphatically signal our reservations about the sweep of Barad's (2007, 132) much-quoted assertion that 'language has been granted too much power':

> The linguistic turn, the semiotic turn, the interpretative turn, the cultural turn: it seems that at every turn lately every 'thing'—even materiality—is turned into a matter of language... Language matters. Discourse matters. Culture matters. There is an important sense in which the only thing that doesn't seem to matter anymore is matter.

This passage clearly exaggerates Barad's position, given that elsewhere in the same text she also asserts: 'Discursive practices and material phenomena do not stand in a relationship of externality to each other' (Barad 2007, 152), and describes the 'exteriority within' material-semiotic phenomena (Barad 2007, 140), a notion that she derives from an in-depth analysis of diffraction patterns and which is also the basis of her conception of the dynamics of *intra-action* and the material-discursive practices of boundary making.

Barad's blunt accusation that 'language has been granted too much power' might be dismissed as a relatively harmless rhetorical overstatement, but it might also encourage distortions of the genealogy of 'new' materialisms, such as we find in Tuck and McKenzie's (2014, 15) assertion that the 'recent new materialist, neo-materialist, or ontological turn has emerged through orientations that object to the hegemony of the linguistic paradigm in poststructuralism'. We interpret Tuck and McKenzie's assertion as an oversimplification—initiating a 'signifying break', great leap or cognitive switch—of positions taken by the many new materialist theorists (including Barad) who in fact acknowledge their continuities with (and indebtedness to) the poststructuralist thought of Jacques Derrida, Deleuze and Michel Foucault, among others (see also Gough 2017). Indeed, as Alaimo and Hekman (2008, 3) state: 'the material force of the work of Deleuze, and especially of Foucault, is often overlooked because of the exclusive focus on the discursive'. Tuck and McKenzie are by no means alone among environmental education researchers in oversimplifying and/or misconstruing poststructuralist contributions to the field (see Gough 2006,

116–117, for another detailed example of environmental education researchers distorting post-structuralism/deconstruction).

We also question the 'newness' of some current emphases on materialism in social inquiry. As Dolphijn and van der Tuin (2012, n.p.) observe: 'For those familiar with the materialism of Walter Benjamin, "new materialism" is ironic for several reasons ... Benjamin rejects the modern fetish of newness and the illusions it presumes'. Benjamin (2002/1982, n.p.) writes: 'Newness is a quality independent of the use value of the commodity', which Dolphijn and van der Tuin interpret as arguing that 'staging a materialism that is "new" makes no sense at all. But of course there is no reason why we should confine ourselves to such a linear modernist idea of History'. Perhaps it is time to wander off the Newtonian path, 'blast out of the continuum of history' (Benjamin 1968, 253). For Benjamin, materialism is bounded up and folded into various moments of time, it flashes up and crystalizes.

Indeed, critical thinking about materiality is apparent in many areas of cultural production and practice relevant to environmental education. Barad (2007) traces what she terms 'the entanglement of matter and meaning' back to Neils Bohr's debates with Albert Einstein, which were indeed a material-discursive series of encounters—a changing of minds and matter—through the materializations of various thought experiments. Such scholarly engagement with 'matter and meaning' is also clearly evident in the long history of the entanglement of literature and nature. For example, one of us (Gough) recently encountered an unexpected example of such thinking during a visit to relatives in Derbyshire, England, where an exhibition devoted to the late poet, novelist and painter Arnold Bennett (1867–1931) was underway at the Potteries Museum and Art Gallery in Stoke-on-Trent to mark the 150th anniversary of his birth. A review of this exhibition by Charlotte Higgins (2017) begins by noting some 'curious ephemera' along-side a display of his water colours: 'his glove stretchers ... shoe horn ... button-hook implements ... grooming brush ... and a case of cut-throat razors'. Higgins (2017) comments:

> One might ... think that such objects are rather too trivial to be included in an exhibition. Bennett himself, though, would disagree. For him, the possession, the object, the thing, was the great clue to the person. A man from an area where livings were in coal, steel and pots, he had an instinctive grasp of the fact that objects ... are the symbols by which we present ourselves to the world. There is something about the Potteries: in its industrial heyday the very earth was mined, pitted and gouged out, so that the landscape itself was made a commodity. Charles Tomlinson, the area's greatest poet, wrote of its 'sedimented air'. For him, even the gases breathed in the Potteries had the weight and substance of a solid object.

Higgins demonstrates the 'thingness' of Bennett's prose by quoting from his novel, *Anna of the Five Towns* (Bennett 1902, 25):

> The room was not a cheerful one in the morning, since the window was small and the aspect westerly. Besides the table and three horse-hair chairs, the furniture consisted of an arm chair, a bent-wood rocking chair, and a sewing machine. A fatigued Brussels carpet covered the floor. Over the mantelpiece was an engraving of 'The Light of the World,' in a frame of polished brown wood. On the other walls were some family photographs in black frames. A two-light chandelier hung from the ceiling, weighed down on one side by a patent gas-saving mantle and a glass shade; over this the ceiling was deeply discoloured.

Bennett's skill in conveying surface detail in his place-conscious prose is a communicative capability of clear importance for environmental educators, perhaps especially for outdoor environmental educators, who must be able to interpret *and verbalise* the material details of landscapes that might be significant not only for learners' education but also for their safety and survival. For example, in Australia, many tree species shed branches spontaneously, so outdoor educators must be able to recognise the material risks of leading learners in forest or woodland environments where such species flourish and be aware of the circumstances in which such risks might be exacerbated by weather conditions, such as high winds or a build-up of ice or snow (see, for example, Brookes 2007, 2018, 143–145).

The extent to which nature has permeated literature might not be apparent to contemporary scholars in disciplines other than literary studies because, as Patrick Murphy (1998, xiii) observes, although nature has always permeated literature, it has 'perhaps received no greater neglect in the Anglo-American intellectual tradition than in the early twentieth-century period of modernism and the early years of contemporary postmodernism'. Following the rise of contemporary conservation movements and growing public awareness of environmental crises, this neglect was to some extent rectified by the late-twentieth-century growth of the academic field of 'ecocriticism', exemplified in the USA by the founding of the Association for the Study of Literature and Environment (ASLE) in 1992 and its journal *ISLE: Interdisciplinary Studies in Literature and Environment*. As Glotfelty (1996, xix), writes: 'Ecocriticism takes as its subject the interconnections between nature and culture, specifically the cultural artifacts of language and literature'. Howarth (1996, 69; italics in original) draws attention to the critical (and political) dimensions of ecocriticism:

> *Ecocriticism* is a name that implies more ecological literacy than its advocates now possess, unless they know what an embattled course ecology has run during its history. *Eco* and *critic* both derive from Greek, *oikos* and *kritis*, and in tandem they mean 'house judge', which might surprise many lovers of green, outdoor writing. A long-winded gloss on *ecocritic* might run as follows: 'a person who judges the merits and faults of writings that depict the effects of culture upon nature, with a view toward celebrating nature, berating its despoilers, and reversing their harm through political action'.

Andrew Jamison (1993, 187) describes a clear example of the 'embattled course ecology has run during its history' in his description of the 'disparate, even contradictory precursors' of contemporary ecology movements, with particular reference to distinctive national versions of systems ecology. Gough (2019) also draws attention to the enduring effects of the US version of systems ecology under the leadership of Odum (1953) whose *Fundamentals of Ecology*, together with its four subsequent editions, portrayed a world of ecosystems tending towards equilibrium and thereby popularised now contested assumptions about the 'balance of nature'. Jamison (1993, 202) observes that by the late 1970s, new evolutionary approaches had become increasingly popular among ecologists, 'so that systems ecology today is only one (and not even the most significant one at that) of a number of competing ecological paradigms'. Gough (2019) cites Australian research demonstrating that teachers frequently identify the idea of 'balance' and presenting students with 'both sides of the climate change argument' as important in climate change education, and speculates that, although conceptions of 'natural' order are pervasive in many disciplines, their persistence in science education might be attributed in large part to the pervasive effects of Odum's 'textbook ecology' received by many undergraduates in US and Australian colleges and universities for more than 50 years.

As noted above, we argue that critical examinations of the linguistic, literary and historical interconnections between readers, writers, texts and the material world are significant foci for environmental education research, yet few environmental education scholars focus on language and literature. Among the exceptions are Andrew Stables' (1996, 1997, 2001) explorations of literary theory vis-á-vis environmental education, Adsit-Morris' (2017) experiments with arts-informed methods and textual strategies, Gough's (1993a, 1993b, 1997, 1998, 2015) demonstrations of the generativity of SF texts as intertextual resources for deconstructing environmental education discourses, and the work of Rousell, Cutter-Mackenzie, and Foster (2017) who position speculative fiction as a mode of creative research that enabled young researchers with whom they collaborated to inhabit possible climate change futures by putting the children's visual and textual representations of these futures into dialogue with concepts derived from Deleuze and Guattari's geophilosophy.

Rich resources for environmental education researchers can be found in other literatures that connect SF with environmental education, especially literature that connects climate change with the material evidence supporting designation of the Anthropocene as a new geophysical epoch. For example, in *The Great Derangement: Climate Change and the Unthinkable*, Ghosh (2016), previously known for novels such as *The Calcutta Chromosome* (Ghosh (1997)), a mystery thriller in the SF sub-genre of alternative history, offers a non-fiction work that interrogates the 'deranged' modes of political and socio-economic organisation impacting human responses to climate change through the themes of literature, history and politics. Recent developments in literary criticism and theory have also sought what Crownshaw (2017, 887) terms 'an interpretive methodology for reading the planetary and the geological in literature'. Focusing in particular on Bacigalupi's (2015) SF novel *The Water Knife*, which imagines a near-future climate-changed American Southwest, Crownshaw (2017, 887) argues that Bacigalupi's future scenarios of climate change, ecological collapse, and near-extinction

> stage cultural memories of the unfolding aetiologies of the conditions imagined in the future but often subject to dissociation in our present. Conceptualising this fiction as the work of speculative memory, this essay finds in such acts of recall a self-reflexiveness as to the mediations of environmental remembrance. That is, in this futural work of cultural memory, the localisation of the planetary particularises the Anthropocene and foregrounds the ways that it is framed.

An example of such 'mediations of environmental remembrance' appears near the beginning of *The Water Knife*. Bacigalupi (2015, 19) describes one of the main characters waking from a dream of rain, which becomes a commentary on the irony of the idea of Manifest Destiny[4], given the ecological realities of the twenty-first century to which she has awoken:

> The dream had seemed so real: the rain pouring down; softness in the air; the smell of plants blossoming. Her clenched pores and the tight clays of the desert all opening wide, welcoming the gift—the land and her body, absorbing the miracle of water that fell from the sky. Godwater, American settlers had once called it as they invaded slowly across the prairies of the Midwest and then pressed into the arid lands beyond the Rocky Mountains.
>
> Godwater.
>
> Water that fell of its own volition, right out of the sky.

As Crownshaw (2017, 903) comments: 'This cultural memory of Manifest Destiny also recalls the fact that the geophysical agency of humanity now affects whether the rain falls, or not… The idea of Manifest Destiny remains a present absence in the text as a means of assessing the reality and tenability of this national narrative'.

In his introduction to a recent Special Issue of *Science Fiction Studies* on SF and the climate crisis, Brent Ryan Bellamy (2018, 417) makes some similar points to Crownshaw's concerning the roles of SF in staging 'cultural memories of the unfolding aetiologies of the conditions imagined in the future but often subject to dissociation in our present'.

> In a form of retroactive continuity, we can now look backward and see the seeds of the current conjuncture and the trajectories that have led to the current climate volatility. These trends may not have been visible before, but they are unmistakable now (despite the fact that some still quibble over their anthropogenic origins). This realization is matched by a turn to our future fortunes… if we begin to behave differently now, how might this change affect the future of the Earth-system?
>
> The gravity of the climate crisis has been steadily pulling genre fiction into its orbit. Global warming unfolds at a spatial scale and temporal rhythm that exceed the capacities of even the most robust literary imagination.

We now turn to more detailed considerations of the particular ways in which creative and critical readings of SF function as ways of performing inquiries in environmental education.

Rupture 2: SF as cognitive estrangement

The publication of Darko Suvin's (1972) 'On the poetics of the science fiction genre'—together with his co-founding of the journal *Science Fiction Studies* in 1973 and the expansion of his 1972 article into the book-length *Metamorphoses of Science Fiction* (Suvin 1979)—is widely regarded by scholars of the genre as a foundational moment for the field of science fiction criticism. Suvin (1972, 372) begins by noting:

> The importance of science fiction (SF) in our time is on the increase. First, there are strong indications that its popularity in the leading industrial nations (USA, USSR, UK, Japan) has risen sharply over the last 100 years, regardless of local and short-range fluctuations. SF has particularly affected some key strata of modern society such as the college graduates, young writers, and general readers appreciative of new sets of values. This is a significant cultural effect which goes beyond any merely quantitative census.

Suvin (1972, 372, his italics) argues for 'a definition of SF as the *literature of cognitive estrangement*'. As Gerry Canavan (2016, xii, his italics) observes, Suvin's influence can be registered in the way his definition 'has become ubiquitous in scholarly introductions to the field, which frequently introduce Suvin's critical work *before* mentioning H.G. Wells, Jules Verne, Mary Shelley, Hugo Gernsback, or any other foundational author associated with the early production of SF texts'. However, Canavan (2016, xii–xiii) adds:

> Of course, SF studies did not begin with Suvin, nor did Suvin solve all its problems in a single move. But what Suvin did was establish a discourse, in the Foucauldian sense, that subsequent SF critics have needed to contend with, whether positively or negatively. Suvin's work established contours for the subdiscipline that have come to structure not only SF criticism but the criticism of related speculative genres like fantasy, fairy tales and horror, the scholarly approaches to which have been defined (usually with significant frustration) by Suvin's totalizing rejection of them in *Metamorphoses*... In a subfield famously devoted to squabbling over definitions and policing generic boundaries, Suvin's definition has become a kind of consensus starting point, a place where we might at least begin to speak to one another.

Suvin's concept of cognitive estrangement derives from the strategy of *defamiliarization* deployed by semioticians to distance themselves from dominant social and linguistic codes. As Chandler (2007, 211–212) explains:

> The notion of cognitive estrangement, or 'making the familiar strange, and the strange familiar' is now a recurrent feature of artistic and photographic manifestos and of creative brainstorming sessions in many fields. The phrase itself has been attributed to the German poet Novalis (1772-1801, aka Friedrich von Hardenberg), who declared that the essence of romanticism was 'to make the familiar strange, and the strange familiar'. The concept is found among other Romantic theorists such as Wordsworth and Coleridge and is also closely associated with Surrealism and Brechtian 'alienation'.

Russian formalist Shklovsky (1917/1965, 13) introduced the concept of defamiliarization (or in Russian, *ostraneniye*, literally 'making strange') to literary theory: 'After we see an object several times, we begin to recognize it. The object is in front of us and we know about it, but we do not see it—hence we cannot say anything significant about it'. That is, over time our perceptions of familiar, everyday situations become stale, blunted, and 'automatized'. Because our everyday perception often is automatic, art and literature employ a variety of defamiliarizing techniques to prolong our perception, attract and hold our attention, and make us look at a familiar object or text with a higher level of awareness. For example, in poetry literary devices—such as word play, deliberately roughened rhythm, or figures of speech—defamiliarize or estrange ordinary speech and force us into a dramatic awareness of the language. Our habitual responses are refreshed, and familiar texts are rendered more perceptible.

The strategy of defamiliarization assumes that the tactic of surprise can serve to diminish distortions and help us to recognize our own preconceptions. However, we need to couple the strategy of 'making the familiar strange' with an awareness that although we might be able to bypass one set of conventions, we are unlikely to ever escape the framing of experience by convention. William A. Reid (1990) introduces his analysis of the history and development of the late twentieth-century institutional categories of schooling in the UK with a quotation from the University of Cambridge, *Statutes of Elizabeth I,* quoted from Costello's (1958, 15) *The Scholastic Curriculum at Early Seventeenth-century Cambridge*:

> From Ash-Wednesday, unto the said Thursday, all the Commencers... are to come to the Schools upon every Monday, Tuesday, Wednesday, Thursday & Friday, at one of the Clock in the afternoon, & to bring

> thither with them every one a Sophister… the said Commencers are there to be ready to define 2 or 3 theses, which they themselves shall make choice of, & deliver unto those Bachelors of Arts, not of the same College, who shall think fit to come thither to reply upon them.

Reid argues that quotations such as the above have more than curiosity value. Rather, they provide an opportunity for improving our understanding of curriculum through the removal of taken-for-granted assumptions:

> The lore of schooling and our familiarity with the world of the classroom can divert our attention from important questions we might be asking about the present functions of curricula and how new functions might be envisaged. One way to raise such questions is to turn away for a while from what is normal and look instead at things and places which strike us as strange. History offers us one medium for achieving this shift of vision. (Reid, 1990, 203)[5]

The tactic of estrangement in the context of environmental education can be exemplified by Le Guin's (1986) novel *Always Coming Home*, in which she tells stories of the Kesh, a people who 'might be going to have lived a long, long time from now in Northern California' and whose stories are written as translations of 'their voices speaking for themselves'.[6] One brief story, 'White Tree' (Le Guin 1986, 273–275) is told by the grandchild of a man named Fairweather. We learn that during Fairweather's adolescence 'he learned arboriculture with his mother's brother, a scholar of the Planting Lodge… and with orchard trees of all kinds'. Fairweather lived in a time and place when 'none of the Valley pears was very good, all were subject to cankers, and most needed irrigation to bear well'. He asked people in the north for help in obtaining different varieties and, by crossbreeding northern seedlings with a pear tree he found growing wild above the oak forests, 'he came upon a strong, small, and drought-hardy tree with excellent fruit… this is the brown pear grown in most orchards and gardens, and people call it the Fairweather pear'. There is much more to this deceptively simple story, which occupies less than two pages of a long novel, than we can examine here. But one of the story's chief delights is Le Guin's (1986, 274–25, her italics) postscript to it:

> TRANSLATOR'S NOTE:
>
> *…he learned arboriculture with his mother's brother… and with orchard trees of all kinds.*
>
> We would be more likely to say that he learned *from* his uncle *about* orchard trees; but this would not be a fair translation of the repeated suffix *oud*, with, together with. To learn *with* an uncle and trees implies that learning is not a transfer of something by someone to someone but is a relationship. Moreover, the relationship is considered to be reciprocal. Such a point of view seems at hopeless odds with the distinction of subject and object considered essential to science. Yet it appears that [Fairweather's] genetic experiments or manipulations were technically skilful, and that he was not ignorant of the theories involved, and it is certain that he achieved precisely what he set out to achieve. And the resulting strain of tree was given his name: a type case, in our vocabulary, of Man's control over Nature. This phrase, however, could not be translated into Kesh, which had no word meaning Nature except *she*, being; and anyhow the Kesh saw the Fairweather pear as the result of a collaboration between a man and some pear trees. The difference of attitude is interesting and the absence of capital letters perhaps not entirely trivial.

The difference of attitude is indeed interesting; moreover, it is the *difference* between the Kesh view of learning—which blurs the boundaries between subject and object—and more conventional understandings that gives the story a critical edge for educators. Fairweather's story, and Le Guin's 'translation' of it, questions the taken-for-grantedness of then-dominant late 20th century conceptions of curriculum and learning and it matters little whether the Kesh exist 'in fact' or that they are a fiction of Le Guin's imagination. The facts of the story's existence and of the reader's critical response to it are more than enough to generate questions for environmental educators. As Le Guin (1986, xi) writes in a preface to *Always Coming Home*:

> The difficulty of translation from a language that doesn't yet exist is considerable, but there's no need to exaggerate it. The past, after all, can be quite as obscure as the future. The ancient Chinese book called *Tao teh ching* has been translated into English dozens of times, and indeed the Chinese have to keep

retranslating it into Chinese at every cycle of Cathay, but no translation can give us the book that Lao Tze (who may not have existed) wrote. All we have is the *Tao teh ching* that is here, now. And so with translations from a literature of the (or a) future. The fact that it hasn't yet been written, the mere absence of a text to translate, doesn't make all that much difference. What was and what may be lie, like children whose faces we cannot see, in the arms of silence. All we ever have is here, now.

Le Guin thus reminds us that the value of a narrative excursion to other times has little to do with its status as historical or scientific 'fact' or speculative fiction. What matters is the wisdom and virtue that may be generated as we (her readers) respond critically and creatively to her story here and now, as it flashes up in the present within which our pasts and futures are enfolded. In another story within *Always Coming Home*, Le Guin (1985, 10–11) offers a useful analogy for the value of speculating about futures: a girl is on a journey from which she makes a short detour to visit her family in a nearby town. She recalls: 'I had been to Madidinou many times, of course, but this time the town looked altogether different, since I was on a journey beyond it'. The alternative futures imagined in the best science fiction have a similar effect: they make our present—and the moral choices and judgments that we perceive within it—look 'altogether different'. This applies as much to the stories we read and/or tell in and about our work as environmental educators as it does to any other aspect of our lives.

Rupture 3: SF as object-orientated philosophy

One of our major reservations about Barad's assertion[7] that 'language has been granted too much power', is that it could be interpreted as calling for a grand leap, or the creation of a 'new' totalising movement (recalling Benjamin's rejection of 'newness', quoted above) that occludes the 'use value' of language (and literature) in fostering *imagination*, and instead leads to a flat ontology that obfuscates the politics of difference, a position that Le Guin (2016a, 4) eloquently explains:

> The imagination is an essential tool of the mind ...
>
> We have to learn to use it, and how to use it ... Young human beings need exercises in imagination as they need exercise in all the basic skills of life ...
>
> When children are taught to hear and learn the central literature of their people, or, in literate cultures, to read and understand it, their imagination is getting a very large part of the exercise it needs.
>
> Nothing else does quite as much for most people, not even the other arts. We are a wordy species. Words are the wings both intellect and imagination fly on ... no art or skill is ever useless learning: but to train the mind to take off from immediate reality and return to it with new understanding and new strength, nothing quite equals poem and story.

The academic curricula of most education systems and institutions in modern, Western, industrial nations (and in systems and institutions modelled on them) tend not to provide 'exercises in imagination' despite the crucial roles that imagination—literally the ability to produce images in one's mind[8]—has played in the development of many disciplines. For example, thought experiments (even in the relatively narrow sense of conceptual experiments performed in circumstances which preclude physical testing procedures) have been particularly significant in the history of physics. However, science education textbooks and curricula rarely foreground the significance of thought experiments and, where they do, tend to diminish their imaginative dimensions (see, e.g. Gilbert and Reiner 2000; Gough 2010).

Notwithstanding Suvin's totalizing exclusion of other speculative genres from his definition of SF, many scholars and critics have embraced broader understandings of the term and we share Haraway's (1989, 5) understandings of what the term 'SF' designates:

> In the late 1960s science fiction anthologist and critic Judith Merril idiosyncratically began using the signifier SF to designate a complex emerging narrative field in which the boundaries between science fiction (conventionally, sf) and fantasy became highly permeable in confusing ways, commercially and linguistically. Her designation, SF, came to be widely adopted as critics, readers, writers, fans, and publishers struggled to comprehend an increasingly heterodox array of writing, reading, and marketing practices indicated by a proliferation of 'sf' phrases: speculative fiction, science fiction, science fantasy, speculative futures, speculative fabulation.

As noted above, Haraway (2016, 2) recently adds 'string figures, speculative feminism, science fact, so far' to her expansion of SF. Her inclusion of science fact as an SF term signals one of the deep continuities between SF and scientific texts, which Damien Broderick (1995, 155) describes as 'attention to the object in preference to the subject'.[9] Gough (1998, 411) argues that this 'object-orientation' of SF—a preoccupation with externalities and a corresponding de-emphasis of fine-grained characterization—marks SF as 'an environmental literature par excellence', because 'the narrative development of SF stories tends to privilege the effects of environments on the actions of characters', rather than the character-driven action of more conventional realist fiction. Much SF speculates on possible human responses to alternative conditions, such as an overpopulated or climate-changed Earth or the strange environs of another planet, and this alternative environment takes priority in the writer's imagination rather than the characters that demonstrate how the imagined conditions might affect humanity in general. In fact, when describing how she writes, Le Guin explains that she creates a world for her characters before she knows what they will do in it (Curry 2018).[10]

The object-orientation of SF clearly converges with a number of relatively recent object-orientated philosophical positions, including Levi Bryant's (2008) 'ontology of immanence', Timothy Morton's (2013) conceptualisation of 'hyperobects', Ian Bogost's (2012) speculations on 'what it's like to be a thing' and Graham Harman's (2018) 'object-orientated ontology' (often abbreviated as 'OOO') which, as Carol Taylor (2016, 202) observes, 'is either an offshoot of speculative realism, its overarching field, or its origin, depending on who you read'. However, Thomas Lemke (2017, 133) argues persuasively that:

> OOO's promise to break once and for all with subject–object dualism results in a revived form of subjectivism. The attempt to go 'back' to the object, to start all over again by exploring the inner depth of things, ends up in a superficial attitude that proclaims the return to the object while erasing any account of situatedness or contextuality.

Noting Iris van der Tuin's (2014, 231) emphasis on 'the androcentrism of much OOO' Lemke also finds the disregard of feminist materialism especially troubling, citing Taylor's (2016, 205) observation that not only are the 'founding fathers' of OOO—Harman, Bogost, Bryant and Morton—all male, but that they also engage with an exclusively male philosophical lineage with Kant, Husserl, and Heidegger as the most prominent figures. This disregard includes feminist theorists' focus on the politics of difference, particularly 'how differences come to matter' (Lemke 2017, 134), which requires an investigation into the situated entanglements of epistemology, ontology and ethics. These practices are predicated on the work of postmodern and poststructuralist thinkers such as Judith Butler, Derrida, Deleuze and Foucault, aimed at uncovering power structures and deconstructing dichotomies in an attempt to disrupt the historical tendency of other movements of thought to reify concepts by swinging, leaping or shifting from one side of the dichotomy to the other.

Drawing on Marilyn Strathern, Haraway (2016) repeatedly reminds us that it matters which concepts think concepts, and what worlds world worlds. Haraway (2014, n.p.) summarizes the importance of such SF practices by stating: 'It matters to destabilize worlds of thinking with other worlds of thinking'. This is because, as Jenny Reardon (2014, n.p.) eloquently explains, relations ground us in particular ways, 'they are not endless and they are not all at once … some relations do matter more than others'.[11] Or to put it another way, it matters which little machine is 'plugged into' each little machine. Le Guin (2016b, 186) reinforces Haraway's conjoining—or plugging into each other—of SF and 'science fact' in her appreciative introduction to a selection of H.G. Wells's short stories:

> Wells wrote science fiction long before it had a name. He called it 'scientific romance'... He didn't distinguish between science fiction and fantasy because nobody did then or for years to come; but he invented a literature because he was the first man to write fiction as a scientist. His imagination was formed and informed by the study of biology, a science in its bright dawn of discovery and expansion.

An object orientation, or what Higgins (2017) calls the 'thingness' of Arnold Bennet's prose, and which Bennett (2010, 2, her italics) calls 'thing-power... to highlight the active role of *non-human* materials in public life', also characterises Wells's stories. According to Le Guin (2016b, 190):

> [Wells's] true mastery... was in that very difficult, underestimated, even maligned element of storytelling, visual description. Wells can make you see what he wants you to see. When this is something that does not in fact exist... his power seems uncanny. He was—literally—a visionary. Perhaps the finest things he wrote are the wonderful lunar morning in *The First Men in the Moon* [1901] and the glimpses of the dying world at the end of *The Time Machine* [1895]... These visions have the authority, in memory, of something seen with one's own eyes.[12]

George Turner (1977, 259) argues that the integrated presentation of an alternative environment as 'a natural development of the narrative' is the 'distinctive technical problem' for authors of SF and 'is the one innovatory technique that science fiction has added to the normal repertoire of the fiction writer'. This problem was not always resolved satisfactorily by some of the genre's pioneers. For example, in *Twenty Thousand Leagues Under the Seas,*[13] Jules Verne (2013/1870) situates speculations about how humans might live for extended periods of time under the world's oceans in the episodic account of the voyage of the submarine *Nautilus,* a nineteenth century precursor of the starship *Enterprise* boldly going where no one had gone before. But the story is frequently—and often awkwardly—interrupted by Professor Aronnax's encyclopaedic catalogues of the sea creatures and other 'wonders' he encounters on the voyage, and by Captain Nemo's proto-environmentalist harangues against the irresponsible exploitation of marine resources. By way of contrast, in *The Island of Dr. Moreau* Wells (1967/1896), relies on allusion and inference, skillfully weaving evolutionary concepts into dialogue and action in ways that interrogate the hitherto uncontested boundary between animals and humans that Charles Darwin problematised. This 'object-orientation' is crucial to exploring how SF materialises our *ontological insecurities* and imaginative possibilities.

Both before and since Wells, many SF authors have interrogated the human/animal boundary. Early examples include Jonathan Swift's (1735) *Gulliver's Travels,* in Part IV of which Gulliver meets the Houyhnhnms, a race of talking horses; and Mary Shelley (1992/1818), whose depiction of the creation of monstrous life in *Frankenstein* both invented and critiqued one of the first great myths of modern industrial society, imagined the creature as being made from both human and non-human animal parts. Subsequent notable examples include Philip K. Dick's (1952) first published story, 'Beyond lies the Wub', and many of Le Guin's stories, such as 'The author of the Acacia seeds and other extracts from the *Journal of the Association of Therolinguistics*' (Le Guin 1984a), which satirises reductionist ethological constructions of animal behaviour, and 'She unnames them' (Le Guin 1987), which speculates that anthropocentric cultural practices of naming erode our sense of community with other organisms. Vint (2010, 444) outlines the contemporary significance of stories like these in her introduction to a symposium on 'Animal studies in the era of biopower':

> The terms of critical currency for the late twentieth and early twenty-first centuries—biopower, biopolitics, posthumanism—and their objects of enquiry—embodiment, subjectivity, the human/animal boundary—have a familiar resonance for readers of sf. Philosophers interrogate the implications of understanding what it means to be human only and always through a splitting off from what is labelled animal, while contemporary science muddles this distinction through research in xenotransplantation and genetic chimeras whose very existence forces us to think about species in terms of connection and continuum

> rather than those of division and differentiation ... This new political relevance of the body forces us to confront our continuity with other animals, and to rethink the nature of governance in a biopolitical era in which power acts upon bodies and forms subjects through this action.

Vint (2011, 161) further expands on the relationship between SF and biopolitics in her editorial introduction to a special issue of *Science Fiction Film and Television:*

> For Foucault, biopower has two interrelated objects of governance: the disciplined body of the individual subject and the managed citizenry, conceived on the aggregate level of the population. In twenty-first-century technoculture, both of these biopolitical objects are thoroughly colonised by subjects which once belonged entirely to the fictional realm. For example, bioethical debates over the status of emergent citizen/subjects, such as embryonic stem cells or 'brain dead' patients, challenge the ideas of what counts as life or death ... At the same time, epidemics and their attendant panics—such as 2005's spread of 'avian flu' and 2008's H1N1, with their images of burning pyres of animals and airports filled with people wearing surgical masks—conflate the management of borders, disease vectors and agriculture trade with speculative fantasies about invader species and zombie plagues. Under biopolitics, life itself becomes the object of political governance, and political governance becomes the practice of steering the biological life of individuals and species. Technoscience, sf speculation and biopolitical practice converge in this context.

In other words, SF and material worlds are now so entwined that they cannot be understood in isolation, as Sarah Franklin (2007) demonstrates in her account of the ways in which Dolly—the now (in)famous cloned sheep produced at Scotland's Roslin Institute in 1996—is situated within a broader genealogy that stretches from the past (where did she come from?) to the future (what does she point toward?). Franklin (2007, 3) emphasises 'the inseparability of the new biologies from the meaning systems they both reproduce and depend upon, such as beliefs about nature, reproduction, scientific progress, or categories such as gender, sex and species'.[14]

Le Guin is especially adept at weaving complex threads of ecology, ethics, language, and criticism into deceptively simple narratives aimed at destabilizing normative worldings. For example, her short story, 'Sur: a summary report of the *Yelcho* expedition to the Antarctic, 1909 – 1910' (Le Guin 1984b), is a seemingly straightforward recount by one member of a group of South American women who explore the Antarctic, and reach the South Pole, several years before Amundsen's and Scott's all-male expeditions. But, as Barr (1993) demonstrates in her reading of 'Sur' as an exemplary humanist and antihumanist text, the story also offers readers the opportunity to explore new articulations of the contradictory positions to be found within and between liberal (modernist) and postmodernist discourses of identity, gender, science, nature, and narrative. Through an accretion of small incidents, rather than the heroic gestures of androcentric adventure stories, Le Guin imagines women of colour attending to nature in very different ways from European men in such matters as their treatment of penguins, their approaches to driving sleds and making camps, and the naming of topographical features.

Turner's (1987) novel, *The Sea and Summer*, is a particularly prescient story of the Earth as it might yet be. Turner imagines the rise and eventual collapse of a 'greenhouse culture' in Australia during the twenty-first century. The effects of global warming on weather and food production are the background to a story of gritty social realism set in a drowning Melbourne. In one particularly insightful chapter, an old woman recalls 'the annual glories of the sea and summer' that she enjoyed as a child and describes the gradual changes in the seasons and neighbourhood gardens as the effects of rising carbon dioxide concentrations and sea levels become increasingly evident. The chapter builds toward a moving conclusion: 'the aging woman has what the child desired—the sea and eternal summer' (Turner 1987, 20), evoking a palpable sense of loss as childhood memories are soured retrospectively by the consequences of climate change.

Where do we go next?

Given that both of us (Adsit-Morris 2017; Gough 2016) are cited in the 'Call for Papers' as part of the rationale for this Special Issue, some readers might be curious as to why we could appear to have taken a sceptical stance. We are not opposed to the research directions inferred by the term 'new materialism' but, rather, vigilant. We have great respect for many of the scholars championed under this label, but we do not regard them as infallible.[15] Taking Le Guin's (2016c) insistence on the materiality of words and imagination to heart, we have explored the generative possibilities of SF for anticipating genuinely 'new' (as distinct from recycled and reconfigured) materialisms. Our premise is that environmental education research in times to come will require apprehension and comprehension of as-yet-unanticipated material-semiotic nodes. We need, as Haraway discussed in a recent public talk, imaginaries for the 'near-term' (i.e. within the next 100–200 years) when humanity will be in the midst of dealing with the effects of climate change and the struggle against capitalism, as opposed to the post-apocalyptic 'doom-and-gloom' dystopias—including catastrophism and accelerationism—prevalent in popular media recently, which do not include imaginaries for grappling with these crises,[16] or as Bellamy (2018, 417) writes, 'the climate crisis forces a cognitive transformation', and SF, 'ever a tool of imagining the (im)possible, has long offered us the chance to imagine retroactive futurities'. We need, as Le Guin (2016a, 4) reminds us, to 'train the mind to take off from immediate reality and return to it with new understanding and new strength' to create new imaginaries for strange and as-yet-unanticipated futures.

We see this approach fitting with an environmental education research agenda informed by conceptualisations of social inquiry as a performative materialist notion of a 'research-assemblage' (see e.g. Malins 2004; Fox and Alldred 2015) comprising researcher, empirical materials, methods and milieux, to which we add an anticipatory dimension. We embrace the multiplicity and variety of theoretical approaches put forward by academics currently labelled as 'new materialists', all of whom engage differently with matter and the material world, amplifying the multiplicity of its manifestations in concrete and imaginative ways—working to make the taken-for-grantedness of material strange again. Indeed, we think that the methods and modes of attention put forth by new materialists contribute significantly to the 'box of tools' available for environmental education researchers. Like Fox and Alldred (2015), our conception of assemblage derives from Deleuze and Guattari (1987, 4), whose work we understand as a collection of machinic concepts that can be plugged into other machines or concepts and made to work. Yet, we believe the task at hand requires not only creating 'new' concepts and assemblages, but just as importantly, it requires 'unblocking attachment sites' (Haraway 2018) to marginalized discourses and cultural resources of popular media and non-western knowings, which tend to be ignored or devalued in mainstream science and philosophy. We suspect that we might be criticised for overquoting SF authors such as Le Guin, to which our response parallels Deleuze and Guattari's: when we write, the question is which other machines the SF machine can be plugged into in order to work. We offer SF stories as 'little machines' than can be plugged into an environmental education research-assemblage to produce new material-semiotic conjunctions and configurations of researchers, data, methods and contexts.

Notes

1. Deleuze refers to a concept as a 'box of tools' in a conversation with Michel Foucault (1997, 207).
2. We prefer to use Norman Denzin's (2013, 355) term 'empirical materials' as an alternative to 'data' although we acknowledge than many of the authors, we cite approvingly use the latter term. Nevertheless, we agree with Denzin who eschews the term data because it brings into play a positivist ontology and an epistemology based on a politics of evidence.
3. Peta Malins (2004, 84) offers a convincing demonstration of the generativity of this conception of machinic assemblage by substituting the concept 'book' with 'drug using body' to suit the purposes of her explorations of the 'ethico-aesthetics of drug use'.
4. Manifest Destiny held that the US was destined—by God, its advocates believed—to expand its dominion and spread democracy and capitalism across the entire North American continent. 'Manifest Destiny', a phrase coined in 1845 by journalist John O'Sullivan, expressed the philosophy that drove 19th-century US territorial expansion (see, for example, The Ohio State University History Institute's lesson plan: https://hti.osu.edu/history-lesson-plans/united-states-history/manifest-destiny-westward-expansion).

5. Reid (1990, 215) adds that other possibilities are comparative studies or futures studies.
6. In fact, the stories and characters in *Always Coming Home* were strongly influenced by Le Guin's father's work as an anthropologist studying indigenous cultures. As director of the University of California Museum of Anthropology, Alfred L. Kroeber was famed for his work with Ishi, an indigenous man known as the 'last wild Indian' (Curry 2018).
7. It should be noted that although Barad is often called a new materialist—e.g. Carol Taylor (2016, 201) refers to Barad's *Meeting the Universe Halfway* (2007), as 'already a "foundational" text for new material feminism'—she is first and foremost a feminist science studies scholar and philosopher. The breadth of academics brought together under the term 'new materialism'—many of whom draw from different theoretical fields—can limit the interpretive potentiality of their work and lead to substantial misinterpretations. As noted in the text, we embrace the multiplicity and variety of theoretical approaches put forth by academics currently labelled as 'new materialists', all of whom engage differently with matter and the material world. We believe that such tensions and contradictions are generative. Yet we caution that the nuances of their theories have often been selectively overlooked or collapsed into each other, leading to misreadings and misunderstandings of particular concepts.
8. One of the anonymous reviewers of this article objected that this 'reads as a reductive definition of imagination' but we stand by it because it is a particularly important characteristic of *scientific* thought experiments, most of which involve visualization of phenomena (Maxwell's daemon, Einstein's train and elevator, Schrodinger's cat, etc.).
9. Haraway (1985, 66) has consistently engaged the material from her earliest work, such as 'A manifesto for cyborgs' in which she states: 'The cyborg is a condensed image of both imagination and material reality', and 'the boundary between science fiction and social reality is an optical illusion'.
10. Which can also be seen as a pointed critique of the bias of using predetermined hypotheses in scientific research.
11. Haraway (2016, 31) explains this another way by stating that: 'Nothing is connected to everything; everything is connected to something.'
12. As a further example of Wells' 'visions' Le Guin (2016, 190) adds: 'A squadron of airplanes over Naples (two years before Kitty Hawk!)', an apparent reference to 'A Dream of Armageddon', a short story by Wells first published in 1901 in the British weekly magazine *Black and White*.
13. Earlier English translations of Verne's (1870) *Vingt mille lieues sous les mers*, inaccurately translated *mers* as the singular 'sea' rather than the plural 'seas'.
14. For further elaboration of biopolitics in SF literature and popular media see Gough (2017).
15. Much of Barad's (2007) *Meeting the Universe Halfway* reflects her training as a high-energy theoretical physicist, and she predicates many of her arguments for 'how and why we must understand in an integral way the roles of human and nonhuman, material and discursive, and natural and cultural factors in scientific and other practices' (Barad 1997, 25) on her interpretation of Neils Bohr's and Werner Heisenberg's quantum mechanics. But, as Silvan Schweber (2008, 881) points out, Barad's reading of quantum physics is selective: she almost completely overlooks the contributions of Wolfgang Pauli (he is mentioned in a footnote) and his extensive correspondence with Bohr and Heisenberg, as well as more recent scholarship on the subject. Schweber (who is also an eminent professor of physics and the history of ideas) disagrees with Barad's assessment of the failure of representational approaches in fundamental physics. He writes: 'advances at the nuclear and subnuclear levels were due to the possibility of a confluence between ontology and representation'. We were also astonished by Barad's (2007, 85) assertion that quantum mechanics is 'the correct theory of nature that applies to all scales'. This ignores Sandra Harding's (1986, 193) call for feminist science scholars to resist 'the longing for "one true story" that has been the psychic motor for [modern] Western science'. Also, as Schweber (2008, 881) points out, although quantum mechanics applies to a wide range of length scales, it is not a 'final theory' or a 'theory of everything', because, for example, it cannot (to date) incorporate gravitational phenomena.
16. On October 13th 2017, Haraway participated in a public dialogue with the SF writer Starhawk at the University of California, Santa Cruz. Sponsored by the Science and Justice Research Center, the talk was titled: 'A Public Conversation with Donna Haraway and Starhawk: Magic, Figuration & Speculative Fiction as Calls to Action'.

References

Adsit-Morris, Chessa. 2017. *Restorying Environmental Education: Figurations, Fictions, Feral Subjectivities*. New York, NY: Palgrave.

Alaimo, Stacy, and Hekman Susan. 2008. "Introduction: Emerging Models of Materiality in Feminist Theory." In *Material Feminisms*, edited by Stacy Alaimo and Susan Hekman, 1–19. Bloomington, IN: Indiana University Press.

Atkinson, Paul, Sara Delamont, and Martin Hammersley. 1988. "Qualitative Research Traditions: A British Response to Jacob." *Review of Educational Research* 58(2): 231–250. doi:10.3102/00346543058002231.

Bacigalupi, Paulo. 2015. *The Water Knife*. London, UK: Little Brown/Orbit Books.

Barad, Karen. 2007. *Meeting the Universe Halfway: Quantum Physics and the Entanglement of Matter and Meaning*. Durham, NC & London, UK: Duke University Press.

Barad, Karen. 2014. "Diffracting Diffraction: Cutting Together-apart." *Parallax* 20(3): 168–187. doi:10.1080/13534645.2014.927623.

Barr, Marlene S. 1993. "Ursula Le Guin's 'Sur' as Exemplary Humanist and Antihumanist Text." In *Lost in Space: Probing Feminist Science Fiction and Beyond*, 154–170. Chapel Hill, NC: The University of North Carolina Press.

Bellamy, Brent Ryan. 2018. "Introduction to Special Issue: Science Fiction and the Climate Crisis." *Science Fiction Studies* 45(3): 417–419. doi:10.5621/sciefictstud.45.3.0417.

Benjamin, Walter. 1968. *Illuminations*. Translated by Harry Zohn. New York, NY: Harcourt, Brace & World.

Benjamin, Walter. 2002/1982. "The Arcades Project." Harvard University Press, accessed 9 November 2017. https://archive.org/stream/BenjaminWalterTheArcadesProject/Benjamin_Walter_The_Arcades_Project_djvu.txt.

Bennett, Arnold. 1902. *Anna of the Five Towns*. London, UK: Chatto & Windus.

Bennett, Jane. 2010. *Vibrant Matter: A Political Ecology of Things*. Durham, NC: Duke University Press.

Bogost, Ian. 2012. *Alien Phenomenology, or What It's like to Be a Thing*. Minneapolis, MN: University of Minnesota Press.

Broderick, Damien. 1995. *Reading by Starlight: Postmodern Science Fiction*. London and New York: Routledge.

Brookes, Andrew. 2007. "Preventing Death and Serious Injury from Falling Trees and Branches." Journal of Outdoor and Environmental Education 11(2): 50–59. doi:10.1007/BF03400857.

Brookes, Andrew. 2018. "Single Fatalities on School or Youth Group Camps and Excursions: Lessons for Prevention from Australia and the UK." In Preventing fatal incidents in school and youth group camps and excursions: Understanding the unthinkable, 127–198. Cham, Switzerland: Springer.

Bryant, Levi. 2008. *Difference and Givenness: Deleuze's Transcendental Empiricism and the Ontology of Immanence*. Evanston, IL: Northwestern University Press.

Burton, James Edward. 2015. *The Philosophy of Science Fiction: Henri Bergson and the Fabulations of Philip K. Dick*. London, UK: Bloomsbury Publishing.

Canavan, Gerry. 2016. "The Suvin event." In *Darko Suvin Metamorphoses of Science Fiction: on the Poetics and History of a Literary Genre*, xi–xxxvi. Bern, Switzerland: Peter Lang.

Caputo, John. 1987. Radical Hermeneutics: Repetition, Deconstruction, and the Hermeneutic Project. Bloomington, IN: Indiana University Press.

Chandler, Daniel. 2007. *Semiotics: The Basics*. 2nd ed. Abingdon, UK: Routledge.

Costello, William T. 1958. The Scholastic Curriculum at Early Seventeenth- century Cambridge. Cambridge, MA: Harvard University Press.

Crownshaw, Richard. 2017. "Speculative Memory, the Planetary and Genre Fiction." *Textual Practice* 31(5): 887–910. doi:10.1080/0950236X.2017.1323484.

Curry, Arwen. 2018. *Worlds of Ursula K. Le Guin*. Grasshopper Film.

Deleuze, Gilles. 1994. *Difference and Repetition*. Translated by Paul Patton. New York, NY: Columbia University Press.

Deleuze, Gilles, and Félix Guattari. 1987. *A Thousand Plateaus: Capitalism and Schizophrenia*. Translated by Brian Massumi. Minneapolis, MN: University of Minnesota Press.

Denzin, Norman K. 2013. "The Death of Data?" *Cultural Studies ↔ Critical Methodologies* 13(4): 353–356. doi:10.1177/1532708613487882.

Dick, Philip K. 1952. "Beyond lies the Wub." *Planet Stories*. New York, NY: Fiction House.

Dolphijn, Rick, and Iris van der Tuin. 2012. "The Transversality of New Materialism." In *New materialisms: Interviews and cartographies*. n.p. Ann Arbor, MI: Open Humanities Press. https://quod.lib.umich.edu/o/ohp/11515701.0001.001/1:5.2/–new-materialism-interviews-cartographies?rgn=div2;view=fulltext

Foucault, Michel. 1997. "Intellectuals and Power: A Conversation between Michel Foucault and Gilles Deleuze." In Language, Counter-Memory, Practice: Selected Essays and Interviews, edited by Donald F. Bouchard, 205–217. New York, NY: Cornell University Press.

Fox, Nick J., and Pam Alldred. 2015. "Inside the Research-Assemblage: New Materialism and the Micropolitics of Social Inquiry." *Sociological Research Online* 20(2): 1. doi:10.5153/sro.3578.

Franklin, Sarah. 2007. Dolly Mixtures: The Remaking of Genealogy. Durham, NC: Duke University Press.

Ghosh, Amitav. 1997. The Calcutta Chromosome: A Novel of Fevers, Delirium and Discovery. New York, NY: Avon Books.

Ghosh, Amitav. 2016. The Great Derangement: Climate Change and the Unthinkable. Chicago, IL: University of Chicago Press.

Gilbert, John, and Miriam Reiner. 2000. "Thought Experiments in Science Education: Potential and Current Realization." *International Journal of Science Education* 22(3): 265–283. doi:10.1080/095006900289877.

Glotfelty, Cheryll. 1996. "Introduction: Literary Studies in an Age of Environmental Crisis." In *The ecocriticism reader: Landmarks in literary ecology*, edited by Cheryll Glotfelty and Harold Fromm, xv–xxxvii. Athens, GA and London, UK: The University of Georgia Press.

Gough, Noel. 1993a. "Environmental Education, Narrative Complexity and Postmodern Science/Fiction." *International Journal of Science Education* 15(5): 607–625. doi:10.1080/0950069930150512.

Gough, Noel. 1993b. "Neuromancing the Stones: Experience, Intertextuality, and Cyberpunk Science Fiction." *Journal of Experiential Education* 16(3): 9–17. doi:10.1177/105382599301600303.

Gough, Noel. 1997. "Weather™ Incorporated: Environmental Education, Postmodern Identities, and Technocultural Constructions of Nature." *Canadian Journal of Environmental Education* 2: 145–162.

Gough, Noel. 1998. "Playing with Wor(l)ds: Science Fiction as Environmental Literature." In *Literature of nature: An international sourcebook*, edited by Patrick D. Murphy, 409–414. Chicago, IL and London, UK: Fitzroy Dearborn.

Gough, Noel. 2006. "Sustainable Development in Learning, Leadership and the Law: A Review Essay." *Australian Journal of Environmental Education* 22(1): 115–121. doi:10.1017/S0814062600001749.

Gough, Noel. 2010. "Performing Imaginative Inquiry: Narrative Experiments and Rhizosemiotic Play." In *Imagination in Educational Theory and Practice: a Many-sided Vision*, edited by Thomas William Nielsen, Rob Fitzgerald and Mark Fettes, 42–60. Newcastle upon Tyne: Cambridge Scholars Publishing.

Gough, Noel. 2015. "Undoing Anthropocentrism in Educational Inquiry: A Phildickian Space Odyssey?" In *Posthumanism and Educational Research*, edited by Nathan J. Snaza and John A. Weaver, 151–166. New York, NY: Routledge.

Gough, Noel. 2016. "Postparadigmatic Materialisms: A 'New Movement of Thought' for Outdoor Environmental Education Research?" *Journal of Outdoor and Environmental Education* 19(2): 51–65. doi:10.1007/BF03400994.

Gough, Noel. 2017. "Specifying a Curriculum for Biopolitical Critical Literacy in Science Teacher Education: Exploring Roles for Science Fiction." *Cultural Studies of Science Education* 12(4): 769–794. doi:10.1007/s11422-017-9834-0.

Gough, Noel. 2019. "Playing within/against Entombed Scholarship: Episodes in an Academic Life." In *Critical voices in science education research: Narratives of hope and struggle*, edited by Jesse Bazzul and Christina Siry, 201–212. New York, NY: Springer.

Haraway, Donna J. 1985. "A Manifesto for Cyborgs: Science, Technology and Socialist Feminism in the 1980s." Australian Feminist Studies 15(2): 65–107. doi:10.1080/08164649.1987.9961538.

Haraway, Donna J. 1989. *Primate Visions: Gender, Race, and Nature in the World of Modern Science*. New York, NY: Routledge.

Haraway, Donna J. 2007. *When Species Meet*. Minneapolis, MN: University of Minnesota Press.

Haraway, Donna J. 2014. "SF: String figures, multispecies muddles, staying with the trouble." [Video file]. Accessed 27 June 2014. https://www.youtube.com/watch?v=Z1uTVnhIHS8.

Haraway, Donna J. 2016. *Staying with the Trouble: Making Kin in the Chthulucene*. Durham, NC: Duke University Press.

Haraway, Donna J. 2018. "Unblocking Attachment Sites." *Rubber Boots Methods for the Anthropocene: A Node-Based Conference*, Aarhus University 26–27 November.

Harman, Graham. 2018. *Object-Oriented Ontology: A New Theory of Everything*. London, UK: Penguin.

Higgins, Charlotte. 2017. "Arnold Bennett: Poet of the Potteries." *The Guardian*, 20 May 2017, Review. https://www.theguardian.com/books/2017/may/20/arnold-bennett-poet-potteries-museum-exhibition.

Howarth, William. 1996. "Some Principles of Ecocriticism." In *The Ecocriticism Reader: Landmarks in Literary Ecology*, edited by Cheryll Glotfelty and Harold Fromm, 69–91. Athens, GA and London, UK: The University of Georgia Press.

Jamison, Andrew. 1993. "National Political Cultures and the Exchange of Knowledge: The Case of Systems Ecology." In *Denationalizing Science: the Contexts of International Scientific Practice*, edited by Elizabeth Crawford, Terry Shinn and Sverker Sörlin, 187–208. Dordrecht, Netherlands: Kluwer.

Kuhn, Thomas S. 1996. *The Structure of Scientific Revolutions*. 3rd ed. Chicago, IL: University of Chicago Press.
Latour, Bruno. 1993. *We Have Never Been Modern*. Translated by Catherine Porter. Cambridge, MA: Harvard University Press.
Le Guin, Ursula K. 1984a. "The Author of the Acacia Seeds and Other Extracts from the Journal of the Association of Therolinguistics." In *The Compass Rose*, 11–19. London, UK: Grafton Books.
Le Guin, Ursula K. 1984b. "The Eye Altering." In *The Compass Rose*, 166–180. London, UK: Grafton Books.
Le Guin, Ursula K. 1986. *Always Coming Home*. London, UK: Victor Gollancz.
Le Guin, Ursula K. 1987. "She Unnames Them." In *Buffalo Gals and Other Animal Presences*., 194–196. New York, NY: New American Library.
Le Guin, Ursula K. 2016a. "The Operating Instructions." In *Words Are My Matter: Writings about Life and Books, 2000-2016 with a Journal of a Writer's Week*, 3–6. Northampton, MA: Small Beer Press.
Le Guin, Ursula K. 2016b. "Wells's Worlds." In *Words Are My Matter: Writings about Life and Books, 2000-2016 with a Journal of a Writer's Week*, 184–190. Northampton, MA: Small Beer Press.
Le Guin, Ursula. 2016c. *Words Are My Matter: Writings about Life and Books, 2000-2016 with a Journal of a Writer's Week*. Northampton, MA: Small Beer Press.
Lemke, Thomas. 2017. "Materialism without Matter: The Recurrence of Subjectivism in Object-Oriented Ontology." *Distinktion: Journal of Social Theory* 18(2): 133–152. doi:10.1080/1600910X.2017.1373686.
Malins, Peta. 2004. "Machinic Assemblages: Deleuze, Guattari and an Ehico-Aesthetics of Drug Use." *Janus Head* 7(1): 84–104.
Morton, Timothy. 2013. *Hyperobjects: Philosophy and Ecology after the End of the World*. Minneapolis, MN: University of Minnesota Press.
Murphy, Patrick D., ed. *Literature of Nature: an International Sourcebook*. Chicago, IL and London, UK: Fitzroy Dearborn. 1998
Odum, Eugene P. 1953. *Fundamentals of Ecology*. Philadelphia, PA: W.B. Saunders.
Reardon, Jennifer E. 2014. "Inhabiting Multispecies Bodies: Panel Discussion with Donna Haraway, Margaret McFall-Ngai, and Jenny Reardon." Presented at the Anthropocene: Arts of Living on a Damaged Planet conference. Filmed May 9, 2014.
Reid, William A. 1990. "Strange curricula: origins and development of the institutional categories of schooling.' Journal Of Curriculum Studies, 22:3, 203–206, DOI:10.1080/0022027900220301
Rousell, David, Amy Cutter-Mackenzie, and Jasmyne Foster. 2017. "Children of an Earth to Come: Speculative Fiction, Geophilosophy and Climate Change Education Research." *Educational Studies* 53(6): 654–669. doi:10.1080/00131946.2017.1369086.
Roy, Kaustuv. 2003. *Teachers in Nomadic Spaces: Deleuze and Curriculum*. New York, NY: Peter Lang.
Schweber, Silvan.S. 2008. "Review of Karen Barad (2007) *Meeting the Universe Halfway: Quantum Physics and the Entanglement of Matter and Meaning*." ISIS 99(4): 879–882. doi:10.1086/597741.
Shelley, Mary. 1992/1818. Frankenstein; or, The Modern Prometheus. London: Penguin.
Shklovsky, Victor. 1917/1965. "Art as Technique (1917)." In *Russian Formalist Criticism: Four Essays*, edited by Lee T. Lemon and Marion J. Reis, 3–24. Lincoln, NE: University of Nebraska Press.
Stables, Andrew. 1996. "Reading the Environment as Text: Literary Theory and Environmental Education." *Environmental Education Research* 2(2): 189–195. doi:10.1080/1350462960020205.
Stables, Andrew. 1997. "The Landscape and the 'Death of the Author." *Canadian Journal of Environmental Education* 2: 104–113.
Stables, Andrew. 2001. "Language and Meaning in Environmental Education: An Overview." *Environmental Education Research* 7(2): 121–128. doi:10.1080/13504620123688.
Suvin, Darko. 1972. "On the Poetics of the Science Fiction Genre." *College English* 34(3): 372–382. doi:10.2307/375141.
Suvin, Darko. 1979. *Metamorphoses of Science Fiction: on the Poetics and History of a Literary Genre*. 1st ed. New Haven, CT: Yale University Press.
Swift, Jonathan. 1735. *Gulliver's Travels, or Travels into Several Remote Nations of the World*. London, UK: George Faulkner. Reprint, Online at Project Gutenberg (1997).
Taylor, Carol A. 2016. "Close Encounters of a Critical Kind: A Diffractive Musing in/between New Material Feminism and Object-Oriented Ontology." *Cultural Studies* 16(2): 201–212. doi:10.1177/1532708616636145.
Todd, Zoe. 2016. "An Indigenous Feminist's Take on the Ontological Turn: 'Ontology' Is Just Another Word for Colonialism." *Journal of Historical Sociology* 29(1): 4–22. doi:10.1111/johs.12124.
Tuck, Eve, and Marcia McKenzie. 2014. *Place in Research: Theory, Methodology, and Methods*. London, UK: Routledge.
Turner, George. 1977. "Science Fiction as Literature." In *The Visual Encyclopedia of Science Fiction*, edited by Brian Ash, 257–262. London, UK and Sydney, Australia: Pan Books.
Turner, George. 1987. *The Sea and Summer*. London, UK: Faber and Faber (US title: *The Drowning Towers*).
van der Tuin, Iris. 2014. "Diffraction as a Methodology for Feminist onto-Epistemology: On Encountering Chantal Chawaf and Posthuman Interpellation." *Parallax* 20(3): 231–244. doi:10.1080/13534645.2014.927631.

Verne, Jules. 2013/1870. *Twenty Thousand Leagues under the Seas*. Translated by William Butcher. Revised ed. London, UK: The Folio Society and Oxford University Press.

Vint, Sherryl. 2010. "Animal Studies in the Era of Biopower." *Science Fiction Studies* 37(3): 444–455.

Vint, Sherryl. 2011. "Introduction: Science Fiction and Biopolitics." *Science Fiction Film and Television* 4(2): 161–172. doi:10.3828/sfftv.2011.11.

Wells, H. G. 1967/1896. The Island of Dr. Moreau. Harmondsworth: Penguin.

Nature matters: diffracting a keystone concept of environmental education research – just for kicks

Jamie Mcphie and David A. G. Clarke

ABSTRACT

As a keystone species the concept 'nature' plays a vital role in shaping our world. In this article, we think with the material turn about the *concept* nature due to its significant performativity in its role within environmental education and research. How nature is conceived is played out on a massive scale as matter itself is morphed through conceptual processes. Therefore, we focus on the matter(ing) of conceptual abstraction, the physical effects – and affects – of thinking a thing into existence. We initiate a pluralistic thought experiment that purposefully diffracts nature into eight performances, to see what it *does*. The concept nature performs ecologically and enacts trophic cascades. This exploration highlights feats of racism, classism, androcentrism, colonialism, homogenization, and mass extinction. What we are proposing is an environmental literacy that attends to what a concept is capable of, what a concept can do, and perhaps even what a concept can prevent, post-nature.

Re-connections

'*A concept is a brick. It can be used to build a courthouse of reason. Or it can be thrown through the window.*' (Massumi in Deleuze and Guattari 2004, xiii)

> In his 61st year on this earth, the man who calls himself Nimblewill Nomad left home and walked a very long way through the mountains [...] Then, he took another, even longer walk. And then another one. And then another. Soon, he had given away almost all of his money and taken to walking almost year-round, roaming the post-industrial wilderness of North America in what he called "a desperate search for peace." (Moore 2017, 1)

Robert Moor, the author of *On Trails: An Exploration*, joined Nimblewill Nomad for a few days of walking, garnering a few glimpses of a post-nature perspective.

> The next night, we slept in a copse of gnarled oaks beside a graveyard, a shady grove carpeted with slender, rippling leaves. It was strangely lovely. Eberhart found them everywhere, these forgotten little shards of wilderness. The problem, he said, was that hikers tended to divide their lives into compartments: wilderness over here, civilization over there. "The walls that exist between each of these compartments are not there naturally," he said. "We create them. The guy that has to stand there and look at Mount Olympus to find peace and quiet and solitude and meaning – life has escaped him totally!"' (Moore 2017, 27)

Initially we might agree with Nimblewill's Whiteheadian insight about the bifurcation of nature – 'wilderness over here, civilization over there'. But look again and we see another slip into the Cartesian trap. It's easy to do. The walls (in this case the wall between nature and culture) *are* there naturally because 'we' invented the concepts of bifurcations and dualisms and we are *of* this world. Human concepts are just as natural as a tree or a crisp packet. Conceptual abstractions are real, material, physical and they perform ecologically.

This article is a playful attempt to unpack the multiplicity of the conceptual abstraction - *nature*. For we must have an idea of what nature 'is' if environmentalists, for example, are to ask people to 're-connect' to 'it'. Though, as will become clear, we are more interested in the *becomings* or *doings* of various natures, rather than their 'beings' – designated by definitions. It is this political aspect of concepts – their acting in the world – that we wish to explore in environmental education research, rather than offering further historical or genealogical explorations of *nature*.

The concept of nature is very much alive – and mattering[1] – in environmental education research. A flick through this journal reveals authors who suggest that there is a 'nature' which is somehow apart from human produce. We don't want to be accused of constructing a straw *man*; we know there are critical and inventive ways in which 'nature' – as well as 'landscape', 'wilderness' and 'place', for example – is being conceived and played within environmental education research and we will discuss some of these presently. However, 'nature' is also often described in simply ecological, biological, geological, or atmospheric terms. Additionally, sometimes the term is not considered past its own use. For example, Braun and Dierkes (2017, 938) admit that straightforward usage is controversial as 'there is no clear, universally accepted definition for the notion of nature connectedness', yet their following literature review does not tackle this controversy. Instead they cite others who have also used 'nature' as axiomatic. Similarly, Typhina (2017) takes nature to be relatively unproblematic, presenting it matter-of-factly, as a thing to be experienced. To pick two instances is perhaps unfair as there are many examples in this and other journals that use the term 'nature' as a destination to be visited, experienced, and connected to. Measuring humans' connectedness or relatedness to a romanticized version of nature is becoming commonplace. For example, there are now 'scales' to objectively measure just how related to nature you are, such as the 'Connectedness to Nature Scale' (CNS) (Mayer and Frantz 2004), the 'Nature Relatedness Scale' and the 'Connectivity to Nature Scale' (Selhub and Logan 2012, 228), and conferences to disseminate the almost always positive results of these scales, such as 'Nature Connections ... an interdisciplinary conference to examine routes to nature connectedness' (Nature Connections 2015). Malone (2016), suggests that the emergence of a 'children in nature' movement, both in academic and educational circles and buoyed by the work of authors such as Richard Louv, relies on an 'adult sentimentality' of urban children's loss of connection to 'nature' (also see Duhn, Malone, and Tesar 2017, 1363). Malone (2016) draws on the work of Rautio (2013) to demonstrate a number of logical problems and anthropocentric assumptions in the view that children need to be more connected to nature: '(1) human societies used to be closer to nature, (2) our current way of life is unnatural or distant from nature, and (3) proximity to nature is a question of learning (and teaching)' (Rautio 2013, 449). We would reassert another fundamental challenge, perhaps read as implicit in Rautio's points by some readers: nature is a concept, not a place or thing in the sense of being an object.

The concept of a distinct nature is also at work when authors discuss ecopedagogy, ecopsychology, ecotherapy, ecojustice, and ecological literacy, for example. Terms that have an 'eco' designator suggest to the reader that what is being discussed is concerned with preserving/conserving/managing/saving/being healed by/learning about/acting for what seems to us to be an idea that *matters* - nature. While we know that 'eco' comes from *oîkos* – Greek for home – and that 'ecology' seems to refer to relationships more than it does to 'nature', there is still a sense in which 'eco' concerns itself with the *other* in the form of the biological sphere drawing upon the hydrosphere, atmosphere and lithosphere – for some reason – however, the anthroposphere still seems largely detached. Ironically 'home' refers to the green/blue planet, rather than the 'artificial' world humans have built and inhabit. What of the ecology that sits between a fridge and its contents? Or the ecology of motorway roadsides with their fusion of cars, carrion and crows? The ecology of the electric grid, coal extraction, nuclear waste storage, and UK/China post-Brexit relations? The ecology of stories? The ecology of percepts, affects and *concepts*? How do we educate for/with these environments? Should we? What does an environmental education without a 'green' version of nature look like? For, what does it *do* to do environmental education research, as we think we must, without the romanticized story of nature? And what does it do not to do an ecology of concepts?

To this last point, we would say that the impact of omitting the mattering of concepts in the cogitations of environmental education research – and ecocriticism more broadly – leaves open the performance of the neglected cogitations themselves (i.e. what do they do? See Affifi, this issue, for a similar consideration of what non/anthropocentrisms do). Concepts perform, whether we pay attention to them or not. The performativity and materiality of concepts need explorers to challenge (un)knowable realms of vital materiality. Conceptual abstractions are not merely philosophical tools for engaging in cerebral flights of fancy, they are made of matter and matter is indeed a conceptual abstraction. 'The meaning of a philosophical concept cannot be reduced to its semantic content, defined in abstraction from this process. There is a transformational aspect to the concept's letting loose, by which it effectively overspills its own definition' (Massumi 2010, 4). Therefore, concepts behave ecologically and produce trophic[2] cascades – although trophic *torrents* might be more apt due to the hierarchical linear cause-and-effect sequencing that 'cascade' implies. '[A] philosophical concept carries a surplus of meaning that is one with the transformative movement of its performative *force*' (Massumi 2010, 4). It is to this *force* of matter, in the guise of concepts, that we must attend as an ethical imperative.

In this article, we think with the material turn[3], unpacking and introducing various romanticized environmental perceptions and queer(y)ing 'critical' positions in environmental education. There is, of course, not *one* critical position – as if critique were something that occurred from an objective stance outside of becoming involved with the problems at hand – what Haraway (1988) calls the 'god trick.' Each one of the natures we explore *is* a critical position, in relation to another position. For instance, the critical position taken up by Arne Naess, sometimes labelled as deep ecology, is deemed critical due to its criticism of shallow ecology for its evident anthropocentrism. However, a number of scholars (Morton, Braidotti, de Vega, Cohen, for example) critique this critique, suggesting it is similarly anthropocentric due to its inherent Cartesian contradictions. These stances themselves can be critiqued from positions that question whether they are ontologically sound – this is not quite the type of critique we are hoping to perform here. There is another manner in which 'critical' is used in educational theory/research; toward uncovering the imbalance of power relations and injustices – again, while there are similarities in our approach, this is not solely the type of critique we are hoping to enact here. What we *are* moving toward is the understanding that we can't step outside of politics (Latour 2004). Our critical position is what Maggie MacLure (2015) calls 'the new materialist critique of critique,' that is the understanding that rationalistic critique 'arrests things – stitches them up, pins them down or closes them down, in the rage to expose error and the rush to pronounce judgement' (101). Rather, new materialist approaches aim for creation and production. MacLure (2015) suggests an 'immanent form of critique' which is

> able to follow, or sense, the multifarious connections and intensities that coalesce in events, rather than sniping from its particular dugout at other disciplines and paradigms. It must be oriented towards eventualities that cannot be foreseen, and where the usual privileges of human agency and the linearity of cause and effect are not in play. (105)

So, while the natures we offer do snipe at each other to a certain extent – and while they do reveal unjust power relations to a certain extent – they also create new, immanent[4] lines of nature, describing versions of nature beyond duality or privileged human agency. Following this immanent mode of critique further, we come to oppose the transcendently[5] *placed*, static environmental concepts – wilderness, environment, landscape, nature – with immanent versions – environ*ing*, landscap*ing*, natur*ing*. For now, it is the juxtaposition of different conceptions of nature which provides the creative lines that co-mingle with the reader (and with the writers) to produce the diffractive patterns. Our hope is that the reader considers their use of 'nature' in the practice of both research and pedagogy, so as to ponder the *effects* of their use of it. Donna Haraway (1997) infers that 'diffraction can be a metaphor for another kind of critical consciousness, [...] one committed to making a difference and not to repeating the Sacred Image of Same' (273). Haraway (1997, 2000) and Barad (2007) point out that thinking with a diffraction pattern is not an attempt to map the appearance of differences but rather the *effects* of where differences appear. And so, by paying attention to nature here, we imagine the diffractive effects of eight versions of nature as spilling over from the page. This is, with the hope that a difference might be made for the reader, in the process that occurs of the reader meeting different conceptions of nature at play in environmental education research and beyond: Nature 1 – scary nature; Nature 2 – scenic nature; Nature 3 – utopian nature; Nature 4 – scarier nature; Nature 5 – artificial nature; Nature 6 – affective nature; Nature 7 – conceptual nature; and Nature 8 – abstract nature. It is this hope – that a difference might be made by exploring difference – that is the political drive of this article. That is, the manner in which ontology, knowledge, and ethics are always already wrapped up with each other – what Barad (2007) calls 'ethico-onto-epistemology.' For us – in the writing of this article – the matter of concern is the political mattering of the ontologically suffused concept *nature*.

Before we meet these eight versions of nature we wish to attend to its appropriation. And as matter is storied, as all things are, we will attempt to weave a narrative similar to that of the ancient Greeks – or Star Wars if you prefer – that always begins *in medias res*[6] ...

Episode IV: the appropriation of nature

A long time ago in a galaxy far, far away....[7]

> The world of nature, it is often said, is what lies 'out there'. [...] Application of this logic forces an insistent dualism, between object and subject, the material and the ideal, operational and cognized, 'etic' and 'emic.' (Ingold 2000, 191)

Nature has evolved conceptually within the realm of science, through the *philosophical* work of Boyle, Bacon, Harvey, Descartes, Hooke and others, who have variously extended the concept of nature from a Greek and medieval organismic agent, to a concept of nature as mechanistic or as a set of laws guiding the operation of clockwork to be observed (Weinert 2004).

Due to the development of quantum physics, the double slit experiment and observations of entanglement and wave-particle duality, the concept of nature in the philosophy of science has moved from this view of the world as deterministic to a view where: 'The basic indeterminacy in the atomic realm leads to a fundamental revision in the concept of Nature: Nature is probabilistic, not just our knowledge of Nature' (Weinert 2004, 59). It is an enacted idea, a concept that is influenced by empirical observations. And yet these observations often have little to say about the fact that the concept itself occurs and that its implications manifest. They perform empirically, even as the concept changes. The concept nature itself literally *matters*.

Human Geography and Environmental History have well explored the conceptual construction of the environment, or nature, establishing its material fallout and the impossibility of considering it an unbiased element of study (Schama 1995; Castree and Braun 2001; Lorimer 2012). We cannot separate scientific observations and claims about nature from the socio-cultural constructions that guide our political interests (and *vice versa*), occasionally serving as 'instruments of power and domination' (Castree and Braun 2001, 9). For instance, in the context of the colonial

European expansion overseas, the existence of a conception of a distinct 'nature' as opposed to civilization was 'easily racialized and, in the guise of scientific racism, provided a rationale for European colonial rule over more 'primitive' cultures and peoples' (Ginn and Demeritt 2008, 303). This perceived inferiority can be linked to myriad 'minority' groups by associating 'them' with that 'thing' – nature (Plumwood 2002). The Occidental bifurcation of nature supported arguments that some cultures remained primitive due to geographical determinism as well as romanticized ideas of the 'noble savage'. Simon Schama (1995) describes how 'The arcadian idyll, for example, seems just another pretty lie told by propertied aristocracies (from slave-owning Athens to slave-owning Virginia) to disguise the ecological [and cultural] consequences of their greed' (12). Similarly the 'received' conception of wilderness, as unpopulated *terra nullius*, has been critiqued as a tool of androcentrism, racism, colonialism, and genocide (Callicott 2000); these are material consequences indeed. The way nature is conceptualized has played a principal role in historical fascism (Cutting 2016).

The appropriation of nature is evident in colonized landscapes all over the world, from Australia and America to India and Kenya. These 'Teletubby Landscapes' (Mcphie 2017) perform homogenizing atrocities, morphing once diverse environments into England's new *Jerusalem*,[8] a 'green and pleasant land' that hides a more sinister Victorian conservatism. Now, as a derivative of the organic 'biological' invention, the concept nature arguably plays the most vital role in shaping our world ... contributing to a(nother) mass extinction event. How nature is conceived is played out on a massive scale. Matter itself is morphed through conceptual processes. The physical effects of thinking a thing into existence has world changing ramifications.

Episode V: ecology without nature

Concepts, then, *matter* in very literal ways. Recognizing the power of concepts, Timothy Morton has suggested that the term environmentalism is curiously similar to both sexism and racism in that it is immediately divisive, setting as it does one group of things against and above another. Next year it will be ten years since Morton (2007) published *Ecology Without Nature* in which he argues:

> the very idea of "nature" which so many hold dear will have to wither away in an "ecological" state of human society. Strange as it may sound, the idea of nature is getting in the way of properly ecological forms of culture, philosophy, politics, and art. (1)

And, we hazard, education. Morton was implying that if nature is abridged to romantic versions of flora, fauna and perhaps a few other phenomena such as weather, water, and earth, then our understanding (and experience) of the world will be less rich, less nuanced, and potentially more damaging due to its monocultural oversimplification and the resulting behavior that it may produce. Since then, more recent versions of ecocriticism – the field of literary study concerned with the depiction of nature and the environment – have begun to move to embrace philosophies that attempt to think beyond the romanticized version of nature, becoming all the more aware of the material effects of the stories we tell. For instance, Cohen's edited collection *Prismatic Ecology: Ecotheory beyond Green* is just such an attempt to expunge the lingering viridescence of (un)critical 'bright green' ecological writing. Drawing from Morton, Cohen (2013, xxii) argues that 'bright green criticism emphasizes balance, the innate, the primal, landscapes with few people, macrosystems, the unrefined' – in short, the supposed 'critical' position in environmental literature studies has reified the very nature that Morton urges us to think beyond. Cohen (2013, xxii) goes on:

> In a green Arcadia what do we make of the airplane, graves, gamma rays, bacteria, invasive bamboo accidently planted as an ornament, inorganic agency, relentless lunar pull, electronic realms, prehistoric flora lingering as plastic refuse, lost supercontinents, parasites, inorganic compounds that act like living creatures, species undergoing sudden change?

Fletcher (2017, 231) suggests that abandoning the conceptual category 'nature' 'can actually provide a stronger basis for critique of environmental degradation than appeals to "nature," which invites counter-critique that such degradation is produced by "natural" human inclinations'. This is similar to a position that we have taken previously, though we would distance ourselves from any statements of 'natural' human inclinations, rather stating that nature as conceived as whole would not be able to be escaped by anything, even by humans and their concepts, such as 'nature' (Clarke and Mcphie 2014). In this sense, the variations and *multiple* natures are 'natural'. However, arguments that we might better care for 'the environment' by dropping the concept of 'nature' leave open the question of environmental degradation itself, as a thing occurring to an environment that is separate from the human, i.e. if 'nature' is the label we apply to the things that should be conserved, upheld or connected to – and if this label is in flux – then environmental damage itself moves into flux. Payne (2016, 175) suggests that 'the universalizing implications of homogenizing and conflating nature and culture, and their environments, and humans and nonhumans, for EE [Environmental Education] and ESD [Education for Sustainable Development] are profound – theoretically, pedagogically, and practically.' Where would the harm be located in an ecology without nature? What would there be to sustain or educate for?

Conceiving the world in terms of material – and materiality – allows researchers to think with different conceptions, different stories that *matter* different consequences. For research is a story and it does *matter* consequences. Like Cohen and Morton, there has occurred a concerted effort by scholars in recent years to 'develop forms of ecological discourse that complicate, critique, historicize, or abandon the concept of nature while taking serious account of the agency of non-human beings and phenomena' (Raine 2014, 103). Rather, there has never been a nature for us to overcome, damage, or even return to. Thinking with MacLure (2015), an *immanent* ecocriticism would aim to be productive of new natures. In what follows we play with nature as a concept to see what varying conceptions of nature *do* as physical phenomena that perform and *matter*, for nature is stubborn.

Episode VI: the diffraction of nature – just for kicks

The following list of natures is not meant to state that there are only eight such versions of nature, we're sure there are many, many more and they can never be simplified to a bounded number. It is merely a pluralistic ploy (for example, Loris Malaguzzi's *Hundred Languages of Children* and Facebooks' 71 gender options) to highlight the shortcomings of reducing the concept 'nature' to just one or two things, objects with binary inferiors or *quiddities* – a thing's *whatness* – when it is clearly many things to many people (or nothing to some). Of course we realize the contradiction inherent in numbering (and naming) them as such, and so for the record we would like to place all of these numbered natures *sous rature*.[9] Although they are conceptual abstractions, the different natures *do* different things. They work in different ways. We accept that this exploration is a little one-sided – toward the *Dionysian* – but there are far more examples of the positive *Apollonian* effects of some versions of nature within environmental education discourse so we thought we'd employ a little positive discrimination, *just for kicks*. This was a term used by Deleuze and Guattari (2004) to demonstrate the importance of minor creative acts as movements toward an ethics of immanence.

> For example, it is relatively easy to stop saying "I," but that does not mean that you have gotten away from the regime of subjectification; conversely, you can keep on saying "I," just for kicks, and already be in another regime in which personal pronouns function only as fictions. (152)

So, for the purposes of this article, we employ the term 'just for kicks' as an experiment – a playfully political transgressive move – because we are already in another regime in which nature functions only as a fiction.

Nature 1 – scary nature: scary, useless or dangerous, inhabited only by wild animals, as in the *wildēornes* of the saga Beowulf. This nature belonged to the pre-picturesque/pre-romantic Western mind. In 1642 James Howell described the Alps as 'uncouth, huge, monstrous Excrescences of Nature, bearing nothing but craggy stones' (Thompson 2010, 20). 'In 1657 a dictionary still describes "forest" as "awful", "gloomy", "desolate", "inhospitable" (Lemaire 1988, 62)' (cited in Egmond 2007, 15-16). Nature 1 destroyed the wolves and bears in the UK and attempted to create a more sterile urban, devoid of more ostensibly mobile diversities, such as wild mammals, insects and germs. This one *sanitized* the land. But Nature 1 still scares people today. Some don't see this nature as at all therapeutic or something *to connect* to, instead wanting to avoid the 'claustrophobia' 'darkness', 'insects' 'anyone lurking', 'getting buried', 'ghost stories', 'pedophiles', 'the Blair Witch' and leading one person to think that because of 'fairytales, I didn't know if there was anything out there and something could come and get me like a monster and stuff like, erm, like a man crossed with something ... sort of an animal or some fairy quality' (Milligan and Bingley 2007, 806-808). Koole and van den Berg (2005) showed how many people have biophobic reactions to (conceptually) 'natural' landscapes, associating 'wilderness' with death, and Milligan and Bingley (2007) found that while some participants found woodland restorative, others felt fearful or were repelled by it. Some environmental educators try to fight this concept of nature.

Nature 2 – scenic nature: an ordered, neat, picturesque, specifically designed (perhaps by Gilpin or Capability Brown) formal nature, framed with a hint of human culture in the scene (a typical Teletubbies landscape). Nature 2 aesthetized (and anesthetized) productive landscapes (from food production to aesthetic production), as only the upper classes possessed unproductive land pre-industrialization. This one *civilized* the land. These elitist picturesque landscape preferences were born out of literature, art and land ownership as they reinforced the personal views and perceptions of its members as *the owners of nature* (Gillespie 2008). Today this nature draws visitors to Blenheim Palace and therapy seekers to Flatford Mill, the site of the Romantic artist Constable's *The Hay Wain* (Barton, Hine and Pretty 2009), a painting 'used to promote a timeless ideal of beauty and social order which belied exploitative labour relations, rural poverty and the political unrest that was sweeping the English countryside at the time [it was] painted' (Hawkins, cited in Howard, Thompson and Waterton 2013, 3–4).

Nature 3 – utopian nature: type in the word 'nature' into any search engine images and you will see a plethora of Disneyfied landscapes. This is a romantically *idealized* nature: Green rolling hills, wild (but never *too* wild), 'particular' mammals, 'particular' flowers, a lake in front of some mountains, green leaved trees (usually deciduous, pictured in the summer), a rainbow, a waterfall, blue sky with white clouds, etc. Yes, that's nature. Along with fresh air, bird song, stars and sweet smelling nectar, this is the sort of nature that people who say 'we must re-connect to nature' generally seem to mean. Cohen (2013) suggests that this sort of romanticized analysis 'often focuses on the destabilizing encroachment of industrialized society into wild spaces, the restorative and even ecstatic powers of unblemished landscapes, and the companionless dignity of nonhuman creatures' in which 'Woodlands, serene waterscapes, sublime vistas, and charismatic megafauna feature prominently' (xix-xx). With Nature 3 there is a 'utopian emphasis on homeostasis, order, and the implicit benevolence of an unexamined force labelled nature' (Cohen 2013, xxii). Morton (2010) suggests that this 'bright green' view peddles nature as 'affirmative, extraverted and masculine [...] sunny, straightforward, ableist, holistic, hearty, and 'healthy'' (16) or as Cohen (2013) posits, 'a purified place to which one travels rather than dwells always within: separate from the human, empty, foundationally pure' (xxi). This is the ecocriticism of deep ecology, for example, one of the aims of which is to 'reconnect humans with nature.'

> The problem with this position is that, in flagrant contradiction with its explicitly stated aims, it promotes full-scale humanization of the environment. This strikes me as a regressive move, reminiscent of the sentimentality of the Romantic phases of European culture. I concur therefore with Val Plumwood's (1993,

> 2003) assessment that deep ecology misreads the earth-cosmos nexus and merely expands the structures of possessive egoism and self-interests to include non-human agents. (Braidotti 2013, 85)

de Vega (n.d.) posits, 'If shallow ecology objectifies nature, deep ecology subjectifies nature, thus it is no less dualistic' (3), which is paradoxical as 'the more we venerate nature as the place we need to appreciate and respect, the more we set up rules and principals [sic] that keep us separated from it' (4). It is not artifice that is the principal driver of mass extinction, it is nature, via this persistent *belief* in separation, the bifurcation of nature. 'Under this conception, nature is thought of as something pristine, pure, wild, and immediate; something that we can look at, sometimes touch, and almost always end up destroying' (de Vega n.d., 4). Ellison (2013) indicates that since the late 19th Century ecology has been dominated by a romantically harmonious notion of landscape which is the wrong sort of nature for an ecologist to study. This nature was born out of Nature 2 (Scenic Nature), the sublime, privileged Grand Tours and oxymoronic literature that combined horror with beauty (influenced by the 1681 book *A Sacred Theory of the Earth* by Thomas Burnet (MacFarlane 2003; Nicolson 1997; Schama 1995; Thompson 2010), which was read by influential progenitors of Nature 3, such as Addison, Dennis, Steele, Wharton, Young, Wordsworth and Coleridge (Nicolson 1997)). Nature 3 romanticized a wild nature. For the Romanticists, it became associated with the 'experience of God's Creation' and was 'the centre of spiritual and religious regeneration' (Egmond 2007, 16). This one *idealized* the land. This is the one that has performed feats of racism, genocide, androcentrism, colonialism, etc. (Callicott 2000, 24). 'For over a century, in countries like Kenya, indigenous peoples have been forcibly removed from, or denied access to, traditional territories because conservationists have argued that segregated 'wildlife parks' are required for species protection' (Castree and Braun 2001, 9). The World Bank estimates that between 1986 and 1996, about 3 million people were forced to move from forests, tribal land and other areas as a result of both development and conservation schemes (Vidal 2001), some of which 'involved the destruction of the resident indigenous peoples, and the problem is now growing more acute as conservationists press harder for governments to set aside 'natural' areas, which in reality have been lived on for generations' (Corry 2011, 211). It is not hard to find examples of environmental education research that promote Nature 3, yet it may also be the nature that co-created the summer camps of the English fascist movement of the 1930s (Cutting 2016).

Nature 4 – scarier nature: dog shit, slime mold, adrenal cancer, earthquakes, strychnine poison, sulphur dioxide, methane, piss, tsunamis, scorpions, rotting cabbage, snot, bile, viruses, the Black Death, phlegm, malaria, weeds, sharks, breast cancer, floods, a flower that smells of rotting meat, rotting meat, puke, forest fires, etc. They're all nature too. This nature is perhaps more reminiscent of Nature 1 (Scary Nature) conceptually conditioned by history and socio-economic status. They are generally labelled as 'natural' or 'nature' but aren't usually thought of when picturing *scenic* landscapes and cuddly mammals. Do you still wish to *re-connect to nature,* re-connect to shit and cancer? Nature 4 is an ill thought of nature and is often discarded, feared, killed, eradicated or 'weeded' out (often in response to Nature 2 and sometimes 3). Arguably 'climate change' is the most cited example of Nature 4 in environmental education literature. Although increasingly viewed as '*man* made' it is generally not conceived in the same way as a car or a house (see Nature 5), or even human waste. It is more like a devastating avalanche encouraged by a skier, for whatever reason.

Nature 5 – artificial nature: a James Bond watch, pickled onion flavour Monster Munch crisps, false teeth, a plastic lawn, a tube of toothpaste, books, computers, stilettos, scissors, electric wire, cars, a guitar, a knife, a plastic flower, etc. They're nature too, only many people in the West, perhaps *especially* environmentalists, ecotherapists and deep ecologists, think not, mostly because these things (what might be labelled as 'artificial') are the produce of humans, even though those same people might argue that humans are nature too. Somewhere along their thought patterns, somehow human produce has become separated from the existing universe. A form of transcendence has invaded the earth. Nature 5 is not thought of as nature to many people, as already mentioned it is of human produce. But if humans are nature too then surely everything we produce is *of* nature (the material, force and energy of the world/universe), so at what point does it become 'not nature' or 'unnatural'? *In* our *heads*? Some environmental education researchers are working to overcome this problem. For instance, Nature 5 might be occurring in Rautio, Hohti, Leinonen and Tammi's (2017) work, where the researcher notes:

> In addition to me and the children 'the urban' and arguably 'the nature' took part and generated each other as green containers, crisscrossing pathways, roads, compost piles, electronics, tyres, cardboard, glass, descending darkness, fluorescent lights, extreme cold, birds and rats and heavy machinery moving around discarded materials from 300,000 people. (1380)

New versions of Nature 5 are co-emerging as we write. YouTube viral videos, twitter storms, digital warfare, real fake news websites and genuine alternative news organizations, online journal access and academic social media sites. The digital age encourages an in-depth access to a topological, intra-relational nature that developed with language, writing, and the printing press, and is now becoming in larger scales and quicker speeds due to the digital era. Like a bat monitor that extends our sensory apparatus to be allowed into a privileged noctillionine epistemology, the internet is a body extended and distributed all over the world (and beyond). This is much deeper than the shallowness of deep ecology. How much 'deeper' into nature can we go while at the same time becoming more diffractive to the inequitable consequences/effects that 'some' of this newly emerging nature seem to enact?

Nature 6 – affective nature: a tear, a frown, a whisper, a tender touch, an annoying cough, the word *cunt* (this word has a particularly strong force of encounter for an English language culture – as it performs ecologically it sets off a trophic torrent of culturally instigated and highly gendered physiological effects… do you feel it?), a cheeky wink, a rousing speech, an uncanny atmosphere, the rush of a blush, the force of an erotic encounter, emotional elation, etc. These are nature too but are harder to define or *capture* as they are an affective nature, one examined under the lens of the *affective turn* perhaps (see Gregg and Seigworth 2010). Nature 6 can lead to marriage or a call to arms. It may be the nature at work in Gannon's (2017) exploration of educational encounters and McKenzie's (2017) contemplation of policy. For Harris (2014) studying affective experiences is more promising than promoting 'environmental' knowledge, though environmental educators have known this for a while (Russell and Oakley 2017).

Nature 7 – conceptual nature: free speech, 56, fascism, yellow, the alphabet, China, nature, culture, place, etc. These are nature too. They exist in the world, a world of (re)cycled materials, forces and energies. When 56 is written on a page it is perhaps ink (that is a material of the world) on paper (from a tree, even the romantic idealists may call this one nature!). If it is uttered from a mouth, that is breath (full of CO_2, etc.). They are percepts as they are things that we can touch or see or hear, for example, but they are not the *idea* of 56, nature or yellow. Alfred North Whitehead (1919) suggested that 'Nature is that which we observe in perception through the senses' (1) but 'Thought about nature is different from the sense-perception of nature.' (2) Yet thought is itself a sense, as we can feel it, just a different kind of sense. What Whitehead is referring to here is what we would call the difference between nature as a concept (still a kind of percept, although less intense) and nature as a percept. The concept can influence the percept (and *vice versa*). Affect may also influence and be influenced by the other two in turn. When 56 is thought but not articulated it is a concept and yet still a percept. We might say it's not *really* real (*virtual*?), yet still real enough to enact a physical presence of some sort (in the *actual*?). Nature 7 allows for many trophic torrents that perform in the world once they've been born. For example, the atomic bomb had a devastating physical consequence, but it was born of and deployed out of abstract concepts - mathematics, and a bunch of numbers. This is how nature works. Although a representational invention, it has physical consequences. But it can be deterritorialized and occupied! Nature 7 would include the 'theory' of any discipline, but theory is always already material both in its presence (concept/percept) and its affects. Environmental education theory and research methodology is thus always also a material practice, and by implication it is also pedagogical. It becomes habit. This realization points to the practical and political work that philosophy does. As Foucault said to Deleuze: 'theory does not express, translate, or serve to apply practice: it is practice. But it is local and regional, as you said, and not totalizing' (Foucault 1977, 208). Just like this article and its iterations of nature, practice is being enacted in your reading of it and whatever affects this may produce.

Nature 8 – abstract nature: a unicorn, pixies, an Orc, fairies, God, Bambi, Shiva, ghosts, ray guns, the USS Enterprise, etc. They're all nature too! As ideas, these ones are abstract concepts, like the thought of 56 from Nature 7 (Conceptual Nature – the not *really* real), only more difficult to prove. For example, we can witness the effects of 56 when applied to atomic physics. Now it gets tricky as they are still percepts. A *picture* of a unicorn is a percept because it is empirical yet can we touch the unicorn itself? Unicorns are concepts, thoughts, and thoughts themselves are physical relational processes of the world of material, force and energy. They are not outside it. They are empirical in some way (electrical impulses, for example) and therefore ecological ... natural. The *thought* of them exists as a percept in the world (a subtle real, a *conceptual percept*) but not the *actual* unicorn. However, unicorn's have the *potential* to impress a physical dent in the world just as perceptual as the hoof prints of a horse. But this is similar to 56 because they all start life as *conceptual percepts*. With the example of 56, we can see its impact in the world, its impression. But then the abstract concept *God* has perhaps made even more of an inscription in the fabric of the world – enacting trophic torrents – due to its performativity. There are degrees of actuality and actualizing (or virtuality and virtualizing if you prefer) rather than the binary actual-virtual as 'Purely actual objects do not exist' (Deleuze, cited in Deleuze and Parnet 2002, 112). Their boundaries are topologically fluid and permeable as they flow and stretch interchangeably. This makes more sense to us due to the allowance of grey bits, dark matter that is perhaps more abundant than what is perceived. An affective realm may be actualized more fully and coherently if we look for *ecotones* – the bits in-between biomes – rather than definite boundaries between virtual and actual planes. The unicorn is also actual in this sense and not merely virtual. It's just not *as* actual as a horse or narwhal which is much more obviously empirical and open to many more sensory apparatuses than the felt presence of a unicorn depending on the particular assemblage (as Jamie knew someone who interacted explicitly with an imaginary 'My Little Pony' – see, hear and talk to it as one might do a dog – whereas Jamie only ever perceives the abstract little pink pony conceptually). Don't forget, it's the 'effects' of conception that 'matter' more intensely in the world. So, for us, we've never actualized a god, it's always been more toward the virtual end of the scale (not that there's an end ... or a scale) whereas for Jamie's auntie, a (lovely) born again Christian, God has been actualized rather explicitly and has achieved a masculinized material status in its actual affects in/on/*of* the world. Nature 8 led to the crusades and various forms of religiously inspired terrorism but also some great speculative fiction. Nature 8 is akin to environmental education approaches that highlight the possibilities of speculation and fiction. In Nature 8, all the natures above (and more) – and their material effects – can be reinforced or new natures created. Nature 8 might be present in the work of Gough (e.g. 1993, 1994, 2008, 2010), Morgan (2010), and Rousell, Cutter-MacKenzie, and Foster (2017).

So, nature has many faces. It 'can refer to stable substrata of brute matter', can signal 'generativity, fecundity, Isis, or Aphrodite' (Bennett 2010, 117). It is infused with bias and informed by propaganda. It can open doors to health or restrict access to livelihoods depending on a variety of cultural (in)equities. Now, what to do with nature?

Episode VII: what to do with nature?

All of the Natures presented here are ecological processes and can indeed perform as they are enacted in the world. This leads to perceptions that force us to assume that either certain ecologically destructive actions won't affect 'us' too much or that we can discard the non-romanticized nature, like pollution, in favour of a pristine wilderness 'untrammeled by *man*'!

Some scholars (Morton, Zizek) disapprove of the term nature altogether and wish to eliminate it. Some scholars (Bryant) think it's the concept 'culture' we must abolish rather than nature. Some (Latour, Haraway) wish to merge it to form 'natureculture'. Cohen (2013) attributes nature with an 'inorganic agency' (xxii) where 'Shadow itself is ecological' (xix). He suggests that if nature was 'refracted through the geological [...] our ethical connectedness to the nonhuman would become more tangible' (Cohen 2015, 12).

Thinking from a flatter, immanent ontology, some scholars (de Vega, Deleuze, Cohen, us) wish to revolutionize it as the concept has its uses. But we must be wary of how we attempt to (re)present it. For how can we justify using a concept that has the potential to perform atrocities? And even if we do excuse the terms of possible oppression, how can we use them as a counter measure to revive or free-up alternative, less problematic meanings? 'How then to write about young people engaging in just such 'sacred spaces', like woods and moorlands, without resorting to reified notions of nature' (Quinn 2013, 738)? Quinn (2013) has 'considered coining the term 'open nature', which could be helpful in conveying a sense of forests and moorlands, but negatively would serve to sub-divide nature in a binary way' and so decides to leave it to other scholars: 'Ultimately, finding a solution to this philosophical problem of naming is not within the scope of this article' (739). But we could wrestle with it a little as there are, as always, alternatives to nature...

Episode VIII: environing

The term 'environment' has become a noun out of the verb 'environ' + ment. The 'Oxford International Dictionary of the English Language' (Little, et al. 1957, 619) provides one definition of 'environ' as 'to surround', 'envelop' or 'enclose.' But to surround or envelop *what* (and indeed *when*)? This definition is undeniably separationist if we assume that it is *we* or other organisms who are surrounded. This bifurcation of nature is the result of a transcendent ontology. Perhaps we must look to non-anthropocentric, flatter ontologies for our conceptual approaches to the nature of our environments. How, then, might it look from an immanent perspective? The 1647 definition, *'to go round in a circle'* (Little, et al. 1957, 619, emphasis added) is perhaps more appropriate to the *condition of a thing* (and is one that best applies to Ingold's (2000, 2007, 2011) lines of living in the world). For example, 'to go round' implies movement *along* rather than an emic-etic split (an *in* and an *out*). The concept 'place', in place-based and place-responsive pedagogy, for example, can also perform romantically if conceived of as a static environment, bound within imaginary borders. Ingold (2011, 148) states that 'we tend to identify traces of the circumambulatory movements that bring a place into being as boundaries that demarcate the place from its surrounding space,' as 'the pathways or trails *along* which movement proceeds are perceived as limits *within* which it is contained [...] turning the 'way through' of the trail into the containment of the place-in-space.'

> [H]uman existence is not fundamentally place-*bound*, as Christopher Tilley (2004: 25) maintains, but place-*binding*. It unfolds not in places but along paths. Proceeding along a path, every inhabitant lays a trail. Where inhabitants meet, trails are entwined, as the life of each becomes bound up with the other. Every entwining is a knot, and the more that lifelines are entwined, the greater the density of the knot. (Ingold 2011, 148)

In this way, and developing Ingold's reasoning further, we would say that we are not *in* spaces or places, nor *on* or *along* paths, rather, we *are* the paths themselves in their continual *environing*. Therefore, the 1603 definition of 'environment' possibly serves our own purposes best when attempting to find a suitable term that describes *what it is* that we are *in* and/or *of*: 'The action of environing' (Little, et al. 1957, 619). Hence, environment is an *action*, something we *do* and are *of* rather than something we are encased *in*. Environment is something that is continually *becoming* and we are *of* that process. One might say, then, that we are living lines of environing. The 1827 version, 'That which environs; *esp.* the conditions or influences under which any person or thing lives or is developed' (Little, et al. 1957, 619) is also useful due to its emphasis on movement and development as opposed to stasis and separation.

Episode IX: landscaping

'the landscape thinks itself in me ... and I am its consciousness.' (Cezanne, cited in Wylie 2007, 2) Landscape is not a fixed scene to be 'gazed upon' by an image capturing spectator (through a lens) that catches and then frames a representation of it. In recent literature landscape has started to become more mobile once again.

> For Wylie in particular, this shift to 'landscaping', [...] turns the word from a noun into a more rhythmic and mobile action verb [...] Body and landscape thus become recursively intertwined, both constitutive and constituting, and always in a process of (re)formation. Indeed, they become, to borrow from Thrift and Dewsbury (2000: 415), extensions of the body and mind, and vice versa. (Waterton 2013, 70)

Rather like Andy Goldsworthy's 'taking a wall for a walk' or Paul Klee's 'taking a line for a walk', Cumbrian poet Norman Nicholson's (1977) poem *'Wall'* emphasizes the animacy of what are normally considered inanimate objects in the landscape, illuminating how 'A wall walks slowly' and 'Is always on the move.' Nicholson had a keen eye for movement and saw the Cumbrian landscape in ways that the romantic poets, such as Wordsworth and Coleridge seemed to omit. For Nicholson's artistic working class gaze, the landscape of the Lake District wasn't merely a romanticized scenic nature for an elite clientele: 'It is futile to assess such country in terms of views. [...] it measures the landscape from the borders of an imaginary picture-frame; it reduces like to a post-card', as for him 'it is also the man-made screes beside the quarries; and whitewash on the Copper Mines Hostel, a stone playing ducks & drakes on Levers-Water, making the black tarn throw up waves like a magicians' steel rings' (Nicholson 1977, 33-34). We might call this more mobile landscape, land*sceppan* (Ingold 2011), landscap*ing* (Wylie 2007), environ*ing* or just *life*.

Episode X: Shelley's romantic (re)visions

In the book 'Romantic Revisions' (Brinkley and Hanley 1992), Brinkley (1992) examined the Romantic poet Shelley's notebook where he drafted a copy of his poem 'Mont Blanc.' Rather than explain the meaning of the finished and polished poem, Brinkley explored the spaces between the words, the omissions and the words placed under erasure by Shelley himself, the words that Shelley changed his mind about including in the finished product. The poem, Brinkley (1992) writes, seemed 'to be structured by the breaks in composition' as 'much of 'Mont Blanc' was inspired as it was composed – by intervals of thwarted writing' (243). In this poem, Shelley originally wrote 'In daylight thoughts, bright or obscure/In day – the stream of various thoughts [eternal] universe of things/Flows thro the mind reflecting rolls & rolls its rapid waves' later changing it to, 'In day – the [eternal] universe of things/Flows thro the mind & rolls its rapid waves' (Brinkley 1992, 247). These reworkings, Brinkley writes, 'articulate a radical epistemology in which *things – and not their representations* – are said to flow through the mind' (1992, 247, emphasis added). Thus, nature is not conceptually separated as a reflection, representation or image. It is neither objectified nor subjectified within this revision. The emic-etic split is dissolved. This is a non-representational philosophy, a philosophy of immanence and vital materiality. It is a radical onto-epistemology which rejects the Freudian *tripartite psyche*, the Cartesian soul which operates a mechanical body from within the confines of the human pineal gland or the Kantian subject as a reflective self. Unfortunately, the romantic poets didn't utilize the same thinking when it came to human produce, falling straight back into the Cartesian well. These *things* – that flow through the mind – are not only majestic mountains and fresh air, they are also crisp packets and mobile phones (not that Shelley would have heard of such things).

As Spirn (1998) clearly points out, 'Humans are not the sole authors of landscape.' (17) The volcanic processes (flows of material, force and energy) that metamorphose rock from one state into another are similar processes to those that metamorphose a mountain into a shopping centre. 'Intention', 'agency', or 'will to act' follow the same complex co-emergences as the rest of the material fluxes that continually transform the planet. Concepts then, such as nature, are agential as well as ecological.

Episode XI: the ecology of nature

Concepts – as physical processes – perform ecologically. Depending on the diversity of their impact, they have the ability to capture matter and slow down energy dissipation over time if they display minimal or low entropy (like an old growth forest where multiple species have evolved with and adapted to each other over time) or the opposite if they display maximal or high entropy (like a newly concreted car park or a volcanic eruption). Similar to capitalism, the concept 'nature' is a keystone species – it has disproportionately large effects in the world considering its seemingly discreet material nature as a conceptual abstraction – but in many of its guises displays maximal entropy. This is a problematic combination due to the trophic torrents that are produced as a result often leading to eutrophic[10] and inequitable homogenized monocultures – although the effects of some other versions of nature can also lead to inequitable homogenized monocultures as a result of oligotrophic[11] exhaustion. We have exampled a few of these torrents in episode VII (we'll let *you* decide which ones are oligotrophic and which are eutrophic).

It may not be an unusual practice to talk about concepts using an ecological patois if discussed metaphorically but our intention here is very literal – we are exploring the physical, nutritional and relational properties of a concept. To answer the question of why it may sound strange describing concepts physically or nutritionally, we must recognize what we've just done. We've *ennatured culture*. We've simply extended what is usually reserved for the field of ecology to include phenomena from human culture - thought. We've dared to imagine that humans and human produce are also *of* this world and as such perform ecologically, just as much as flora, fauna and myriad other physical processes. We could just as easily have *encultured nature* by personifying the environment[12] – as employed by animist ontologies. Both are practices of rectifying the bifurcation of nature and making more permeable the Cartesian barrier that was extended during the Enlightenment. Weaving these non-dualistic perspectives together could be an example of what Bartlett, Marshall and Marshall (2012) calls 'two-eyed seeing' – the integration of 'indigenous and mainstream knowledges within science educational curricula' (331). This is a pluralistic experiment that could prove beneficial regarding socio-environmental equity.

Episode XII: Post-Nature – natur*ing* or environ*ing* education

Let us be clear, we are no fans of mass extinction due to an (perhaps misplaced) empathy with the current epoch. We are certainly no fans of multispecies suffering. And we are no fans of desertification, inequity and homogeneity, of loss and misery. But we feel stultified by simplistic, or romantic environmental discourse and research, especially when there is a world of (immanent) critical environmental thought occurring just a journal away. Rather than suggesting anyone 'reconnect' with nature, as an endeavour of environmental education, we instead propose the idea of playing with nature, as a concept with learners. This can be conceived as a process of natur*ing* or as environ*ing* education. We cannot hope for people to reconnect to nature – as there is no such ideal state – but we can hope that people consider nature as a material concept that can be experimented with in the process of concept creation. de Freitas and Palmer (2016, 1220) suggest thinking of concepts as playmates:

> this way of thinking of concepts as flexible and amusing playmates participating in children's explorative investigations is not essentialist or normative, but is rather an ongoing intervention that is never the same but differs according to the situation. The concept operates differently and becomes different depending on the circumstances of each event.

In this way we can experiment with how concepts, namely nature, come to matter. Rather than attempting to 'mush' together culture and nature, or overcome Cartesian dualism with a piece of sticky tape holding the subject and object together in 'interaction' or 'connection', we can instead realize that concepts are performative through our specific intra-actions and storytelling of the world. Karen Barad (2003) suggests that the realization that concepts come to matter has paradigm changing implications for the dominant view of perceiving humans as subjects outside of their objects of observation. Boundaries, according to Barad, are created in the agential intra-actions of material-discursive events. So, there is a sense in which the process of natur*ing*, of creating new concepts of nature, is in effect a mode of becoming *of* the world. In this respect it just might be the connection that many wish to attain.

In ethical terms, Deleuze and Guattari argue that concept making is a political and revolutionary act, arguing that reflection and communication only get us so far:

> We do not lack communication. On the contrary, we have too much of it. We lack creation. *We lack resistance to the present*. The creation of concepts in itself calls for a future form, for a new earth and people that do not yet exist. Europeanization does not constitute a becoming but merely the history of capitalism, which prevents the becoming of subjected peoples. (Deleuze and Guattari 1994, 108, cited in Peters 2004, 224)

And so, according to Peters (2004), the future of philosophy 'is the role of the philosopher as physician, as the physician of culture, 'an inventor of new immanent modes of existence' (Deleuze and Guattari 1994, 113). In philosophy of education these categories, these becomings have an easy resonance' (224).

We are interested in the type of experiments that allow us, and our learners, to become philosopher physicians, critically playing with the everyday concepts we pick up in literature and our daily lives to literally create new material-conceptual worlds. But we should also take care when conceptualizing nature. As Cutting (2016, 112) notes, educational approaches that claim to 'encourage a deeper emotional engagement with the natural environment may not necessarily promote liberal thought, because while there is nothing wrong with promoting a love of the countryside, how students come to conceptualize this relationship is critical'. With this article we hope we have shown how critical the endeavor of creating concepts of nature can be. We have played with nature as a concept, to see what different iterations do. Nature 1 – scary nature – sanitized the land; Nature 2 – scenic nature – civilized the land; Nature 3 – utopian nature – idealized the land and created a romanticized, harmonious, Apollonian and privileged green environment; Nature 4 – scarier nature – weeded out unwanted nature and turned a blind eye to the Dionysian inaesthetic; Nature 5 – artificial nature – created a conceptual nature-culture bifurcation; Nature 6 – affective nature – emotionalized the world; Nature 7 – conceptual nature – and Nature 8 – abstract nature – ecologized thought to create monsters.

We have exposed the agential ecology of a concept and followed the trophic torrents left in its wake. What we are proposing is an environmental literacy that attends to what a concept is capable of, what a concept can do, and perhaps even what a concept can prevent, *post-nature*.

Notes

1. The use of the term 'mattering' is twofold; firstly to denote importance and secondly to highlight the physical and ecological performance of the concept nature.

2. 'Trophic' usually refers to the nutritive relations between organisms – in other words, the matter that is passed between things.
3. We do not *explain* the material turn as that is already attempted in the editorial of this special issue. However, it is worth noting that by thinking *with* the material turn we are embracing some of its key themes, such as a move away from dominant enlightenment epistemologies and anthropocentrism and a move towards distributed conceptions of agency and a focus on materiality.
4. Inherently 'of' the world, always processual and inseparable.
5. Imagined to be detached, separate from and independent of other 'things'.
6. We imagine many of the readers of this article will have seen the film *Star Wars* (Lucas, 1977) at some point in their life. It begins in the middle, with episode IV. The Greeks did this and called it *in media res*. The purpose of this is to distort the linearity of the storyline, to fit more closely to the non-linear nature of temporality. It encourages the observer/reader to think beyond the now normalized representational trends that exist within academe. It also reminds the reader that this is just another story (in the Brechtian tradition) and most certainly not 'the truth'. This emphasis on storied matter sits well with both new materialisms and material ecocriticisms.
7. Quote taken from the opening credits of *Star Wars*, used here to highlight the romanticized and transcendent appropriation of nature.
8. The English hymn *Jerusalem* was composed by Hubert Parry in 1916 but was originally written by William Blake as a poetic preface to *Milton: A Poem in Two Books,* where Blake imagined Jesus Christ walking upon England's 'green and pleasant land'.
9. *Sous rature* (under erasure) is a (Heideggarian/Derridean) ploy to problematize a concept whilst keeping it in play.
10. High nutrient content and species proliferation which can eventually lead to a reduced oxygen supply and species depletion if one species dominates or the rate of decay increases.
11. Low nutrient content.
12. As opposed to anthropomorphizing or Disneyfying the environment which simply reinforces the Romanticized perception of nature.

Acknowledgments

We'd like to thank the reviewers for their time and careful readings of the article and Nigel Dykes for his advice on ecological processes.

Disclosure statement

No potential conflict of interest was reported by the authors.

ORCID

Jamie Mcphie http://orcid.org/0000-0001-5290-1685
David A. G. Clarke http://orcid.org/0000-0002-4330-0818

References

Barad, K. 2003. "Posthumanist Performativity: Toward an Understanding of How Matter Comes to Matter." *Signs: Journal of Women in Culture and Society* 28 (3):801–831.

Barad, K. 2007. *Meeting The Universe Halfway: Quantum Physics and the Entanglement of matter and meaning*. Durham and London: Duke University Press.

Bartlett, C., M. Marshall, and A. Marshall. 2012. "Two-Eyed Seeing and Other Lessons Learned within a co-Learning Journey of Bringing Together Indigenous and Mainstream Knowledges and Ways of Knowing." *Journal of Environmental Studies and Sciences* 2 (4):331–340. doi:10.1007/s13412-012-0086-8.

Barton, J., R. Hine, and J. Pretty. 2009. "The Health Benefits of Walking in Greenspaces of High Natural and Heritage Value." *Journal of Integrative Environmental Sciences* 6 (4):261–278.

Bennett, J. 2010. *Vibrant Matter: A Political Ecology of Things*. Durham and London: Duke University Press.

Braidotti, R. 2013. *The Posthuman*. Cambridge, UK and Malden, US: Polity Press.

Braun, T., and P. Dierkes. 2017. "Connecting Students to Nature–How Intensity of Nature Experience and Student Age Influence the Success of Outdoor Education Programs." *Environmental Education Research* 23 (7):937–49.

Brinkley, R. 1992. "Spaces Between Words: Writing *Mont Blanc*." In *Romantic Revisions*, edited by R. Brinkley and K. Hanley, 243–267. Cambridge, UK: Cambridge University Press.

Brinkley, R., and K. Hanley. 1992. *Romantic revisions*. Cambridge, UK: Cambridge University Press.

Callicott, J. B. 2000. "Contemporary Criticisms of the Received Wilderness Idea." In *Wilderness Science in a Time of Change, Vol 1: Changing Perspectives and Future Directions*, edited by D. N. Cole, S. F. McCool, W. A. Freimund, and J. O'Loughlin, 24–31. Ogden, UT: U.S. Department of Agriculture, Forest Service, Rocky Mountain Research Station.

Castree, N., and B. Braun. 2001. *Social Nature Theory, Practice, and Politics*. London: Blackwell.

Clarke, D. A., and J. Mcphie. 2014. "Becoming Animate in Education: Immanent Materiality and Outdoor Learning for Sustainability." *Journal of Adventure Education & Outdoor Learning* 14 (3):198–216.

Cohen, J. J. 2013. "Introduction: Ecology's Rainbow." In *Prismatic Ecology: Ecotheory Beyond Green*, edited by J. J. Cohen, xv–xxxv. Minneapolis and London: University of Minnesota Press.

Cohen, J. J. 2015. *Stone: An Ecology of the Inhuman*. Minneapolis and London: University of Minnesota Press.

Corry, S. 2011. *Tribal Peoples for Tomorrow's World*. Cookhill, Worcestershire, UK: Freeman Press.

Cutting, R. 2016. "Reflections on Outdoor Education and English 'Indigenous Organic Fascism' in the 1930s." *Journal of Adventure Education and Outdoor Learning* 16 (2):105–116.

de Freitas, E., and A. Palmer. 2016. "How Scientific Concepts Come to Matter in Early Childhood Curriculum: rethinking the Concept of Force." *Cultural Studies of Science Education* 11 (4):1201–1222.

de Vega, E. P. n.d. Thinking the Ecological Present. http://www.academia.edu/2964186/Thinking_The_Ecological_Present

Deleuze, G., and F. Guattari. 1994. *What is Philosophy*? (Trans. G. Burchell and H. Tomlinson). London, UK: Verso.

Deleuze, G., and F. Guattari. 2004. *A Thousand Plateaus: Capitalism and Schizophrenia* (Trans. B. Massumi). London, UK: Continuum.

Deleuze, G., and C. Parnet. 2002. *Dialogues II* (Trans. H. Tomlinson and B. Habberjam). New York: Columbia University Press.

Duhn, I., K. Malone, and M. Tesar. 2017. "Troubling the Intersections of Urban/Nature/Childhood in Environmental Education." *Environmental Education Research* 23 (10):1357–1368. doi:10.1080/13504622.2017.1390884

Egmond, T. V. 2007. *Understanding Western tourists in developing countries. Oxfordshire.*, UK and Cambridge, USA: CAB International.

Ellison, A. M. 2013. "The Suffocating Embrace of Landscape and the Picturesque Conditioning of Ecology." *Landscape Journal* 32 (1):79–94. doi:10.3368/lj.32.1.79.

Fletcher, R. 2017. "Connection with Nature Is an Oxymoron: A Political Ecology of "Nature-Deficit Disorder." *The Journal of Environmental Education* 48 (4):226–233.

Foucault, M. 1977. "Intellectuals and Power: A Conversation Between Michel Foucault and Giles Deleuze." In *Language, Counter-Memory, Practice: Selected Essays and Interviews by Michel Foucault*, edited by D. Bouchard. Oxford: Basil Blackwell.

Gannon, S. 2017. "Saving Squawk? Animal and Human Entanglement at the Edge of the Lagoon." *Environmental Education Research* 23 (1):91–110.

Gillespie, G. 2008. *Hunting for Empire: Narratives of Sport in Rupert's Land*, 1840–70. UBC Press.

Ginn, F., and D. Demeritt. 2010. "Nature." In *Key Concepts in Geography*, edited by S. L. Holloway, S. P. Rice, G. Valentine, and N. Clifford, (2nd ed., 300–311). London: Sage.

Gough, N. 1993. "Environmental Education, Narrative Complexity and Postmodern Science/Fiction." *International Journal of Science Education* 15 (5):607–625.

Gough, N. 1994. "Playing at Catastrophe: Ecopolitical Education after Poststructuralism." *Educational Theory* 44 (2): 189–210.

Gough, N. 2008. "Narrative Experiments and Imaginative Inquiry." *South African Journal of Education* 28 (3):335–349.

Gough, N. 2010. *Performing imaginative inquiry: narrative experiments and rhizosemiotic play. Imagination in educational theory and practice: A many-sided vision*, 42–60.

Gregg, M., and Seigworth, G. (eds.). 2010. *The Affect Theory Reader*. Durham, NC: Duke University Press.

Haraway, D. J. 1988. "Situated Knowledges: The Science Question in Feminism and the Privilege of Partial Perspective." *Feminist Studies* 14 (3):575–599.

Haraway, D. J. 1997. *Modest_Witness@Second_Millennium.FemaleMan_Meets_OncoMouse: Feminism and Technoscience*. New York: Routledge.

Haraway, D. J. 2000. *How Like a Leaf: An Interview with Thyrza Nichols Goodeve*. London and New York: Routledge.

Harris, L. M. 2014. "Imaginative Geographies of Green: difference, Postcoloniality, and Affect in Environmental Narratives in Contemporary Turkey." *Annals of the Association of American Geographers* 104 (4):801–15.

Howard, P., Thompson, I. and Waterton, E. (Eds.). 2013. *The Routledge Companion to Landscape Studies*. London and New York: Routledge.

Ingold, T. 2000. *The Perception of the Environment: Essays on Livelihood, Dwelling and Skill*. Oxon: Routledge.

Ingold, T. 2007. *Lines: A Brief History*. Oxon: Routledge.

Ingold, T. 2011. *Being Alive: Essays on Movement, Knowledge and Description*. Oxon: Routledge.

Latour, B. 2004. "Why Has Critique Run out of Steam? from Matters of Fact to Matters of Concern." *Critical Inquiry* 30 (2):225–248.

Little, W., H. W. Fowler, J. Coulson, and C. T. Onion. 1957. *The Oxford International Dictionary of the English Language (Unabridged)*. Toronto: Leland Publishing Company, Ltd.

Lorimer, J. 2012. "Multinatural Geographies for the Anthropocene." *Progress in Human Geography* 36 (5):593–612.

MacFarlane, R. 2003. *Mountains of the Mind: A History of a Fascination*. London: Granta Publications.

MacLure, M. 2015. "The New Materialisms: A Thorn in the Flesh of Critical Qualitative Inquiry?" In *Critical Qualitative Inquiry*, edited by G. S. Cannella, M. S. Pérez, and P. A. Pasque, 93–112. Walnut Creek, CA: Left Coast Press.

Malone, K. 2016. "Reconsidering Children's Encounters with Nature and Place Using Posthumanism." *Australian Journal of Environmental Education* 32 (01):42–56.

Massumi, B. 2010. "What Concepts Do: Preface to the Chinese Translation of a Thousand Plateaus." *Deleuze Studies* 4 (1):1–15.

Mayer, F. S., and C. M. Frantz. 2004. "The Connectedness to Nature Scale: A Measure of Individuals' Feeling in Community with Nature." *Journal of Environmental Psychology* 24 (4):503–515.

McKenzie, M. 2017. "Affect Theory and Policy Mobility: challenges and Possibilities for Critical Policy Research." *Critical Studies in Education* 58 (2):187–204.

Mcphie, J. 2017. "Embodied Walls and Extended Skins: Exploring the Distribution of Mental Health Through Tataus and Graffiti." In *Street Art of Resistance*, edited by S. Awad and B. Wagoner. Palgrave Macmillan Series on Culture and Creativity.

Milligan, C., and A. Bingley. 2007. "Restorative Places or Scary Spaces? The Impact of Woodland on the Mental Well-Being of Young Adults." *Health & Place* 13 (4):799–811.

Moore, R. 2017. "The Man Who Went on a Hike – and Never Stopped Walking." *The Guardian* July 3. https://www.theguardian.com/world/2017/jul/03/hiking-walking-nimblewill-nomad-mj-eberhart

Morgan, A. 2010. "The Lord of the Rings – A Mythos Applicable in Unsustainable Times?." *Environmental Education Research* 16 (3–4):383–399.

Morton, T. 2007. *Ecology Without Nature: Rethinking Environmental Aesthetics*. London, UK: Harvard University Press.

Nature Connections. 2015. *Nature Connections 2015: An Interdisciplinary Conference to examine routes to nature connectedness. College of Life and Natural Sciences*. University of Derby, March 26, 2015. http://www.derby.ac.uk/enterprisecentre/events/nature-connections/

Nicholson, N. 1977. *The Lake District: An Anthology Compiled by Norman Nicholson*. Harmondsworth: Penguin.

Nicolson, M. H. 1997. *Mountain Gloom and Mountain Glory: The Development of the Aesthetics of the Infinite*. Seattle and London: University of Washington Press.

Payne, P. G. 2016. "What Next? Post-Critical Materialisms in Environmental Education." *The Journal of Environmental Education* 47 (2):169–178. doi:10.1080/00958964.2015.1127201

Peters, M. A. 2004. "Geophilosophy, Education and the Pedagogy of the Concept." *Educational Philosophy and Theory* 36 (3):217–226.

Plumwood, V. 2002. *Feminism and the Mastery of Nature*. Oxon: Routledge.

Quinn, J. 2013. "Theorising Learning and Nature: post-Human Possibilities and Problems." *Gender and Education* 25 (6):738–753.

Raine, A. 2014. "Ecocriticism and Modernism." In *The Oxford Handbook of Ecocriticism*, edited by G. Garrard, 98–117. Oxford: Oxford University Press.

Rautio, P. 2013. "Being Nature: Interspecies Articulation as a Species-Specific Practice of Relating to Environment." *Environmental Education Research* 19 (4):445–457.

Rautio, P., R. Hohti, R. M. Leinonen, and T. Tammi. 2017. "Reconfiguring Urban Environmental Education with 'Shitgull'and a 'Shop." *Environmental Education Research* 23 (10):1379–1390. 10.1080/13504622.2017.1325446

Rousell, D., A. Cutter-Mackenzie, and J. Foster. 2017. "Children of an Earth to Come: Speculative Fiction, Geophilosophy and Climate Change Education Research." *Educational Studies* 53 (6):654–669.

Russell, C., and J. Oakley. 2017. "Engaging the Emotional Dimensions of Environmental Education." *Canadian Journal of Environmental Education (CJEE)*, 21:13–22.

Schama, S. 1995. *Landscape and Memory*. London: Harper Collins Publishers.

Selhub, E. M., and A. C. Logan. 2012. *Your Brain on Nature: the Science of Nature's Influence on Your Health, Happiness, and Vitality*. Canada: John Wiley and Sons.

Spirn, A. W. 1998. *The Language of Landscape*. New Haven and London: Yale University Press.

Lucas, G. (Writer and Director). 1977. *Star Wars [film]*. United States: Lucasfilm and Twentieth Century Fox.

Thompson, I. 2010. *The English Lakes: A History*. London, New York and Berlin: Bloomsbury Publishing.

Typhina, E. 2017. "Urban Park Design + Love for Nature: Interventions for Visitor Experiences and Social Networking." *Environmental Education Research* 23 (8):1169–1181.

Vidal, J. 2001. *A Great White Lie*. http://www.theguardian.com/education/2001/dec/01/highereducation.books

Waterton, E. 2013. "Landscape and Non-Representational Theories." In *The Routledge Companion to Landscape Studies*, edited by P. Howard, I. Thompson and E. Waterton, 66–75. Oxford, UK: Routledge.

Weinert, F. 2004. *The Scientist as Philosopher: Philosophical Consequences of Great Scientific Discoveries*. Berlin, Germany: Springer.

Whitehead, A. N. 1919. *The Concept of Nature. The Tarner lectures, delivered in Trinity college, November, 1919*. http://archives.library.illinois.edu/erec/University%20Archives/1515022/OriginalFiles/LITERATURE/WHITEHEAD/Concept%20of%20Nature%20Whitehead.pdf

Wylie, J. 2007. *Landscape*. London: Routledge.

Index

Page numbers in **bold** refer to tables and those in *italic* refer to figures.

New Materialisms and Environmental Education

'New materialisms' refers to a broad, contemporary, and significant movement of thought across the social sciences and cultural studies which attempts to (re)turn to, renew, or create alternative philosophies of matter. Such philosophies spring from multiple sources but are in general an attempt to bring the indissolubility of the social and environmental more forcefully into our analytical frames and modes of inquiry and tackle a perceived over-reliance on discourse and language in the so-called postmodern era of philosophy and social science. This movement in thought is underlaid by, and meets up with, the climate and biodiversity crises and the nature of the human condition (and modes of learning or becoming), within the field of environmental education. This volume brings together academics working at differing intersections of environmental education and new materialisms, highlighting tensions, knots, and lines of flight across and for research, practice, and theory. As such this collection draws on multiple interpretations and streams of thought within new materialisms and demonstrates their significance for those engaging with environmental education policy, practice, and research. This book was originally published as a special issue of the journal *Environmental Education Research*.

David A. G. Clarke lectures in Environmental Education at The University of Edinburgh (UK). He is a member of the University's Centre for Creative-Relational Inquiry (CCRI), and the Sustainability in Education Research Group (SIERG). His academic interests traverse education, creative inquiry, life experience, and ethics in the Anthropocene.

Jamie Mcphie's work traverses Health, Environmental Humanities, and Experiential Education. He is a co-theme lead for one of the Learning, Education and Development Research Centre themes based at the University of Cumbria (UK). His research interests include therapeutic landscapes, environmental ethics, contemporary animisms, posthumanism, and psychogeography. He recently authored the book Mental Health and Wellbeing in the Anthropocene: A Posthuman Inquiry (2019).

New Materialisms and Environmental Education